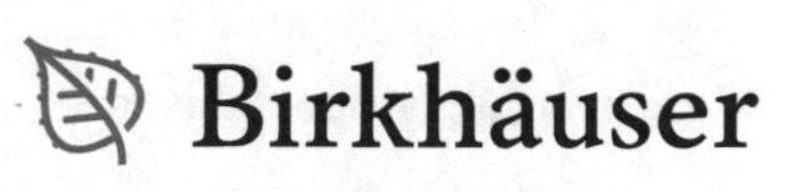
Birkhäuser

Operator Theory: Advances and Applications

Volume 287

Founded in 1979 by Israel Gohberg

More information about this series at https://link.springer.com/bookseries/4850

Christian Seifert • Sascha Trostorff •
Marcus Waurick

Evolutionary Equations

Picard's Theorem for Partial Differential Equations, and Applications

Christian Seifert
Institut für Mathematik
Technische Universität Hamburg
Hamburg, Germany

Sascha Trostorff
Mathematisches Seminar
Christian-Albrechts-Universität zu Kiel
Kiel, Germany

Marcus Waurick
Institut für Angewandte Analysis
TU Bergakademie Freiberg
Freiberg, Germany

ISSN 0255-0156 ISSN 2296-4878 (electronic)
Operator Theory: Advances and Applications
ISBN 978-3-030-89396-5 ISBN 978-3-030-89397-2 (eBook)
https://doi.org/10.1007/978-3-030-89397-2

This book is published under the imprint Birkhäuser, www.birkhauser-science.com, by the registered company Springer Nature Switzerland AG.
The registered company address is: Gewerbestrasse 11, 6330 Cham, Switzerland

Preface

The theory of evolutionary equations has its origins in the seminal paper [82] by Rainer Picard, working at the Technische Universität Dresden, Germany. All three of us were students at this university at the time. Thus, we were lucky enough to learn the theory of evolutionary equations from its early days on. We took and still take the opportunity to be part of the continuously growing group of people actively developing the theory further. In fact, both the PhD and the habilitation theses of S.T. and M.W. are concerned with generalisations of the initial theory as well as opening up new directions of research. It is also an aim of these lecture notes to present some of these latest results in a coherent text.

In general terms, the theory of evolutionary equations provides a Hilbert space method to understand differential equations. It comprises a unified approach to solving both ordinary and partial differential equations as well as to show general well-posedness results for both stationary and nonstationary, that is, time-dependent problems. Besides well-posedness theorems for large classes of differential equations (including nonlinear problems), the theory addresses quantitative and qualitative questions related to exponential stability, homogenisation and regularity. This list is bound to get longer in future. The general approach, furthermore, allows for either a comparison or unification (depending on the context) of approaches initially tailored for particular types of equations, such as parabolic, hyperbolic or elliptic. In particular, mixed type equations can be considered and understood with the presented perspective. Thus, many fundamental equations of mathematical physics such as the heat equation, wave equation, Maxwell's equations and the equations of elasticity theory can be treated using this method.

The abovementioned equations fitting into a general solution theory posed a surprising fact (at least for us). Even more so as the general problem class of evolutionary equations bases on four rather elementary observations being shortly summarised as follows:

- the (distributional, time) derivative can be realised as a boundedly invertible, normal operator in exponentially weighted L_2-spaces,

- many equations of mathematical physics (including the above) can be written as a sum of two unbounded operators: one of them involving first order differential operators in space and the other one a first order differential operator in time,
- the introduction of abstract so-called 'material laws' or 'material law operators' as coefficients of the time derivative describes both heterogeneous media and coupling effects,
- the solution mechanism is based on monotonicity of both the sum of the mentioned unbounded operators together with its adjoint computed in the space-time Hilbert space; in many cases, this monotonicity readily follows from the one of the time derivatives multiplied with the material law operator.

The last observation is particularly striking in as much as the monotonicity of the time derivative multiplied with the material law operator is rather easily obtained in many applications. This provides a well-posedness criterion that is both elementary and general, often leading to generalisations of known solution criteria for particular situations. From an applied perspective, these criteria can often be verified without diving into the intricacies of more involved solution methods and, thus, the existing numerical methods for evolutionary equations can be used to numerically solve the considered equation at hand.

In the context of time-dependent equations and related topics, there is a well-established format of introducing various subjects to advanced master or diploma students as well as PhD students, namely the Internet Seminar on Evolution Equations. Since 1997, it has been organised by various groups from Germany, Hungary, Italy, the UK and the Netherlands, providing virtual lectures as well as supervised student projects. In the academic year 2019–2020, we organised the Internet Seminar focussing on evolutionary equations. The present book is an extended version of the lecture notes for the virtual lectures. As such, it presents a thorough introduction to the theory of evolutionary equations and the corresponding solution theory and provides many properties, different classes of examples and properties of solutions, taking the reader from the early beginning of Picard's theorem to (almost) the state-of-the-art in this theory.

As the text is based on weekly virtual lectures, each chapter of the book is intended to (roughly) comprise a selection of material that covers 4 h of lectures and 2 h of exercise classes. Hence, this book covers material for one or two semesters. It is intended for master or diploma students as well as PhD students and researchers and requires only basic knowledge on functional analysis, foundations in Hilbert space theory and complex analysis in one variable. The needed amount of these is similar to the ones provided in basic courses on these topics. Apart from these prerequisites, the material of the book is self-contained. At the end of each chapter, we appended 7 exercises of varying difficulties from easy to challenging and also we commented on further reading and/or on the wider context of the contents of the chapter.

We are indebted to Rainer Picard for introducing this theory to us more than a decade ago and for his past and ongoing support in many areas. We are very grateful to the participants of the 23rd Internet Seminar for reading the manuscript, working with the material and thus checking large parts of the present text. In particular, we cordially thank Jürgen Voigt, Hendrik Vogt and Michael Doherty for their valuable comments, which led to many improvements. M.W. thanks Jussi Behrndt for the invitation on a guest professorship at the TU Graz at the end of 2020 and the beginning of 2021. This guest appointment led to the presentation of the course at TU Graz with many interested students, in particular, Julia Hauser, Peter Schlosser, Georg Stenzel and Raphael Watschinger, studying the material and providing useful feedback that helped to profoundly improve the text. We thank the anonymous referees for their comments that led to further improvements. All the remaining mistakes are our own.

We thank Christiane Tretter, Editor of the *Operator Theory* series, for her encouragement and guidance. Moreover, we thank Dorothy Mazlum for her support during the earlier stages of the manuscript (and its submission) as well as Daniel Jagadisan for the completion and final submission process. Last but not the least, we thank the TU Bergakademie Freiberg for providing the open access costs for this manuscript, thus making the final version of these lecture notes easily available around the world without further costs.

Hamburg, Germany — Christian Seifert
Kiel, Germany — Sascha Trostorff
Freiberg, Germany — Marcus Waurick
August 2021

Contents

Chapter 1
Introduction

This chapter is intended to give a brief introduction as well as a summary of the present text. We shall highlight some of the main ideas and methods behind the theory and will also aim to provide some background on the main concept in the manuscript: the notion of so-called

Evolutionary Equations

dating back to Picard in the seminal paper [82]; see also [84, Chapter 6].

Another expression used to describe the same thing (and in order to distinguish the concept from *evolution equations*) is that of *evo-systems*. Before going into detail on what we think of when using the term evolutionary equations, we provide some wider context to (some) solution methods of partial differential equations.

1.1 From ODEs to PDEs

In order to study and understand partial differential equations (PDEs) in general people have started out looking for methods known from the theory of ordinary differential equations (ODEs) to apply these to PDEs. The process of getting from a PDE to some ODE is by no means unique nor 'canonical'. That is to say there might be more than one way of reformulating a PDE into an (generalised) ODE setting (if at all).

The benefits of such a strategy, if it works, are obvious: Since for ODEs solution methods are well-known and well understood, some intuition from ODEs may be passed onto the solution process for PDEs. One way of directly apply ODE-methods to PDEs can be carried out for transport type equations, where the method of characteristics uses the fact that—using the implicit function theorem—some solutions of PDEs correspond to solutions of ODEs. In this section we shall not

C. Seifert et al., *Evolutionary Equations*, Operator Theory: Advances and Applications 287, https://doi.org/10.1007/978-3-030-89397-2_1

delve into this direction of PDE theory but refer to the standard literature such as [39] instead.

Another way of using ODE theory for PDEs is summarised by what might be called infinite-dimensional generalisations. In a nutshell instead of solving a PDE directly, one solves (infinitely many) ODEs instead. For some equations this strategy can be applied by the separation of variables ansatz. Somewhat similarly, one can generalise linear ODEs into an infinite-dimensional setting under the umbrella term evolution equation to signify differential equations involving time. In order to provide some more detail to this strategy we shortly recall how to solve linear ODEs: Let us consider an $n \times n$-matrix A with entries from the field $\mathbb{K}$ of complex or real numbers, $\mathbb{C}$ or $\mathbb{R}$, and address the system of ordinary differential equations

$$\begin{cases} u'(t) = Au(t), & t > 0, \\ u(0) = u_0 \end{cases}$$

for some given initial datum, $u_0 \in \mathbb{K}^n$. This solution can be computed with the help of the matrix exponential

$$\mathrm{e}^{tA} = \sum_{k=0}^{\infty} \frac{(tA)^k}{k!} \in \mathbb{K}^{n \times n}$$

in the form

$$u(t) = \mathrm{e}^{tA} u_0.$$

As it turns out, this u is continuously differentiable and u satisfies the above equation. We note in particular that $\mathrm{e}^{tA}u_0 \to \mathrm{e}^{0A}u_0 = u_0$ as $t \to 0+$ and that $\mathrm{e}^{(t+s)A} = \mathrm{e}^{tA}\mathrm{e}^{sA}$. In a way, to obtain the solution for the system of ordinary differential equations we need to construct $(\mathrm{e}^{tA})_{t \geqslant 0}$, the so-called fundamental solution.

In order to have a particular example for the infinite-dimensional generalisation in mind, let us have a look at the heat equation next. This is the prototypical example for an (infinite-dimensional) evolution equation: Let $\Omega \subseteq \mathbb{R}^d$ be open. Then consider

$$\begin{cases} \partial_t \theta(t,x) = \Delta\theta(t,x), & (t,x) \in (0,\infty) \times \Omega, \\ \theta(0,x) = \theta_0(x), & x \in \Omega, \end{cases}$$

where $\Delta = \sum_{j=1}^{d} \partial_j^2$ is the usual Laplacian carried out with respect to the 'x-variables' or 'spatial variables', and θ_0 is a given initial heat distribution and θ is the unknown (scalar-valued) heat distribution. The above heat equation is also accompanied with some boundary conditions for $\theta(t,x)$ which are required to be valid for all $t > 0$ and $x \in \partial\Omega$. For definiteness, we consider homogeneous Dirichlet boundary conditions, that is, $\theta(t,x) = 0$ for all $t > 0$ and $x \in \partial\Omega$, in the following.

In order to mark the considered boundary conditions we shall write Δ_D instead of just Δ and look at the heat equation in the form

$$u' = \Delta_D u, \quad u(0) = u_0$$

with the understanding that u is considered to be a vector-valued function assigning each time $t \geqslant 0$ to a function space X of functions $\Omega \to \mathbb{K}$; here we choose $X = L_2(\Omega)$. If Ω is bounded, it is possible to diagonalise Δ_D and the corresponding eigenvector expansion leads to infinitely many ODEs of the form

$$u_k' = \lambda_k u_k, \quad u_k(0) = u_{0,k}$$

for suitable scalars λ_k, $k \in \mathbb{N}$. The solution sequence $(u_k)_k$ for these ODEs is the sequence of coefficients of the eigenvector expansion of u.

A different infinite-dimensional generalisation of the finite-dimensional setting leads to a solution method valid for all Ω.

This generalisation does not consist in changing the PDE to many ODEs but only to a single one with an infinite-dimensional state space. The method is described best by looking at the fundamental solution in the ODE setting rather than the equation. The idea is to find a fundamental solution with state space X so that we replace the family $(e^{tA})_{t\geqslant 0}$ of matrices acting on $\mathbb{K}^n$ by a family $(T(t))_{t\geqslant 0}$ of linear operators in X. This leads to the notion of so-called C_0-semigroups and the fundamental solution of the heat equation is then the (appropriately interpreted) family $(e^{t\Delta_D})_{t\geqslant 0}$, see [38, 48, 81] for some standard references. More precisely, for $X = L_2(\Omega)$ and $\theta_0 \in L_2(\Omega)$, the function $\theta\colon t \mapsto e^{t\Delta_D}\theta_0 \in L_2(\Omega)$ satisfies the above heat equation in a certain *generalised* sense.

In general, for equations written in the form $u' = Au$ for appropriate A, a solution theory, that is, the proof for existence, uniqueness and continuous dependence on the data, is then contained in the construction of the fundamental solution (e.g., C_0-semigroup) in terms of the ingredients of the equation. This infinite-dimensional generalisation from the ODE case proves to be versatile and has been applied to many different particular PDEs of the form $u' = Au$.

Albeit quite successful there are also some drawbacks in the application of the abovementioned theories. For particular PDEs either the considered methods are not applicable or their application necessitates more or less involved workarounds.

In the next section, we describe a particular problem for which invoking for instance semigroup theory would seem unnatural let alone not at all straightforward. It follows, however, the general scheme of looking at fundamental solutions in an infinite-dimensional context.

1.2 Time-independent Problems

The construction of fundamental solutions is also a valuable method for obtaining a solution for time-independent problems, see, e.g., [39]. To see this, let us consider Poisson's equation in $\mathbb{R}^3$: Given $f \in C_c^\infty(\mathbb{R}^3)$ we want to find a function $u\colon \mathbb{R}^3 \to \mathbb{R}$ with the property that

$$-\Delta u(x) = f(x) \quad (x \in \mathbb{R}^3).$$

It can be shown that u given by

$$u(x) = \frac{1}{4\pi} \int_{\mathbb{R}^3} \frac{1}{|x-y|} f(y)\, \mathrm{d}y$$

is well-defined, twice continuously differentiable and satisfies Poisson's equation; cf. Exercise 1.3. Note that $x \mapsto \frac{1}{4\pi|x|}$ is also referred to as the *fundamental solution* or *Green's function* for Poisson's equation. The formula presented for u is the *convolution* with the fundamental solution. The formula used to define u also works for f being merely bounded and measurable with compact support. In this case, however, the pointwise formula of Poisson's equation cannot be expected to hold anymore, since changing f on a set of measure 0 does not influence the values of u. Thus, only a posteriori estimates under additional conditions on f render u to be twice continuously differentiable (say) with Poisson's equation holding for all $x \in \mathbb{R}^3$. However, similar to the semigroup setting, it is possible to *generalise* the meaning of $-\Delta u = f$. Then, again, the fundamental solution can be used to construct a solution for Poisson's equation for more general f.

The situation becomes different when we consider a boundary value problem instead of the problem above. More precisely, let $\Omega \subseteq \mathbb{R}^3$ be an open set and let $f \in L_2(\Omega)$. We then ask whether there exists $u \in L_2(\Omega)$ such that

$$\begin{cases} -\Delta u = f, & \text{on } \Omega, \\ u = 0, & \text{on } \partial\Omega. \end{cases}$$

Notice that the task of just (mathematically) formulating this equation, let alone establishing a solution theory, is something that needs to be addressed. Indeed, we emphasise that it is unclear as to what Δu is supposed to mean if $u \in L_2(\Omega)$, only. It turns out that the problem described is not well-posed in general. In particular—depending on the shape of Ω and the norms involved—it might, for instance, lack continuous dependence on the data, f.

In any case, the solution formula that we have used for the case when $\Omega = \mathbb{R}^3$ does not work anymore. Indeed, only particular shapes of Ω permit to explicitly construct a fundamental solution; see [39, Section 2.2]. Despite this, when Ω is merely bounded, it is still possible to construct a solution, u, for the above problem. There are two key ingredients required for this approach. One is a clever application of Riesz's representation theorem for functionals in Hilbert spaces and the other one involves inventing 'suitable' interpretations of Δu in Ω and $u = 0$ on $\partial\Omega$. Thus, the method of 'solving' Poisson's equation amounts to posing the correct question, which then can be addressed *without* invoking the fundamental solution. With this in mind, one could argue that the *setting* makes the problem solvable.

1.3 Evolution*ary* Equations

The central aim for evolutionary equations is to combine the rationales from both the C_0-semigroup theory and that from the time-independent case. That is to say, we wish to establish a setting that treats time-independent problems as well as time-dependent problems. At the same time we need to *generalise* solution concepts. We shall not aim to construct the fundamental solution in either the spatial or the temporal directions. The problem class will comprise of problems that can be written in the form

$$(\partial_t M(\partial_t) + A)\, U = F$$

where U is the unknown and F the known right-hand side. Furthermore, A is an (unbounded, skew-selfadjoint) operator acting in some Hilbert space that is thought of as modelling spatial coordinates; ∂_t is a realisation of the (time-)derivative operator and $M(\partial_t)$ is an analytic, bounded operator-valued function M, which is evaluated at the time derivative. In the course of the next chapters, we shall specify the definitions and how standard problems fit into this problem class. In particular, we will specify the Hilbert spaces modelling space-time in which the above equation is considered.

Before going into greater depth on this approach, we would like to emphasise the key differences and similarities which arise when compared to the derivation of more traditional solution theories that we outlined above.

Since the solution theory for evolutionary equations will also encapsulate time-independent problems, we predominantly focus on inhomogeneous problems. In fact, the choice of Hilbert spaces implies implicit homogeneous initial conditions at $t = -\infty$. However, inhomogeneous initial values at $t = 0$ will also be considered in this manuscript. In fact, it turns out that these initial value problems can be recast into problems of the above type.

In any case, as we do not want to require the existence of any fundamental solution we will also need to introduce a *generalisation* of the concept of a solution. Moreover, we shall see that both ∂_t and A are *unbounded* operators whereas $M(\partial_t)$ is a bounded operator. Thus, we need to make sense of the operator sum of the two unbounded operators $\partial_t M(\partial_t)$ and A, which, in general, cannot be realised as being onto but rather as having dense range, only.

A post-processing procedure will then ensure that for more regular right-hand sides, F, the solution U will also be more regular. In some cases this will, for instance, amount to U being continuous in the time variable. We shall entirely confine ourselves within the Hilbert space case though. In this sense, the solution theory to be presented will be, in essence, an application of the projection theorem applied in a Hilbert space that combines both spatial and temporal variables.

The operator $M(\partial_t)$ is thought of as carrying all the 'complexity' of the model. What we mean by complexity will become more apparent when we discuss some examples.

Finally, let us stress that A being 'skew-selfadjoint' is a way of implementing first order systems in our abstract setting. In fact, we shall focus on first order equations in *both* time *and* space. This is also another change in perspective when compared to classical approaches. As classical treatments might emphasise the importance of the Laplacian (and hence Poisson's equation) and variants thereof, evolutionary equations rather emphasise *Maxwell's equations* as the prototypical PDE. This change of point of view will be illustrated in the following section, where we address some classical examples.

1.4 Particular Examples and the Change of Perspective

Here we will focus on three examples. These examples will also be the first to be readdressed when we discuss the solution theory of evolutionary equations in a later chapter. In order to simplify the current presentation we will not consider boundary value problems but solely concentrate on problems posed on $\Omega = \mathbb{R}^3$. Furthermore, we shall dispose of any initial conditions. For a more detailed account on the derivation of these equations, we refer to the appendix of this manuscript.

Maxwell's Equations

The prototypical evolutionary equation is the system provided by Maxwell's equations. Maxwell's equations consist of two equations describing an electro-magnetic field, (E, H), subject to a given certain external current, j,

$$\begin{aligned} \partial_t \varepsilon E + \sigma E - \operatorname{curl} H &= j, \\ \partial_t \mu H + \operatorname{curl} E &= 0. \end{aligned}$$

We shall detail the properties of the material parameters ε, μ, and σ later on; for a definition of curl see Sect. 6.1. For the time being it is safe to assume that they are non-negative real numbers and that they additionally satisfy that $\mu(\varepsilon + \sigma) > 0$. Now, in the setting of evolutionary equations, we gather the electro-magnetic field into one column vector and obtain

$$\left(\partial_t \begin{pmatrix} \varepsilon & 0 \\ 0 & \mu \end{pmatrix} + \begin{pmatrix} \sigma & 0 \\ 0 & 0 \end{pmatrix} + \begin{pmatrix} 0 & -\operatorname{curl} \\ \operatorname{curl} & 0 \end{pmatrix} \right) \begin{pmatrix} E \\ H \end{pmatrix} = \begin{pmatrix} j \\ 0 \end{pmatrix}.$$

We shall see later that we obtain an evolutionary equation by setting

$$M(\partial_t) := \begin{pmatrix} \varepsilon & 0 \\ 0 & \mu \end{pmatrix} + \partial_t^{-1} \begin{pmatrix} \sigma & 0 \\ 0 & 0 \end{pmatrix} \text{ and } A := \begin{pmatrix} 0 & -\operatorname{curl} \\ \operatorname{curl} & 0 \end{pmatrix}.$$

A formulation that fits well into an infinite-dimensional ODE-setting would be, for example,

$$\partial_t \begin{pmatrix} E \\ H \end{pmatrix} = \begin{pmatrix} \varepsilon & 0 \\ 0 & \mu \end{pmatrix}^{-1} \begin{pmatrix} -\sigma & \operatorname{curl} \\ -\operatorname{curl} & 0 \end{pmatrix} \begin{pmatrix} E \\ H \end{pmatrix} + \begin{pmatrix} \varepsilon & 0 \\ 0 & \mu \end{pmatrix}^{-1} \begin{pmatrix} j \\ 0 \end{pmatrix},$$

provided that $\varepsilon > 0$. The inhomogeneous right-hand side $(\frac{1}{\varepsilon} j, 0)$ can then be dealt with by means of the variation of constants formula, which is the incarnation of the convolution of $(\frac{1}{\varepsilon} j, 0)$ with the fundamental solution in this time-dependent situation. Thus, in order to apply for example semigroup theory, the main task lies in showing that

$$\widetilde{A} := \begin{pmatrix} -\frac{1}{\varepsilon}\sigma & \frac{1}{\varepsilon}\operatorname{curl} \\ -\frac{1}{\mu}\operatorname{curl} & 0 \end{pmatrix}$$

gives rise to a suitable interpretation of $(\mathrm{e}^{t\widetilde{A}})_{t\geqslant 0}$.

A different formulation needs to be put in place if $\varepsilon = 0$ everywhere. The situation becomes even more complicated if ε and σ are bounded, non-negative, measurable functions of the spatial variable such that $\varepsilon + \sigma \geqslant c$ for some $c > 0$. In the setting of evolutionary equations, this problem, however, *can* be dealt with. Note that then one cannot expect E to be continuous with respect to the temporal variable unless j is smooth enough.

Wave Equation

We shall discuss the scalar wave equation in a medium where the wave propagation speed is inhomogeneous in different directions of space. This is modelled by finding $u: \mathbb{R} \times \mathbb{R}^3 \to \mathbb{R}$ such that, given a suitable forcing term $f: \mathbb{R} \times \mathbb{R}^3 \to \mathbb{R}$ (again we skip initial values here), we have

$$\partial_t^2 u - \operatorname{div} a \operatorname{grad} u = f,$$

where $a = a^\top \in \mathbb{R}^{3\times 3}$ is positive definite; that is, $\langle \xi, a\xi \rangle_{\mathbb{R}^3} > 0$ for all $\xi \in \mathbb{R}^3 \setminus \{0\}$. In the context of evolutionary equations, we rewrite this as a first order problem in time *and* space. For this, we introduce $v := \partial_t u$ and $q := -a \operatorname{grad} u$ and obtain that

$$\left(\partial_t \begin{pmatrix} 1 & 0 \\ 0 & a^{-1} \end{pmatrix} + \begin{pmatrix} 0 & \operatorname{div} \\ \operatorname{grad} & 0 \end{pmatrix}\right) \begin{pmatrix} v \\ q \end{pmatrix} = \begin{pmatrix} f \\ 0 \end{pmatrix}.$$

Thus,

$$M(\partial_t) := \begin{pmatrix} 1 & 0 \\ 0 & a^{-1} \end{pmatrix} \text{ and } A := \begin{pmatrix} 0 & \operatorname{div} \\ \operatorname{grad} & 0 \end{pmatrix}$$

render the wave equation as an evolutionary equation.

Let us mention briefly that it is also possible to rewrite the wave equation as a first order system in time only. For this, a standard ODE trick is used: one simply sticks with the additional variable $v = \partial_t u$ and obtains that

$$\partial_t \begin{pmatrix} u \\ v \end{pmatrix} = \begin{pmatrix} 0 & 1 \\ \operatorname{div} a \operatorname{grad} & 0 \end{pmatrix} \begin{pmatrix} u \\ v \end{pmatrix} + \begin{pmatrix} 0 \\ f \end{pmatrix}.$$

In this formulation the ‘complexity’ of the model is contained in the operator

$$\begin{pmatrix} 0 & 1 \\ \operatorname{div} a \operatorname{grad} & 0 \end{pmatrix}.$$

Heat Equation

We have already formulated classical approaches to the heat equation

$$\partial_t \theta - \operatorname{div} a \operatorname{grad} \theta = Q,$$

in which we have added a heat source Q and a conductivity $a = a^\top \in \mathbb{R}^{3\times 3}$ being positive definite. Here, however, we reformulate the heat equation as a first order system in time and space to end up (again setting $q := -a \operatorname{grad} \theta$) with

$$\left(\partial_t \begin{pmatrix} 1 & 0 \\ 0 & 0 \end{pmatrix} + \begin{pmatrix} 0 & 0 \\ 0 & a^{-1} \end{pmatrix} + \begin{pmatrix} 0 & \operatorname{div} \\ \operatorname{grad} & 0 \end{pmatrix} \right) \begin{pmatrix} \theta \\ q \end{pmatrix} = \begin{pmatrix} Q \\ 0 \end{pmatrix}.$$

In the context of evolutionary equations we then have that

$$M(\partial_t) := \begin{pmatrix} 1 & 0 \\ 0 & 0 \end{pmatrix} + \partial_t^{-1} \begin{pmatrix} 0 & 0 \\ 0 & a^{-1} \end{pmatrix} \text{ and } A := \begin{pmatrix} 0 & \operatorname{div} \\ \operatorname{grad} & 0 \end{pmatrix}.$$

The advantage of this reformulation is that it becomes easily comparable to the first order formulation of the wave equation outlined above. For instance it is now possible to easily consider mixed type problems of the form

$$\left(\partial_t \begin{pmatrix} 1 & 0 \\ 0 & (1-s)a^{-1} \end{pmatrix} + \begin{pmatrix} 0 & 0 \\ 0 & sa^{-1} \end{pmatrix} + \begin{pmatrix} 0 & \operatorname{div} \\ \operatorname{grad} & 0 \end{pmatrix} \right) \begin{pmatrix} \theta \\ q \end{pmatrix} = \begin{pmatrix} Q \\ 0 \end{pmatrix},$$

with $s \colon \mathbb{R}^3 \to [0, 1]$ being an arbitrary measurable function. In fact, in the solution theory for evolutionary equations, this does not amount to any additional complication of the problem. Models of this type are particularly interesting in the context of so-called solid-fluid interaction, where the relations of a solid body and a flow of fluid surrounding it are addressed.

1.5 A Brief Outline of the Course

We now present an overview of the contents of the following chapters.

Basics
In order to properly set the stage, we shall begin with some background of operator theory in Banach and Hilbert spaces. We assume the reader to be acquainted with some knowledge on bounded linear operators, such as the uniform boundedness principle, and basic concepts in the topology of metric spaces, such as density and closure. The most important new material will be the adjoint of an operator, which needs not be bounded anymore. In order to deal with this notion, we will consider relations rather than operators as they provide the natural setting for *unbounded* operators. Having finished this brief detour on operator theory, we will turn to a generalisation of Lebesgue spaces. More precisely, we will survey ideas from Lebesgue's integration theory for functions attaining values in an infinite-dimensional Banach space.

The Time Derivative
Banach space-valued (or rather Hilbert space-valued) integration theory will play a fundamental role in defining the time derivative as an unbounded, continuously invertible operator in a suitable Hilbert space. In order to obtain continuous invertibility, we have to introduce an exponential weighting function, which is akin to the exponential weight introduced in the space of continuous functions for a proof of the Picard–Lindelöf theorem; that is, the unique existence theorem for solutions for ODEs. It is therefore natural to discuss the application of this operator to ODEs. Hence, in passing, we will present a Hilbert space solution theory for ordinary differential equations. Here, we will also have the opportunity to discuss ordinary differential equations with delay and memory. After this short detour, we will turn back to the time derivative operator and describe its spectrum. For this we introduce the so-called Fourier–Laplace transformation which transforms the time derivative into a multiplication operator. This unitary transformation will additionally serve to define (analytic and bounded) functions of the time derivative. This is absolutely essential for the formulation of evolutionary equations.

Evolutionary Equations
Having finished the necessary preliminary work, we will then be in a position to provide the proper justification of the formulation and solution theory for evolutionary equations. We will accompany this solution theory not only with the three leading examples from above, but also with some more sophisticated equations. Amazingly, the considered space-time setting will allow us to discuss

(time-)fractional differential equations, partial differential equations with delay terms and even a class of integro-differential equations. Withdrawing the focus on regularity with respect to the temporal variable, we are en passant able to generalise well-posedness conditions from the classical literature. However, we shall stick to the treatment of analytic operator-valued functions M only. Therefore, we will also include some arguments as to why this assumption seems to be *physically* meaningful. It will turn out that analyticity and causality are intimately related via both the so-called Paley–Wiener theorem and a representation theorem for time translation invariant causal operators.

Initial Value Problems for Evolutionary Equations
As it has been outlined above, the focus of evolutionary equations is on inhomogeneous right-hand sides rather than on initial value problems. However, there is also the possibility to treat initial value problems with the approach discussed here. For this, we need to introduce extrapolation spaces. This then enables us to formulate initial value problems as inhomogeneous equations. We have to make a concession on the structure of the problem, however. In fact, we will focus on the case when $M(\partial_t) = M_0 + \partial_t^{-1} M_1$ for some bounded linear operators M_0, M_1 acting in the spatial variables alone. The initial condition will then read as $(M_0 U)(0+) = M_0 U_0$. Hence, one might argue that the initial condition $U(0+) = U_0$ is only assumed in a rather generalised sense. This is due to the fact that M_0 might be zero. However, for the case $A = 0$ we will also discuss the initial condition $U(0+) = U_0$, which amounts to a treatment of so-called differential-algebraic equations in both finite- and inifinite-dimensional state spaces.

Properties of Solutions and Inhomogeneous Boundary Value Problems
Turning back to the case when $A \neq 0$ we will discuss qualitative properties of solutions of evolutionary equations. One of which will be exponential decay. We will identify a subclass of evolutionary equations where it is comparatively easy to show that if the right-hand side decays exponentially then so too must the solution. If the right-hand side is smooth enough we obtain that $U(t)$, the solution of the evolutionary equation at time t, decays exponentially if $t \to \infty$. Furthermore, we will frame inhomogeneous boundary value problems in the setting of evolutionary equations. The method will require a bit more on the regularity theory for evolutionary equations and a definition of suitable boundary values. In particular, we shall present a way of formulating classical inhomogeneous boundary value problems for domains without any boundary regularity.

Properties of the Solution Operator and Extensions
In the final part, we shall have another look at the advantages of the problem formulation. In fact, we will have a look at the notion of homogenisation of differential equations. In the problem formulation presented here, we shall analyse the continuity properties of the solution operator with respect to weak operator topology convergence of the operator $M(\partial_t)$. We will address an example for ordinary differential equations (when $A = 0$) and one for partial differential

equations (when $A \neq 0$). It will turn out that the respective continuity properties are profoundly different from one another.

Furthermore, we have the occasion to address the notion of 'maximal regularity' in the context of evolutionary equations. Maximal regularity has initially been coined for parabolic-type problems like the heat equation. It turns out that evolutionary equations have a property similar to maximal regularity if one assumes the block structure of $M(\partial_t)$ and A to satisfy certain requirements. These requirements lead to a subclass of evolutionary equations containing classical parabolic type equations. We conclude the body of the text with two extensions of Picard's theorem. The first of which addresses non-autonomous problems and the second non-linear evolutionary inclusions.

1.6 Comments

The focus presented here on the main notions behind evolutionary equations is mostly in order to properly motivate the theory and highlight the most striking differences in the philosophy. There are other solution concepts (and corresponding general settings) developed for partial differential equations; either time-dependent or without involving time.

There is an abundance of examples and additional concepts for C_0-semigroups for which we refer to the aforementioned standard treatments again. There is also a generalisation to problems that are second order in time, e.g., $u'' = Au$, where $u(0)$ and $u'(0)$ are given. This gives rise to cosine families of bounded linear operators which is another way of generalising the fundamental solution concept, see, for example, [107].

The main focus of all of these equations is to address *initial value problems*, where the (first/second) time derivative of the unknown is explicit.

Another way of writing many PDEs from mathematical physics into a common form uses the notion of Friedrichs systems, see [43, 44]. However, the main focus of Friedrichs systems is on static, that is, time-independent partial differential equations. A time-dependent variant of constant coefficient Friedrichs systems are so-called symmetric-hyperbolic systems, see e.g. [12]. In these cases, whether the authors treat constant coefficients or not, the framework of evolutionary equations adds a profound amount of additional complexity by including the operator $M(\partial_t)$.

The treatment of time-dependent problems in space-time settings and addressing corresponding well-posedness properties of a sum of two unbounded operators has also been considered in [26] with elaborate conditions on the operators involved. In their studies, the flexibility introduced by the operator $M(\partial_t)$ in our setting is missing, thus the time derivative operator is not thought of having any variable coefficients attached to it.

Exercises

Exercise 1.1 Let $\phi \in C(\mathbb{R}, \mathbb{R})$. Assume that $\phi(t+s) = \phi(t)\phi(s)$ for all $t, s \in \mathbb{R}$, $\phi(0) = 1$. Show that $\phi(t) = \mathrm{e}^{\alpha t}$ $(t \in \mathbb{R})$ for some $\alpha \in \mathbb{R}$.

Exercise 1.2 Let $n \in \mathbb{N}$, $T : \mathbb{R} \to \mathbb{R}^{n \times n}$ continuously differentiable such that $T(t+s) = T(t)T(s)$ for all $t, s \in \mathbb{R}$, $T(0) = I$. Show that there exists $A \in \mathbb{R}^{n \times n}$ with the property that $T(t) = \mathrm{e}^{tA}$ $(t \in \mathbb{R})$.

Exercise 1.3 Show that $x \mapsto u(x) = \frac{1}{4\pi} \int_{\mathbb{R}^3} \frac{1}{|x-y|} f(y)\,\mathrm{d}y$ satisfies Poisson's equation, given $f \in C_c^\infty(\mathbb{R}^3)$.

Exercise 1.4 Let $f \in C_c^\infty(\mathbb{R})$. Define $u(t, x) := f(x+t)$ for $x, t \in \mathbb{R}$. Show that u satisfies the differential equation $\partial_t u = \partial_x u$ and $u(0, x) = f(x)$ for all $x \in \mathbb{R}$.

Exercise 1.5 Let X, Y be Banach spaces, $(T_n)_{n \in \mathbb{N}}$ be a sequence in $L(X, Y)$, the set of bounded linear operators. If $\sup\{\|T_n\| \,;\, n \in \mathbb{N}\} = \infty$, show that there is $x \in X$ and a strictly increasing sequence $(n_k)_{k \in \mathbb{N}}$ in $\mathbb{N}$ such that $\|T_{n_k} x\| \to \infty$.

Exercise 1.6 Let $n \in \mathbb{N}$. Denote by $\mathrm{GL}(n; \mathbb{K})$ the set of continuously invertible $n \times n$ matrices. Show that $\mathrm{GL}(n; \mathbb{K}) \subseteq \mathbb{K}^{n \times n}$ is open.

Exercise 1.7 Let $n \in \mathbb{N}$. Show that $\Phi \colon \mathrm{GL}(n; \mathbb{K}) \ni A \mapsto A^{-1} \in \mathbb{K}^{n \times n}$ is continuously differentiable. Compute Φ'.

References

12. S. Benzoni-Gavage, D. Serre, *Multidimensional Hyperbolic Partial Differential Equations*. Oxford Mathematical Monographs. First-order systems and applications (The Clarendon Press, Oxford University Press, Oxford, 2007)
26. G. Da Prato, P. Grisvard, Sommes d'opérateurs linéaires et équations différentielles opérationnelles. J. Math. Pures Appl. (9) **54**(3), 305–387 (1975)
38. K.-J. Engel, R. Nagel, *One-Parameter Semigroups for Linear Evolution Equations*, vol. 194. Graduate Texts in Mathematics. With contributions by S. Brendle, M. Campiti, T. Hahn, G. Metafune, G. Nickel, D. Pallara, C. Perazzoli, A. Rhandi, S. Romanelli, R. Schnaubelt (Springer, New York, 2000)
39. L.C. Evans, *Partial Differential Equations*, vol. 19. Graduate Studies in Mathematics (American Mathematical Society, Providence, RI, 1998)
43. K.O. Friedrichs, Symmetric hyperbolic linear differential equations. Commun. Pure Appl. Math. **7**, 345–392 (1954)
44. K.O. Friedrichs, Symmetric positive linear differential equations. Commun. Pure Appl. Math. **11**, 333–418 (1958)
48. E. Hille, R.S. Phillips, *Functional Analysis and Semi-Groups*. American Mathematical Society Colloquium Publications, vol. 31, rev. edn. (American Mathematical Society, Providence, RI, 1957)
81. A. Pazy, *Semigroups of Linear Operators and Applications to Partial Differential Equations*, vol. 44. Applied Mathematical Sciences (Springer, New York, 1983)
82. R. Picard, A structural observation for linear material laws in classical mathematical physics. Math. Methods Appl. Sci. **32**, 1768–1803 (2009)

84. R. Picard, D. McGhee, *Partial Differential Equations: A Unified Hilbert Space Approach*, vol. 55. Expositions in Mathematics (DeGruyter, Berlin, 2011)
107. M. Sova, Cosine operator functions. Rozprawy Mat. **49**, 47 (1966)

Chapter 2
Unbounded Operators

We will gather some information on operators in Banach and Hilbert spaces. Throughout this chapter let X_0, X_1, and X_2 be Banach spaces and H_0, H_1, and H_2 be Hilbert spaces over the field $\mathbb{K} \in \{\mathbb{R}, \mathbb{C}\}$.

2.1 Operators in Banach Spaces

We define the set of continuous linear operators

$$L(X_0, X_1) := \left\{ B \colon X_0 \to X_1 \,;\, B \text{ linear}, \ \|B\| := \sup_{x \in X_0 \setminus \{0\}} \frac{\|Bx\|}{\|x\|} < \infty \right\}$$

with the usual abbreviation $L(X_0) := L(X_0, X_0)$. In contrast to a bounded linear operator, a discontinuous or unbounded linear operator only needs to be defined on a proper albeit possibly dense subset of X_0. In order to define unbounded linear operators, we will first take a more general point of view and introduce (linear) relations. This perspective will turn out to be the natural setting later on.

Definition A subset $A \subseteq X_0 \times X_1$ is called a *relation in* X_0 *and* X_1. We define the *domain*, *range* and *kernel of* A as follows

$$\begin{aligned}
\operatorname{dom}(A) &:= \{x \in X_0 \,;\, \exists y \in X_1 \colon (x, y) \in A\}, \\
\operatorname{ran}(A) &:= \{y \in X_1 \,;\, \exists x \in X_0 \colon (x, y) \in A\}, \\
\ker(A) &:= \{x \in X_0 \,;\, (x, 0) \in A\}.
\end{aligned}$$

C. Seifert et al., *Evolutionary Equations*, Operator Theory: Advances and Applications 287, https://doi.org/10.1007/978-3-030-89397-2_2

The *image*, $A[M]$, *of a set* $M \subseteq X_0$ *under* A is given by

$$A[M] := \{y \in X_1\,;\ \exists x \in M\colon (x, y) \in A\}\,.$$

A relation A is called *bounded* if for all bounded $M \subseteq X_0$ the set $A[M] \subseteq X_1$ is bounded. For a given relation A we define the *inverse relation*

$$A^{-1} := \{(y, x) \in X_1 \times X_0\,;\ (x, y) \in A\}\,.$$

A relation A is called *linear* if $A \subseteq X_0 \times X_1$ is a linear subspace. A linear relation A is called *linear operator* or just *operator from* X_0 *to* X_1 if

$$A[\{0\}] = \{y \in X_1\,;\ (0, y) \in A\} = \{0\}.$$

In this case, we also write

$$A\colon \operatorname{dom}(A) \subseteq X_0 \to X_1$$

to denote a linear operator from X_0 to X_1. Moreover, we shall write $Ax = y$ instead of $(x, y) \in A$ in this case. A linear operator A, which is not bounded, is called *unbounded*.

For completeness, we also define the sum, scalar multiples, and composition of relations.

Definition Let $A \subseteq X_0 \times X_1$, $B \subseteq X_0 \times X_1$ and $C \subseteq X_1 \times X_2$ be relations, $\lambda \in \mathbb{K}$. Then we define

$$\begin{aligned}
A + B &:= \{(x, y + w) \in X_0 \times X_1\,;\ (x, y) \in A, (x, w) \in B\}\,,\\
\lambda A &:= \{(x, \lambda y) \in X_0 \times X_1\,;\ (x, y) \in A\}\,,\\
CA &:= \{(x, z) \in X_0 \times X_2\,;\ \exists y \in X_1\colon (x, y) \in A, (y, z) \in C\}\,.
\end{aligned}$$

For a relation $A \subseteq X_0 \times X_1$ we will use the abbreviation $-A := -1A$ (so that the minus sign only acts on the second component). We now proceed with topological notions for relations.

Definition Let $A \subseteq X_0 \times X_1$ be a relation. A is called *densely defined* if $\operatorname{dom}(A)$ is dense in X_0. We call A *closed* if A is a closed subset of the direct sum of the Banach spaces X_0 and X_1. If A is a linear operator then we will call A *closable*, whenever $\overline{A} \subseteq X_0 \times X_1$ is a linear operator.

Proposition 2.1.1 *Let $A \subseteq X_0 \times X_1$ be a relation, $C \in L(X_2, X_0)$ and $B \in L(X_0, X_1)$. Then the following statements hold.*

(a) *A is closed if and only if A^{-1} is closed. Moreover, we have $(\overline{A})^{-1} = \overline{A^{-1}}$.*
(b) *A is closed if and only if $A + B$ is closed.*
(c) *If A is closed, then AC is closed.*

Proof Statement (a) follows upon realising that $X_0 \times X_1 \ni (x, y) \mapsto (y, x) \in X_1 \times X_0$ is an isomorphism.

For statement (b), it suffices to show that the closedness of A implies the same for $A + B$. Let $((x_n, y_n))_n$ be a sequence in $A + B$ convergent in $X_0 \times X_1$ to some (x, y). Since $B \in L(X_0, X_1)$, it follows that $((x_n, y_n - Bx_n))_n$ in A is convergent to $(x, y - Bx)$ in $X_0 \times X_1$. Since A is closed, $(x, y - Bx) \in A$. Thus, $(x, y) \in A + B$.

For statement (c), let $((w_n, y_n))_n$ be a sequence in AC convergent in $X_2 \times X_1$ to some (w, y). Since C is continuous, $(Cw_n)_n$ converges to Cw. Hence, $(Cw_n, y_n) \to (Cw, y)$ in $X_0 \times X_1$ and since $(Cw_n, y_n) \in A$ and A is closed, it follows that $(Cw, y) \in A$. Equivalently, $(w, y) \in AC$, which yields closedness of AC. □

We shall gather some other elementary facts about closed operators in the following. We will make use of the following notion.

Definition Let $A\colon \operatorname{dom}(A) \subseteq X_0 \to X_1$ be a linear operator. Then the *graph norm* of A is defined by $\operatorname{dom}(A) \ni x \mapsto \|x\|_A := \sqrt{\|x\|^2 + \|Ax\|^2}$.

Lemma 2.1.2 *Let $A\colon \operatorname{dom}(A) \subseteq X_0 \to X_1$ be a linear operator. Then the following statements are equivalent:*

(i) *A is closed.*
(ii) $\operatorname{dom}(A)$ *equipped with the graph norm is a Banach space.*
(iii) *For all $(x_n)_n$ in* $\operatorname{dom}(A)$ *convergent in X_0 such that $(Ax_n)_n$ is convergent in X_1 we have* $\lim_{n\to\infty} x_n \in \operatorname{dom}(A)$ *and* $A \lim_{n\to\infty} x_n = \lim_{n\to\infty} Ax_n$.

Proof For the equivalence (i)⇔(ii), it suffices to observe that $\operatorname{dom}(A) \ni x \mapsto (x, Ax) \in A$, where $\operatorname{dom}(A)$ is endowed with the graph norm, is an isomorphism. The equivalence (i)⇔(iii) is an easy reformulation of the definition of closedness of $A \subseteq X_0 \times X_1$. □

Unless explicitly stated otherwise (e.g. in the form $\operatorname{dom}(A) \subseteq X_0$, where we regard $\operatorname{dom}(A)$ as a subspace of X_0), for closed operators A we always consider $\operatorname{dom}(A)$ as a Banach space in its own right; that is, we shall regard it as being endowed with the graph norm.

Lemma 2.1.3 *Let $A\colon \operatorname{dom}(A) \subseteq X_0 \to X_1$ be a closed linear operator. Then A is bounded if and only if* $\operatorname{dom}(A) \subseteq X_0$ *is closed.*

Proof First of all note that boundedness of A is equivalent to the fact that the graph norm and the X_0-norm on $\operatorname{dom}(A)$ are equivalent. Hence, the closedness and boundedness of A implies that $\operatorname{dom}(A) \subseteq X_0$ is closed. On the other hand, the embedding

$$\iota\colon (\operatorname{dom}(A), \|\cdot\|_A) \hookrightarrow (\operatorname{dom}(A), \|\cdot\|_{X_0})$$

is continuous and bijective. Since the range is closed, the open mapping theorem implies that ι^{-1} is continuous. This yields the equivalence of the graph norm and the X_0-norm and, thus, the boundedness of A. □

For unbounded operators, obtaining a precise description of the domain may be difficult. However, there may be a subset of the domain which essentially (or approximately) describes the operator. This gives rise to the following notion of a core.

Definition Let $A \subseteq X_0 \times X_1$. A set $D \subseteq \operatorname{dom}(A)$ is called a *core for A* provided $\overline{A \cap (D \times X_1)} = \overline{A}$.

Proposition 2.1.4 *Let $A \in L(X_0, X_1)$, and $D \subseteq X_0$ a dense linear subspace. Then D is a core for A.*

Corollary 2.1.5 *Let $A\colon \operatorname{dom}(A) \subseteq X_0 \to X_1$ be a densely defined, bounded linear operator. Then there exists a unique $B \in L(X_0, X_1)$ with $B \supseteq A$. In particular, we have $B = \overline{A}$ and*

$$\|B\| = \sup_{x \in \operatorname{dom}(A), x \neq 0} \frac{\|Ax\|}{\|x\|}.$$

The proofs of Proposition 2.1.4 and Corollary 2.1.5 are asked for in Exercise 2.2.

2.2 Operators in Hilbert Spaces

Let us now focus on operators on Hilbert spaces. In this setting, we can additionally make use of scalar products $\langle\cdot, \cdot\rangle$, which in this course are considered to be linear in the second argument (and anti-linear in the first, in the case when $\mathbb{K} = \mathbb{C}$).

For a linear operator $A\colon \operatorname{dom}(A) \subseteq H_0 \to H_1$ the graph norm of A is induced by the scalar product

$$(x, y) \mapsto \langle x, y\rangle + \langle Ax, Ay\rangle,$$

known as the *graph scalar product of* A. If A is closed then $\operatorname{dom}(A)$ (equipped with the graph norm) is a Hilbert space.

Of course, no presentation of operators in Hilbert spaces would be complete without the central notion of the adjoint operator. We wish to pose the adjoint within the relational framework just established. The definition is as follows.

Definition For a relation $A \subseteq H_0 \times H_1$ we define the *adjoint relation* A^* by

$$A^* := -\left(\left(A^{-1}\right)^{\perp}\right) \subseteq H_1 \times H_0,$$

where the orthogonal complement is computed in the direct sum of the Hilbert spaces H_1 and H_0; that is, the set $H_1 \times H_0$ endowed with the scalar product $\big((x, y), (u, v)\big) \mapsto \langle x, u\rangle_{H_1} + \langle y, v\rangle_{H_0}$.

Remark 2.2.1 Let $A \subseteq H_0 \times H_1$. Then we have

$$A^* = \left\{(u, v) \in H_1 \times H_0\,;\ \forall (x, y) \in A : \langle u, y\rangle_{H_1} = \langle v, x\rangle_{H_0}\right\}.$$

In particular, if A is a linear operator, we have

$$A^* = \left\{(u, v) \in H_1 \times H_0\,;\ \forall x \in \operatorname{dom}(A) : \langle u, Ax\rangle_{H_1} = \langle v, x\rangle_{H_0}\right\}.$$

Lemma 2.2.2 *Let $A \subseteq H_0 \times H_1$ be a relation. Then A^* is a linear relation. Moreover, we have*

$$A^* = -\left(\left(A^{\perp}\right)^{-1}\right) = \left((-A)^{-1}\right)^{\perp} = \left(-\left(A^{-1}\right)\right)^{\perp} = \left((-A)^{\perp}\right)^{-1} = \left(-\left(A^{\perp}\right)\right)^{-1}.$$

The proof of this lemma is left as Exercise 2.3.

Remark 2.2.3 Let $A \subseteq H_0 \times H_1$. Since A^* is the orthogonal complement of $-A^{-1}$, it follows immediately that A^* is closed. Moreover, $A^* = \left(\overline{A}\right)^*$ since $A^{\perp} = \left(\overline{A}\right)^{\perp}$.

Lemma 2.2.4 *Let $A \subseteq H_0 \times H_1$ be a linear relation. Then*

$$A^{**} := \left(A^*\right)^* = \overline{A}.$$

Proof We compute using Lemma 2.2.2

$$A^{**} = \left(\left(-\left(A^*\right)\right)^{-1}\right)^{\perp} = \left(\left(-\left(-\left(\left(A^{\perp}\right)^{-1}\right)\right)\right)^{-1}\right)^{\perp} = \left(A^{\perp}\right)^{\perp} = \overline{A}.$$ □

Theorem 2.2.5 *Let $A \subseteq H_0 \times H_1$ be a linear relation. Then*

$$\operatorname{ran}(A)^{\perp} = \ker(A^*) \quad \textit{and} \quad \overline{\operatorname{ran}}(A^*) = \ker(\overline{A})^{\perp}.$$

Proof Let $u \in \ker(A^*)$ and let $y \in \operatorname{ran}(A)$. Then we find $x \in \operatorname{dom}(A)$ such that $(x, y) \in A$. Moreover, note that $(u, 0) \in A^*$. Then, we compute

$$\langle u, y\rangle_{H_1} = \langle 0, x\rangle_{H_0} = 0.$$

This equality shows that $\operatorname{ran}(A)^\perp \supseteq \ker(A^*)$. If on the other hand, $u \in \operatorname{ran}(A)^\perp$ then for all $(x, y) \in A$ we have that

$$0 = \langle u, y\rangle_{H_1},$$

which implies $(u, 0) \in A^*$ and hence $u \in \ker(A^*)$. The remaining equation follows from Lemma 2.2.4 together with the first equation applied to A^*. □

The following decomposition result is immediate from the latter theorem and will be used frequently throughout the text.

Corollary 2.2.6 *Let $A \subseteq H_0 \times H_1$ be a closed linear relation. Then*

$$H_1 = \overline{\operatorname{ran}}(A) \oplus \ker(A^*) \quad \textit{and} \quad H_0 = \ker(A) \oplus \overline{\operatorname{ran}}(A^*).$$

We will now turn to the case where the adjoint relation is actually a linear operator.

Lemma 2.2.7 *Let $A \subseteq H_0 \times H_1$ be a linear relation. Then A^* is a linear operator if and only if A is densely defined. If, in addition, A is a linear operator, then A is closable if and only if A^* is densely defined.*

Proof For the first equivalence, it suffices to observe that

$$A^*[\{0\}] = \operatorname{dom}(A)^\perp. \tag{2.1}$$

Indeed, A being densely defined is equivalent to having $\operatorname{dom}(A)^\perp = \{0\}$. Moreover, A^* is an operator if and only if $A^*[\{0\}] = \{0\}$. Next, we show (2.1). For this, apply Theorem 2.2.5 to the linear relation A^{-1}. One obtains $(\operatorname{ran} A^{-1})^\perp = \ker(A^{-1})^*$. Hence, $(\operatorname{dom}(A))^\perp = \ker(A^*)^{-1} = A^*[\{0\}]$, which is (2.1). For the remaining equivalence, we need to characterise $\overline{A}$ being an operator. Using Lemma 2.2.4 and the first equivalence, we deduce that $\overline{A} = (A^*)^*$ is a linear operator if and only if A^* is densely defined. □

Remark 2.2.8 Note that the statement "A^* is an operator if A is densely defined" asserted in Lemma 2.2.7 is also true for *any* relation. For this, it suffices to observe that (2.1) is true for any relation $A \subseteq H_0 \times H_1$. Indeed, let $A \subseteq H_0 \times H_1$ be a relation; define $B := \operatorname{lin} A$. Then $\operatorname{dom}(B) = \operatorname{lin}\operatorname{dom}(A)$. Also, we have

$$A^* = -(A^\perp)^{-1} = -(B^\perp)^{-1} = B^*.$$

With these preparations, we can write

$$\operatorname{dom}(A)^{\perp} = (\operatorname{lin}\operatorname{dom}(A))^{\perp} = \operatorname{dom}(B)^{\perp} = B^*[\{0\}] = A^*[\{0\}],$$

where we used that (2.1) holds for linear relations.

Lemma 2.2.9 *Let $A \subseteq H_0 \times H_1$ be a linear relation. Then $\overline{A} \in L(H_0, H_1)$ if and only if $A^* \in L(H_1, H_0)$. In either case, $\|A^*\| = \|\overline{A}\|$.*

Proof Note that $\overline{A} \in L(H_0, H_1)$ implies that A is closable and densely defined. Thus, by Lemma 2.2.7, A^* is a densely defined, closed linear operator. For $u \in \operatorname{dom}(A^*)$ we compute using Lemma 2.2.4

$$\left\|A^*u\right\| = \sup_{x \in H_0 \setminus \{0\}} \frac{|\langle A^*u, x\rangle|}{\|x\|} = \sup_{x \in H_0 \setminus \{0\}} \frac{\left|\langle u, \overline{A}x\rangle\right|}{\|x\|} \leqslant \left\|\overline{A}\right\| \|u\|,$$

yielding $\|A^*\| \leqslant \left\|\overline{A}\right\|$. On the one hand, this implies that A^* is bounded, and on the other, since A^* is densely defined we deduce $A^* \in L(H_1, H_0)$ by Lemma 2.1.3. The other implication (and the other inequality) follows from the first one applied to A^* instead of A using $A^{**} = \overline{A}$. □

We end this section by defining some special classes of relations and operators.

Definition Let H be a Hilbert space and $A \subseteq H \times H$ a linear relation. We call A *(skew-)Hermitian* if $A \subseteq A^*$ ($A \subseteq -A^*$). We say that A is *(skew-)symmetric* if A is (skew-)Hermitian and densely defined (so that A^* is a linear operator), and A is called *(skew-)selfadjoint* if $A = A^*$ ($A = -A^*$). Additionally, if A is densely defined, then we say that A is *normal* if $AA^* = A^*A$.

2.3 Computing the Adjoint

In general it is a very difficult task to compute the adjoint of a given (unbounded) operator. There are, however, cases, where the adjoint of a sum or the product can be computed more readily. We start with the most basic case of bounded linear operators.

Proposition 2.3.1 *Let $A, B \in L(H_0, H_1), C \in L(H_2, H_0)$. Then $(A + B)^* = A^* + B^*$ and $(AC)^* = C^*A^*$.*

The latter results are special cases of more general statements to follow.

Theorem 2.3.2 *Let $A, B \subseteq H_0 \times H_1$ be relations. Then $A^* + B^* \subseteq (A + B)^*$. If, in addition, $B \in L(H_0, H_1)$, then $(A + B)^* = A^* + B^*$.*

Proof In order to show the claimed inclusion, let $(u, r) \in A^* + B^*$. By definition of the sum of relations, we find $v, w \in H_0$, $r = v + w$, with $(u, v) \in A^*$ and

$(u, w) \in B^*$. We compute for all $(x, s) \in A + B$, that is, $(x, y) \in A$ and $(x, z) \in B$ for some $y, z \in H_1$ with $s = y + z$

$$\begin{aligned}\langle x, r\rangle_{H_0} &= \langle x, v + w\rangle_{H_0} = \langle x, v\rangle_{H_0} + \langle x, w\rangle_{H_0} \\ &= \langle y, u\rangle_{H_1} + \langle z, u\rangle_{H_1} = \langle y + z, u\rangle_{H_1} = \langle s, u\rangle_{H_1} .\end{aligned}$$

This shows the desired inclusion. Next, we assume in addition that $B \in L(H_0, H_1)$. For the equality, it remains to show that $(A + B)^* \subseteq A^* + B^*$, which in conjunction with the above follows if $\mathrm{dom}((A+B)^*) \subseteq \mathrm{dom}(A^* + B^*) = \mathrm{dom}(A^*) \cap \mathrm{dom}(B^*)$. By Lemma 2.2.9, we have $\mathrm{dom}(B^*) = H_1$. Hence, it suffices to show that $\mathrm{dom}((A+B)^*) \subseteq \mathrm{dom}(A^*)$. For this, let $(u, v) \in (A + B)^*$. Then we compute for all $(x, y) \in A$ using Lemma 2.2.9 again

$$\langle x, v\rangle_{H_0} = \langle y + Bx, u\rangle_{H_1} = \langle y, u\rangle_{H_1} + \left\langle x, B^*u\right\rangle_{H_0} .$$

Thus, $\langle x, v - B^*u\rangle_{H_0} = \langle y, u\rangle_{H_1}$, which yields $(u, v - B^*u) \in A^*$; whence, $u \in \mathrm{dom}(A^*)$ as desired. □

Corollary 2.3.3 *Let $A \subseteq H_0 \times H_1$, $B \in L(H_0, H_1)$. If A is densely defined, then $A^* + B^*$ is an operator and $(A + B)^* = A^* + B^*$.*

Theorem 2.3.4 *Let $A \subseteq H_0 \times H_1$ and $C \subseteq H_2 \times H_0$. Then $\overline{C^*A^*} \subseteq (AC)^*$. If, in addition, $A \subseteq H_0 \times H_1$ is closed and linear as well as $C \in L(H_2, H_0)$, then $(AC)^* = \overline{C^*A^*}$.*

Proof For the first inclusion, let $(u, w) \in C^*A^*$. Thus, we find $v \in H_0$ such that $(u, v) \in A^*$ and $(v, w) \in C^*$. Next, let $(r, y) \in AC$. Then we find $x \in H_0$ such that $(r, x) \in C$ and $(x, y) \in A$. We compute

$$\langle y, u\rangle_{H_1} = \langle x, v\rangle_{H_0} = \langle r, w\rangle_{H_2} .$$

Since $(r, y) \in AC$ were chosen arbitrarily, we infer $C^*A^* \subseteq (AC)^*$. As every adjoint is closed, we obtain $\overline{C^*A^*} \subseteq (AC)^*$.

Next, we assume that A is closed and linear as well as that C is bounded and linear. Then, by what we have just shown, we obtain $AC \subseteq (C^*A^*)^*$. Next, let $(w, y) \in (C^*A^*)^*$. Then for all $(u, v) \in A^*$ and $z = C^*v$ we obtain

$$\langle u, y\rangle_{H_1} = \langle z, w\rangle_{H_2} = \left\langle C^*v, w\right\rangle_{H_2} = \langle v, Cw\rangle_{H_0} .$$

Thus, we obtain $(Cw, y) \in A^{**} = \overline{A} = A$. Thus, $(w, y) \in AC$. Hence,

$$AC = \left(C^*A^*\right)^* ,$$

which yields the assertion by adjoining this equation. □

Corollary 2.3.5 *Let $A \subseteq H_0 \times H_1$ be a linear relation and $C \in L(H_2, H_0)$. Then* $\left(\overline{A}C\right)^* = \overline{C^*A^*}$.

Proof The result follows upon realising that $A^* = A^{***} = \left(\overline{A}\right)^*$. □

Corollary 2.3.6 *Let $A \subseteq H_0 \times H_1$ be a linear relation and $C \in L(H_2, H_0)$. If $\overline{A}C$ is densely defined, then C^*A^* is a closable linear operator with $\overline{C^*A^*} = \left(\overline{A}C\right)^*$.*

Remark 2.3.7 Let us comment on the equalities in the prevoius statements.

(a) Note that if $B \in L(H_1, H_2)$ and $A \subseteq H_0 \times H_1$ is linear, then $\left(B\overline{A}\right)^* = A^*B^*$. Indeed, this follows from Theorem 2.3.4 applied to A^* and B instead of A and C^*, respectively, since then we obtain $(A^*B^*)^* = \overline{B^{**}A^{**}} = \overline{B\overline{A}}$. Computing adjoints on both sides again and using that A^*B^* is closed by Proposition 2.1.1, we get the assertion.
(b) We note here that in Corollary 2.3.5 and Corollary 2.3.6 $\overline{A}C$ cannot be replaced by $\overline{AC}$ and encourage the reader to find a counterexample for A being a closable linear operator. We also refer to [94] for a counterexample due to J. Epperlein.

We have already seen that $A^* = \overline{A}^*$. We can even restrict A to a core and still obtain the same adjoint.

Proposition 2.3.8 *Let $A \subseteq H_0 \times H_1$ be a linear relation, $D \subseteq \operatorname{dom}(A)$ a linear subspace. Then D is a core for A if and only if $(A \cap (D \times H_1))^* = A^*$.*

Proof We set $A|_D := A \cap (D \times H_1)$. Then

$$D \text{ core} \iff \overline{A|_D} = \overline{A} \iff \overline{A|_D}^{\perp} = \overline{A}^{\perp} \iff A|_D{}^{\perp} = A^{\perp} \iff A|_D^* = A^*. \quad \square$$

2.4 The Spectrum and Resolvent Set

In this section, we focus on operators acting on a single Banach space. As such, throughout this section let X be a Banach space over $\mathbb{K} \in \{\mathbb{R}, \mathbb{C}\}$ and let $A\colon \operatorname{dom}(A) \subseteq X \to X$ be a closed linear operator.

Definition The set

$$\rho(A) := \left\{\lambda \in \mathbb{K};\ (\lambda - A)^{-1} \in L(X)\right\}$$

is called the *resolvent set* of A. We define

$$\sigma(A) := \mathbb{K} \setminus \rho(A)$$

to be the *spectrum* of A.

We state and prove some elementary properties of the spectrum and the resolvent set. We shall see natural examples for A which satisfy that $\sigma(A) = \mathbb{K}$ or $\sigma(A) = \varnothing$ later on.

For a metric space (X, d), we will write $B(x, r) = \{y \in X\,;\ d(x, y) < r\}$ for the open ball around x of radius r and $B[x, r] = \{y \in X\,;\ d(x, y) \leqslant r\}$ for the closed ball.

Proposition 2.4.1 *If $\lambda, \mu \in \rho(A)$, then the* resolvent identity *holds. That is*

$$(\lambda - A)^{-1} - (\mu - A)^{-1} = (\mu - \lambda)\,(\lambda - A)^{-1}\,(\mu - A)^{-1}.$$

Moreover, the set $\rho(A)$ is open. More precisely, if $\lambda \in \rho(A)$ then $B\left(\lambda, 1/\left\|(\lambda - A)^{-1}\right\|\right) \subseteq \rho(A)$ and for $\mu \in B\left(\lambda, 1/\left\|(\lambda - A)^{-1}\right\|\right)$ we have

$$(\mu - A)^{-1} = \sum_{k=0}^{\infty} (\lambda - \mu)^k \left((\lambda - A)^{-1}\right)^{k+1}$$

as well as

$$\left\|(\mu - A)^{-1}\right\| \leqslant \frac{\left\|(\lambda - A)^{-1}\right\|}{1 - |\lambda - \mu|\,\left\|(\lambda - A)^{-1}\right\|}.$$

The mapping $\rho(A) \ni \lambda \mapsto (\lambda - A)^{-1} \in L(X)$ is analytic.

Proof For the first assertion, we let $\lambda, \mu \in \rho(A)$ and compute

$$\begin{aligned}(\lambda - A)^{-1} - (\mu - A)^{-1} &= (\lambda - A)^{-1}\big((\mu - A) - (\lambda - A)\big)(\mu - A)^{-1} \\ &= (\lambda - A)^{-1}(\mu - \lambda)(\mu - A)^{-1} \\ &= (\mu - \lambda)(\lambda - A)^{-1}(\mu - A)^{-1}.\end{aligned}$$

Next, let $\lambda \in \rho(A)$ and $\mu \in B\left(\lambda, 1/\left\|(\lambda - A)^{-1}\right\|\right)$. Then

$$\left\|(\lambda - \mu)(\lambda - A)^{-1}\right\| < 1.$$

Hence, $1 - (\lambda - \mu)(\lambda - A)^{-1}$ admits an inverse in $L(X)$ satisfying

$$\left(1 - (\lambda - \mu)(\lambda - A)^{-1}\right)^{-1} = \sum_{k=0}^{\infty} \left((\lambda - \mu)(\lambda - A)^{-1}\right)^k. \tag{2.2}$$

We claim that $\mu \in \rho(A)$. For this, we compute

$$\mu - A = \lambda - A - (\lambda - \mu) = (\lambda - A)\left(1 - (\lambda - \mu)(\lambda - A)^{-1}\right).$$

Since $\left(1-(\lambda-\mu)(\lambda-A)^{-1}\right)$ is an isomorphism in $L(X)$, we deduce that the right-hand side admits a continuous inverse if and only if the left-hand side does. As $\lambda \in \rho(A)$, we thus infer $\mu \in \rho(A)$. The estimate follows from (2.2). Indeed, we have

$$\begin{aligned}
\left\|(\mu-A)^{-1}\right\| &\leqslant \left\|(\lambda-A)^{-1}\right\| \left\|\sum_{k=0}^{\infty}\left((\lambda-\mu)(\lambda-A)^{-1}\right)^k\right\| \\
&\leqslant \left\|(\lambda-A)^{-1}\right\| \sum_{k=0}^{\infty}\left\|(\lambda-\mu)(\lambda-A)^{-1}\right\|^k = \frac{\left\|(\lambda-A)^{-1}\right\|}{1-\left\|(\lambda-\mu)(\lambda-A)^{-1}\right\|}.
\end{aligned}$$

For the final claim of the present proposition, we observe that

$$\begin{aligned}
(\mu-A)^{-1} &= \left(1-(\lambda-\mu)(\lambda-A)^{-1}\right)^{-1}(\lambda-A)^{-1} \\
&= \sum_{k=0}^{\infty}(\lambda-\mu)^k\left((\lambda-A)^{-1}\right)^{k+1},
\end{aligned}$$

which is an operator norm convergent power series expression for the resolvent at μ about λ. Thus, analyticity follows. □

For a given measure space (Ω, Σ, μ) we shall consider multiplication operators in $L_2(\mu)$ next. For a measurable function $V\colon \Omega \to \mathbb{R}$ we will use the notation $[V \leqslant c] := V^{-1}\left[(-\infty, c]\right]$ for some constant $c \in \mathbb{R}$ (and similarly for other relational symbols).

Remark 2.4.2 Before we turn to more general multiplication operators, we like to reason our notation for them by illustrating the example case of multiplication operators in $L_2(\mathbb{R})$. A multiplication operator that immediately comes to mind is the so-called multiplication-by-the-argument operator on $L_2(\mathbb{R})$, which we shall denote by m. Expressed differently, let

$$\mathrm{m}\colon \operatorname{dom}(\mathrm{m}) \subseteq L_2(\mathbb{R}) \to L_2(\mathbb{R}),\ f \mapsto (x \mapsto xf(x)),$$

where $\operatorname{dom}(\mathrm{m})$ consists of all those $L_2(\mathbb{R})$-functions f such that $(x \mapsto xf(x)) \in L_2(\mathbb{R})$. Being a multiplication operator, m admits what is called a 'functional calculus': It is possible to define functions of m, which will turn out to be operators themselves. Thus, if $V\colon \mathbb{R} \to \mathbb{C}$ is measurable, we can define $V(\mathrm{m})$ to denote an operator in $L_2(\mathbb{R})$ acting as follows

$$(V(\mathrm{m})f)(x) := V(x)f(x)$$

for suitable f. To apply V to m turns out to be the same as the operator of multiplication by V. This correspondence serves to justify the notation of multiplication operators acting on $L_2(\mu)$ for some measure space (Ω, Σ, μ). We will re-use the notation $V(\mathrm{m})$ to denote the operator of multiplication-by-V, even in cases where there is no well-defined multiplication-by-argument-operator m in $L_2(\mu)$.

Theorem 2.4.3 *Let (Ω, Σ, μ) be a measure space and $V\colon \Omega \to \mathbb{K}$ a measurable function. Then the operator*

$$V(\mathrm{m})\colon \operatorname{dom}(V(\mathrm{m})) \subseteq L_2(\mu) \to L_2(\mu)$$
$$f \mapsto \big(\omega \mapsto V(\omega) f(\omega)\big),$$

with $\operatorname{dom}(V(\mathrm{m})) := \big\{f \in L_2(\mu)\,;\, \big(\omega \mapsto V(\omega)f(\omega)\big) \in L_2(\mu)\big\}$ *satisfies the following properties:*

(a) *$V(\mathrm{m})$ is densely defined and closed.*
(b) *$(V(\mathrm{m}))^* = V^*(\mathrm{m})$ where $V^*(\omega) = V(\omega)^*$ for all $\omega \in \Omega$ (here $V(\omega)^*$ denotes the complex conjugate of $V(\omega)$).*
(c) *If V is μ-almost everywhere bounded, then $V(\mathrm{m})$ is continuous. Moreover, we have $\|V(\mathrm{m})\|_{L(L_2(\mu))} \leqslant \|V\|_{L_\infty(\mu)}$.*
(d) *If $V \neq 0$ μ-a.e. then $V(\mathrm{m})$ is injective and $V(\mathrm{m})^{-1} = \frac{1}{V}(\mathrm{m})$, where*

$$\frac{1}{V}(\omega) := \begin{cases} \frac{1}{V(\omega)}, & V(\omega) \neq 0, \\ 0, & V(\omega) = 0, \end{cases}$$

for all $\omega \in \Omega$.

Proof For the whole proof we let $\Omega_n := [|V| \leqslant n]$ and put $\mathbb{1}_n := \mathbb{1}_{\Omega_n}$.

(a) We first show that $V(\mathrm{m})$ is densely defined. Let $f \in L_2(\mu)$. Then, we have for all $n \in \mathbb{N}$ that $\mathbb{1}_n f \in \operatorname{dom}(V(\mathrm{m}))$. From $\Omega = \bigcup_n \Omega_n$ and $\Omega_n \subseteq \Omega_{n+1}$ it follows that $\mathbb{1}_n f \to f$ in $L_2(\mu)$ as $n \to \infty$.
Next, we confirm that $V(\mathrm{m})$ is closed. Let $(f_k)_k$ in $\operatorname{dom}(V(\mathrm{m}))$ convergent in $L_2(\mu)$ with $(V(\mathrm{m})f_k)_k$ be convergent in $L_2(\mu)$. Denote the respective limits by f and g. It is clear that for all $n \in \mathbb{N}$ we have $\mathbb{1}_n f_k \to \mathbb{1}_n f$ as $k \to \infty$. Also, we have

$$\mathbb{1}_n g = \lim_{k\to\infty} \mathbb{1}_n V(\mathrm{m}) f_k = \lim_{k\to\infty} V(\mathrm{m})(\mathbb{1}_n f_k) = V(\mathrm{m})(\mathbb{1}_n f) = \mathbb{1}_n V f.$$

Hence, $g = Vf$ μ-almost everywhere and since $g \in L_2(\mu)$, we have that $f \in \operatorname{dom}(V(\mathrm{m}))$.
(b) It is easy to see that $V^*(\mathrm{m}) \subseteq V(\mathrm{m})^*$. For the other inclusion, we let $u \in \operatorname{dom}(V(\mathrm{m})^*)$. Then, for all $f \in L_2(\mu)$ and $n \in \mathbb{N}$ we have $\mathbb{1}_n f \in \operatorname{dom}(V(\mathrm{m}))$ and, hence,

$$\begin{aligned}\langle f, \mathbb{1}_n V^* u\rangle &= \textstyle\int_{\Omega_n} f^* V^* u \,\mathrm{d}\mu = \langle V(\mathrm{m})(\mathbb{1}_n f), u\rangle = \langle \mathbb{1}_n f, V(\mathrm{m})^* u\rangle \\ &= \langle f, \mathbb{1}_n V(\mathrm{m})^* u\rangle\,.\end{aligned}$$

It follows that $\mathbb{1}_n V^* u = \mathbb{1}_n V(\mathrm{m})^* u$ for all $n \in \mathbb{N}$. Thus, $\Omega = \bigcup_n \Omega_n$ implies $V^* u = V(\mathrm{m})^* u$ and therefore $u \in \operatorname{dom}(V^*(\mathrm{m}))$ and $V^*(\mathrm{m})u = V(\mathrm{m})^* u$.

(c) If $|V| \leqslant \kappa$ μ-almost everywhere for some $\kappa \geqslant 0$, then for all $f \in L_2(\mu)$ we have $|V(\omega)f(\omega)| \leqslant \kappa |f(\omega)|$ for μ-almost every $\omega \in \Omega$. Squaring and integrating this inequality yields boundedness of $V(\mathrm{m})$ and the asserted inequality.

(d) Assume that $V \neq 0$ μ-a.e. and $V(\mathrm{m})f = 0$. Then, $f(\omega) = 0$ for μ-a.e. $\omega \in \Omega$, which implies $f = 0$ in $L_2(\mu)$. Moreover, if $V(\mathrm{m})f = g$ for $f, g \in L_2(\mu)$, then for μ-a.e. $\omega \in \Omega$ we deduce that $f(\omega) = \frac{1}{V}(\omega)g(\omega)$, which shows $\frac{1}{V}(\mathrm{m}) \supseteq V(\mathrm{m})^{-1}$. If on the other hand $g \in \operatorname{dom}\left(\frac{1}{V}(\mathrm{m})\right)$, then a similar computation reveals that $\frac{1}{V}(\mathrm{m})g \in \operatorname{dom}(V(\mathrm{m}))$ and $V(\mathrm{m})\frac{1}{V}(\mathrm{m})g = g$. □

The spectrum of $V(\mathrm{m})$ from the latter example can be computed once we consider a less general class of measure spaces. We provide a characterisation of these measure spaces first.

Proposition 2.4.4 *Let (Ω, Σ, μ) be a measure space. Then the following statements are equivalent:*

(i) (Ω, Σ, μ) *is* semi-finite, *that is, for every $A \in \Sigma$ with $\mu(A) = \infty$, there exists $B \in \Sigma$ with $0 < \mu(B) < \infty$ such that $B \subseteq A$.*

(ii) *For all measurable $V : \Omega \to \mathbb{K}$ with $V(\mathrm{m}) \in L(L_2(\mu))$, we have $V \in L_\infty(\mu)$ and $\|V\|_{L_\infty(\mu)} \leqslant \|V(\mathrm{m})\|_{L(L_2(\mu))}$.*

Proof (i)⇒(ii): Let $\varepsilon > 0$ and $A_\varepsilon := [|V| \geqslant \|V(\mathrm{m})\|_{L(L_2(\mu))} + \varepsilon]$. Assume that $\mu(A_\varepsilon) > 0$. Since (Ω, Σ, μ) is semi-finite we find $B_\varepsilon \subseteq A_\varepsilon$ such that $0 < \mu(B_\varepsilon) < \infty$. Define $f := \mu(B_\varepsilon)^{-1/2}\mathbb{1}_{B_\varepsilon} \in L_2(\mu)$ with $\|f\|_{L_2(\mu)} = 1$. Consequently, we obtain

$$\|V(\mathrm{m})\|_{L(L_2(\mu))} \geqslant \|V(\mathrm{m})f\|_{L_2(\mu)} \geqslant \|V(\mathrm{m})\|_{L(L_2(\mu))} + \varepsilon,$$

which yields a contradiction, and hence (ii).
(ii)⇒(i): Assume that (Ω, Σ, μ) is not semi-finite. Then we find $A \in \Sigma$ with $\mu(A) = \infty$ such that for each $B \subseteq A$ measurable, we have $\mu(B) \in \{0, \infty\}$. Then $V := \mathbb{1}_A$ is bounded and measurable with $\|V\|_{L_\infty(\mu)} = 1$. However, $V(\mathrm{m}) = 0$. Indeed, if $f \in L_2(\mu)$ then $[f \neq 0] = \bigcup_{n \in \mathbb{N}} [|f|^2 \geqslant n^{-1}]$. Thus,

$$[V(\mathrm{m})f \neq 0] = [f \neq 0] \cap A = \bigcup_{n \in \mathbb{N}} [|f|^2 \geqslant n^{-1}] \cap A.$$

Since $\mu([|f|^2 \geqslant n^{-1}]) < \infty$ as $f \in L_2(\mu)$, we infer $\mu([|f|^2 \geqslant n^{-1}] \cap A) = 0$ by the property assumed for A. Thus, $\mu([V(\mathrm{m})f \neq 0]) = 0$ implying $V(\mathrm{m}) = 0$. Hence, $\|V(\mathrm{m})\|_{L(L_2(\mu))} = 0 < 1 = \|V\|_{L_\infty(\mu)}$. □

Remark 2.4.5 Any σ-finite measure space is semi-finite. Indeed, let (Ω, Σ, μ) be σ-finite and $A \in \Sigma$ with $\mu(A) = \infty$. We find a sequence $(G_n)_n$ of pairwise disjoint, measurable sets with finite measure satisfying $\bigcup_n G_n = \Omega$. Hence, $\mu(G_n \cap A) \leqslant \mu(G_n) < \infty$. If $\mu(G_n \cap A) = 0$ for all n, then $\mu(A) = 0$ by the σ-additivity of μ. Thus, as $\mu(A) \neq 0$, we find n such that $0 < \mu(G_n \cap A) < \infty$ and (Ω, Σ, μ) is semi-finite.

A straightforward consequence of Theorem 2.4.3 (c) and Proposition 2.4.4 is the following.

Proposition 2.4.6 *Let (Ω, Σ, μ) be a semi-finite measure space, $V: \Omega \to \mathbb{K}$ measurable and bounded. Then $\|V(\mathrm{m})\|_{L(L_2(\mu))} = \|V\|_{L_\infty(\mu)}$.*

Theorem 2.4.7 *Let (Ω, Σ, μ) be a semi-finite measure space and let $V: \Omega \to \mathbb{K}$ be measurable. Then*

$$\sigma(V(\mathrm{m})) = \text{ess-ran } V := \{\lambda \in \mathbb{K};\ \forall \varepsilon > 0\colon \mu([|\lambda - V| < \varepsilon]) > 0\}.$$

Proof Let $\lambda \in$ ess-ran V. For all $n \in \mathbb{N}$ we find $B_n \in \Sigma$ with non-zero, but finite measure such that $B_n \subseteq \left[|\lambda - V| < \frac{1}{n}\right]$. We define $f_n := \sqrt{\frac{1}{\mu(B_n)}}\mathbb{1}_{B_n} \in L_2(\mu)$. Then $\|f_n\|_{L_2(\mu)} = 1$ and

$$|V(\omega) f_n(\omega)| \leqslant |V(\omega) - \lambda|\,|f_n(\omega)| + |\lambda|\,|f_n(\omega)| \leqslant \left(\frac{1}{n} + |\lambda|\right)|f_n(\omega)|$$

for $\omega \in \Omega$, which shows that $(f_n)_n$ is in $\mathrm{dom}(V(\mathrm{m}))$. A similar estimate, on the other hand, shows that

$$\|(V(\mathrm{m}) - \lambda) f_n\|_{L_2(\mu)} \to 0 \quad (n \to \infty).$$

Thus, $(V(\mathrm{m}) - \lambda)^{-1}$ cannot be continuous as $\|f_n\|_{L_2(\mu)} = 1$ for all $n \in \mathbb{N}$.

Let now $\lambda \in \mathbb{K}\setminus$ess-ran V. Then there exists $\varepsilon > 0$ such that $N := [|\lambda - V| < \varepsilon]$ is a μ-nullset. In particular, $\lambda - V \neq 0$ μ-a.e. Hence, $(\lambda - V(\mathrm{m}))^{-1} = \frac{1}{\lambda - V}(\mathrm{m})$ is a linear operator. Since, $\left|\frac{1}{\lambda - V}\right| \leqslant 1/\varepsilon$ μ-almost everywhere, we deduce that $(\lambda - V(\mathrm{m}))^{-1} \in L(L_2(\mu))$ and hence, $\lambda \in \rho(V(\mathrm{m}))$. □

We conclude this chapter by sketching that multiplication operators as discussed in Theorem 2.4.3, Propositions 2.4.4, 2.4.6, and Theorem 2.4.7 are *the* prototypical example for normal operators. In fact it can be shown that normal operators are unitarily equivalent to multiplication operators on some $L_2(\mu)$. This fact is also known as the 'spectral theorem'. It is also important to note that, as we have seen in Theorem 2.4.3, a multiplication operator in $L_2(\mu)$ is self-adjoint if and only if V assumes values in the real numbers, only.

2.5 Comments

The material presented in this chapter is basic textbook knowledge. We shall thus refer to the monographs [54, 139]. Note that spectral theory for self-adjoint operators is a classical topic in functional analysis. For a glimpse on further theory of linear relations we exemplarily refer to [7, 14, 25]. The restriction in Proposition 2.4.6 and Theorem 2.4.7 to semi-finite measure spaces is not very severe. In fact, if (Ω, Σ, μ) was not semi-finite, it is possible to construct a semi-finite measure space $(\Omega_{\mathrm{loc}}, \Sigma_{\mathrm{loc}}, \mu_{\mathrm{loc}})$ such that $L_p(\mu)$ is isometrically isomorphic to $L_p(\mu_{\mathrm{loc}})$, see [129, Section 2].

Exercises

Exercise 2.1 Let $A \subseteq X_0 \times X_1$ be an unbounded linear operator. Show that for every linear operator $B \subseteq X_0 \times X_1$ with $B \supseteq A$ and $\mathrm{dom}(B) = X_0$, we have that B is not closed.

Exercise 2.2 Prove Proposition 2.1.4 and Corollary 2.1.5. Hint: One might use that bounded linear relations are always operators.

Exercise 2.3 Prove Lemma 2.2.2.

Exercise 2.4 Let $A\colon \mathrm{dom}(A) \subseteq H_0 \to H_0$ be a closed and densely defined linear operator. Show that for all $\lambda \in \mathbb{K}$ we have

$$\lambda \in \rho(A) \iff \lambda^* \in \rho(A^*).$$

Exercise 2.5 Let $U \subseteq H_0 \times H_1$ satisfy $U^{-1} = U^*$. Show that $U \in L(H_0, H_1)$ and that U is *unitary*, that is, U is onto and for all $x \in H_0$ we have $\|Ux\|_{H_1} = \|x\|_{H_0}$.

Exercise 2.6 Let $\delta\colon C[0,1] \subseteq L_2(0,1) \to \mathbb{K}$, $f \mapsto f(0)$, where $C[0,1]$ denotes the set of $\mathbb{K}$-valued continuous functions on $[0,1]$. Show that δ is not closable. Compute $\overline{\delta}$.

Exercise 2.7 Let $C \subseteq \mathbb{C}$ be closed. Provide a Hilbert space H and a densely defined closed linear operator A on H such that $\sigma(A) = C$.

References

7. R. Arens, Operational calculus of linear relations. Pac. J. Math. **11**, 9–23 (1961)
14. T. Berger, C. Trunk, H. Winkler, Linear relations and the Kronecker canonical form. Linear Algebra Appl. **488**, 13–44 (2016)
25. R. Cross, *Multivalued Linear Operators*, vol. 213. Monographs and Textbooks in Pure and Applied Mathematics (Marcel Dekker, Inc., New York, 1998)

54. T. Kato, *Perturbation Theory for Linear Operators*. Classics in Mathematics. Reprint of the 1980 edition (Springer, Berlin, 1995)
94. R. Picard, Mother operators and their descendants. J. Math. Anal. Appl. **403**(1), 54–62 (2013). With an extension by S. Trostorff, M. Waurick. arXiv:1203.6762
129. H. Vogt, J. Voigt, Bands in L_p-spaces. Math. Nachr. **290**(4), 632–638 (2017)
139. J. Weidmann, *Linear Operators in Hilbert Spaces*, vol. 68. Graduate Texts in Mathematics. Translated from the German by Joseph Szücs (Springer, New York, Berlin, 1980)

Chapter 3
The Time Derivative

It is the aim of this chapter to define a derivative operator on a suitable L_2-space, which will be used as the derivative with respect to the temporal variable in our applications. As we want to deal with Hilbert space-valued functions, we start by introducing the concept of Bochner–Lebesgue spaces, which generalises the classical scalar-valued L_p-spaces to the Banach space-valued case.

3.1 Bochner–Lebesgue Spaces

Throughout, let (Ω, Σ, μ) be a σ-finite measure space and X a Banach space over the field $\mathbb{K} \in \{\mathbb{R}, \mathbb{C}\}$. We are aiming to define the spaces $L_p(\mu; X)$ for $1 \leqslant p \leqslant \infty$. This is the space of (equivalence classes of) measurable functions attaining values in X, which are p-integrable (if $p < \infty$), or essentially bounded (if $p = \infty$) with respect to the measure μ. We begin by defining the space of simple functions on Ω with values in X and the notion of Bochner-measurability.

Definition For a function $f : \Omega \to X$ and $x \in X$ we set

$$A_{f,x} := f^{-1}[\{x\}].$$

A function $f : \Omega \to X$ is called *simple* if $f[\Omega]$ is finite and for each $x \in X \setminus \{0\}$ the set $A_{f,x}$ belongs to Σ and has finite measure. We denote the set of simple functions by $S(\mu; X)$. A function $f : \Omega \to X$ is called *Bochner-measurable* if there exists a sequence $(f_n)_{n\in\mathbb{N}}$ in $S(\mu; X)$ such that

$$f_n(\omega) \to f(\omega) \quad (n \to \infty)$$

for μ-a.e. $\omega \in \Omega$.

C. Seifert et al., *Evolutionary Equations*, Operator Theory: Advances and Applications 287, https://doi.org/10.1007/978-3-030-89397-2_3

Remark 3.1.1 Let us comment on the definition of Bochner-measurability.

(a) For a simple function f we have

$$f = \sum_{x \in X} x \cdot \mathbb{1}_{A_{f,x}},$$

where the sum is actually finite, since $\mathbb{1}_{A_{f,x}} = 0$ for all $x \notin f[\Omega]$.

(b) If $X = \mathbb{K}$, then a function is Bochner-measurable if and only if it has a μ-measurable representative. Indeed, if f is Bochner-measurable, we find a sequence $(f_n)_n$ in $S(\mu; \mathbb{K})$ such that $f_n \to f$ pointwise μ-a.e. Hence, we find a μ-nullset $N \in \Sigma$ such that $g_n := \mathbb{1}_{\Omega \setminus N} f_n \to \mathbb{1}_{\Omega \setminus N} f =: g$ pointwise on all of Ω. Since g_n is μ-measurable and μ-measurable functions are stable under pointwise limits, g is μ-measurable itself. Since $f = g$ except for a μ-nullset, f has a μ-measurable representative. If, on the other hand, f has a μ-measurable representative, let g be this representative. Approximating real and imaginary parts separately, it suffices to treat the case $\mathbb{K} = \mathbb{R}$. Then consider for $n \in \mathbb{N}$

$$s_n := \sum_{k \in \mathbb{Z}} \frac{k+1}{n} \mathbb{1}_{M_n^k},$$

where $M_n^k := g^{-1}[(\frac{k}{n}, \frac{k+1}{n}]]$. It is easy to see that $\sup_{\omega \in \Omega} |s_n(\omega) - g(\omega)| \leqslant 1/n$ for all $\omega \in \Omega$. Hence,

$$\widetilde{s}_n := \sum_{k \in \mathbb{Z}, |k| \leqslant 2^n} \frac{k+1}{n} \mathbb{1}_{M_n^k} \in S(\mu; \mathbb{R})$$

converges pointwise everywhere to g. In consequence, f is Bochner-measurable.

(c) It is easy to check that $S(\mu; X)$ is a vector space and an $S(\mu; \mathbb{K})$-module; that is, for $f \in S(\mu; X)$ and $g \in S(\mu; \mathbb{K})$ we have $g \cdot f \in S(\mu; X)$.

(d) If $f : \Omega \to X$ is Bochner-measurable, then $\|f(\cdot)\|_X : \Omega \to \mathbb{R}$ is Bochner-measurable. Indeed, since

$$\|f(\cdot)\|_X = \lim_{n \to \infty} \|f_n(\cdot)\|_X$$

μ-a.e. and a sequence $(f_n)_{n \in \mathbb{N}}$ in $S(\mu; X)$, it suffices to show that $\|f_n(\cdot)\|_X$ is simple for all $n \in \mathbb{N}$. The latter follows since $A_{f_n,x} \cap A_{f_n,y} = \varnothing$ for $x \neq y$ and thus

$$\|f_n(\cdot)\|_X = \sum_{x \in f_n[\Omega]} \|x\|_X \cdot \mathbb{1}_{A_{f_n,x}}$$

is a real-valued simple function.

(e) If one deals with arbitrary measure spaces, the definition of simple functions has to be weakened by allowing the sets $A_{f,x}$ to have infinite measure. However, since in the applications to follow we only work with weighted Lebesgue measures, we restrict ourselves to σ-finite measure spaces.

Definition (Bochner–Lebesgue Spaces) For $p \in [1, \infty]$ we define

$$\mathcal{L}_p(\mu; X) := \left\{ f : \Omega \to X \, ; \, f \text{ Bochner-measurable}, \; \|f(\cdot)\|_X \in \mathcal{L}_p(\mu) \right\},$$

as well as

$$L_p(\mu; X) := \mathcal{L}_p(\mu; X)/_{\sim},$$

where $\sim$ denotes the usual equivalence relation of equality μ-almost everywhere. We equip $L_p(\mu; X)$ with the norm

$$\|f\|_p := \begin{cases} \left(\int_\Omega \|f(\omega)\|_X^p \, \mathrm{d}\mu(\omega) \right)^{\frac{1}{p}}, & \text{if } p < \infty, \\ \operatorname{ess-sup}_{\omega \in \Omega} \|f(\omega)\|_X, & \text{if } p = \infty \end{cases} \qquad (f \in L_p(\mu; X)).$$

We first prove a density result.

Lemma 3.1.2 *The space $S(\mu; X)$ is dense in $L_p(\mu; X)$ for $p \in [1, \infty)$.*

Proof Let $f \in L_p(\mu; X)$. Then there exists a sequence $(f_n)_{n \in \mathbb{N}}$ in $S(\mu; X)$ such that $f_n(\omega) \to f(\omega)$ for all $\omega \in \Omega \setminus N$ for some nullset $N \subseteq \Omega$. W.l.o.g. we may assume that $\|f_n(\cdot)\|_X$ and $\|f(\cdot)\|_X$ are μ-measurable on $\Omega \setminus N$ for each $n \in \mathbb{N}$. For $n \in \mathbb{N}$ we define the set

$$I_n := \left\{ \omega \in \Omega \setminus N \, ; \, \|f_n(\omega)\|_X \leqslant 2 \, \|f(\omega)\|_X \right\} \in \Sigma,$$

and set $\widetilde{f_n} := f_n \mathbb{1}_{I_n}$. Then $\widetilde{f_n} \in S(\mu; X)$ and we claim that $\widetilde{f_n}(\omega) \to f(\omega)$ for all $\omega \in \Omega \setminus N$. Indeed, if $f(\omega) = 0$ then $\widetilde{f_n}(\omega) = 0$ and the claim follows. If $f(\omega) \neq 0$, then there is some $n_0 \in \mathbb{N}$ such that $\|f_n(\omega)\|_X \leqslant 2 \, \|f(\omega)\|_X$ for $n \geq n_0$, and hence $\omega \in \bigcap_{n \geqslant n_0} I_n$. Consequently $\widetilde{f_n}(\omega) = f_n(\omega) \to f(\omega)$. By dominated convergence, it now follows that

$$\int_\Omega \left\| \widetilde{f_n}(\omega) - f(\omega) \right\|_X^p \mathrm{d}\mu(\omega) \to 0 \quad (n \to \infty),$$

which proves the claim. □

As a consequence of the latter lemma, we can show that Bochner-measurability is preserved by pointwise convergence almost everywhere.

Proposition 3.1.3 *Let $f_n, f : \Omega \to X$ for $n \in \mathbb{N}$. Moreover, assume that f_n is Bochner-measurable for each $n \in \mathbb{N}$ and $f_n(\omega) \to f(\omega)$ as $n \to \infty$ for μ-almost every $\omega \in \Omega$. Then f is Bochner-measurable.*

Proof Since $f_n \to f$ almost everywhere, we have $[f \neq 0]\backslash N \subseteq \bigcup_{n\in\mathbb{N}}[f_n \neq 0]\backslash N$ for some nullset $N \subseteq \Omega$. Moreover, since f_n is Bochner-measurable, the definition of simple functions yields that $\bigcup_{n\in\mathbb{N}}[f_n \neq 0] \subseteq \bigcup_{n\in\mathbb{N}} B_n$, where, for all $n \in \mathbb{N}$, B_n is measurable with $\mu(B_n) < \infty$. The latter implies that there exists a sequence of measurable sets $(A_n)_{n\in\mathbb{N}}$ such that $A_n \subseteq A_{n+1}$, $\mu(A_n) < \infty$ for all $n \in \mathbb{N}$ and

$$[f \neq 0] \setminus N \subseteq \bigcup_{n\in\mathbb{N}} A_n.$$

For $n \in \mathbb{N}$ we set $g_n := \mathbb{1}_{A_n \cap [\widetilde{f_n} \leqslant n]} f_n$, where $\widetilde{f_n} : \Omega \to \mathbb{R}$ is measurable and equals $\|f_n(\cdot)\|_X$ μ-almost everywhere (cp. Remark 3.1.1(d) and (b)). In this way we obtain a sequence of Bochner-measurable functions with $g_n \to f$ μ-almost everywhere. Moreover, $g_n \in L_1(\mu; X)$ for each $n \in \mathbb{N}$ and thus, for each $n \in \mathbb{N}$ we find a simple function h_n with $\|g_n - h_n\|_1 \leqslant 2^{-n}$ by Lemma 3.1.2. Then

$$\int_\Omega \sum_{n\in\mathbb{N}} \|g_n(\omega) - h_n(\omega)\|_X \, \mathrm{d}\mu(\omega) < \infty$$

and hence, $\sum_{n\in\mathbb{N}} \|g_n(\omega) - h_n(\omega)\|_X < \infty$ for μ-almost every $\omega \in \Omega$, which particularily implies $g_n - h_n \to 0$ μ-almost everywhere. Hence, $h_n \to f$ μ-almost everywhere, which shows the Bochner-measurability of f. □

We can now prove that the spaces $L_p(\mu; X)$ are actually Banach spaces.

Proposition 3.1.4 *Let $p \in [1, \infty]$. Then $(L_p(\mu; X), \|\cdot\|_p)$ is a Banach space and if $X = H$ is a Hilbert space, then so too is $L_2(\mu; H)$ with the scalar product given by*

$$\langle f, g\rangle_2 := \int_\Omega \langle f(\omega), g(\omega)\rangle_H \, \mathrm{d}\mu(\omega) \quad (f, g \in L_2(\mu; H)).$$

Proof We just show the completeness of $L_p(\mu; X)$. Let $(f_n)_{n\in\mathbb{N}}$ be a sequence in $L_p(\mu; X)$ such that $\sum_{n=1}^{\infty} \|f_n\|_p < \infty$. We set

$$g_n(\omega) := \|f_n(\omega)\|_X \quad (n \in \mathbb{N}, \omega \in \Omega).$$

Then $(g_n)_{n\in\mathbb{N}}$ is a sequence in $L_p(\mu)$ such that $\sum_{n=1}^{\infty} \|g_n\|_p < \infty$. By the completeness of $L_p(\mu)$ we infer that

$$g := \sum_{n=1}^{\infty} g_n$$

exists and is an element in $L_p(\mu)$. In particular, $g(\omega) < \infty$ for μ-a.e. $\omega \in \Omega$ and thus,

$$\sum_{n=1}^{\infty} \|f_n(\omega)\|_X = \sum_{n=1}^{\infty} g_n(\omega) < \infty$$

for μ-a.e. $\omega \in \Omega$. By the completeness of X we can define

$$f(\omega) := \sum_{n=1}^{\infty} f_n(\omega)$$

for μ-a.e. $\omega \in \Omega$. Note that f is Bochner-measurable by Proposition 3.1.3. We need to prove that $f \in L_p(\mu; X)$ and that $\sum_{n=1}^{k} f_n \to f$ in $L_p(\mu; X)$ as $k \to \infty$. For this, it suffices to prove that

$$\sum_{n=k}^{\infty} f_n \in L_p(\mu; X) \text{ and } \sum_{n=k}^{\infty} f_n \to 0 \text{ in } L_p(\mu; X) \text{ as } k \to \infty. \tag{3.1}$$

Indeed, this would imply both $f - \sum_{n=1}^{k} f_n \in L_p(\mu; X)$ and the desired convergence result. We prove (3.1) for $p < \infty$ and $p = \infty$ separately.

First, let $p = \infty$. For each $n \in \mathbb{N}$ we have $f_n \in L_\infty(\mu; X)$ and thus $\|f_n(\omega)\|_X \leqslant \|f_n\|_\infty$ for all $\omega \in \Omega \setminus N_n$ and some nullset $N_n \subseteq \Omega$. We set $N := \bigcup_{n=1}^{\infty} N_n$, which is again a nullset. For $k \in \mathbb{N}$ and $\omega \in \Omega \setminus N$ we then estimate

$$\left\| \sum_{n=k}^{\infty} f_n(\omega) \right\|_X \leqslant \sum_{n=k}^{\infty} \|f_n(\omega)\|_X \leqslant \sum_{n=k}^{\infty} \|f_n\|_\infty ,$$

which yields (3.1).

Now, let $p < \infty$. For $k \in \mathbb{N}$ we estimate

$$\begin{aligned} \left(\int_\Omega \left(\left\| \sum_{n=k}^{\infty} f_n(\omega) \right\|_X \right)^p \mathrm{d}\mu(\omega) \right)^{\frac{1}{p}} &\leqslant \left(\int_\Omega \left(\sum_{n=k}^{\infty} \|f_n(\omega)\|_X \right)^p \mathrm{d}\mu(\omega) \right)^{\frac{1}{p}} \\ &= \left(\int_\Omega \lim_{m\to\infty} \left(\sum_{n=k}^{m} \|f_n(\omega)\|_X \right)^p \mathrm{d}\mu(\omega) \right)^{\frac{1}{p}} \end{aligned}$$

$$= \lim_{m\to\infty} \left(\int_\Omega \left(\sum_{n=k}^{m} \| f_n(\omega)\|_X \right)^p \, \mathrm{d}\mu(\omega) \right)^{\frac{1}{p}}$$

$$\leqslant \lim_{m\to\infty} \sum_{n=k}^{m} \| f_n\|_p = \sum_{n=k}^{\infty} \| f_n\|_p ,$$

where we have used monotone convergence in the third line. This estimate yields (3.1). □

We now want to define an X-valued integral for functions in $L_1(\mu; X)$; the so-called Bochner-integral.

Proposition 3.1.5 *The mapping*[1]

$$\int_\Omega \mathrm{d}\mu \colon S(\mu; X) \subseteq L_1(\mu; X) \to X$$

$$f \mapsto \sum_{x\in X} x \cdot \mu(A_{f,x})$$

is linear and continuous, and thus has a unique continuous linear extension to $L_1(\mu; X)$*, called the* Bochner-integral. *Moreover,*

$$\left\| \int_\Omega f \, \mathrm{d}\mu \right\|_X \leqslant \|f\|_1 \quad (f \in L_1(\mu; X)),$$

and for $A \in \Sigma$, $f \in L_1(\mu; X)$ *we set*

$$\int_A f \, \mathrm{d}\mu := \int_\Omega f \cdot \mathbb{1}_A \, \mathrm{d}\mu.$$

Proof We first show linearity. Let $f, g \in S(\mu; X)$ and $\lambda \in \mathbb{K}$. Then, for $x \in X$ we have

$$A_{\lambda f+g,x} = (\lambda f+g)^{-1}[\{x\}] = \bigcup_{y\in X} \left(f^{-1}[\{y\}]\cap g^{-1}[\{x-\lambda y\}]\right) = \bigcup_{y\in X} A_{f,y}\cap A_{g,x-\lambda y},$$

[1] Note that the sum is indeed finite and all summands are well-defined if we set $0_X \cdot \infty := 0_X$.

and therefore $\mu(A_{\lambda f+g,x}) = \sum_{y\in X} \mu(A_{f,y} \cap A_{g,x-\lambda y})$. Thus, we compute

$$
\begin{aligned}
\int_\Omega (\lambda f + g)\,\mathrm{d}\mu &= \sum_{x\in X} x\cdot \mu(A_{\lambda f+g,x}) = \sum_{x\in X}\sum_{y\in X} x\cdot\mu(A_{f,y}\cap A_{g,x-\lambda y}) \\
&= \sum_{y\in X}\sum_{x\in X} \lambda y\cdot \mu(A_{f,y}\cap A_{g,x-\lambda y}) \\
&\quad + \sum_{y\in X}\sum_{x\in X} (x-\lambda y)\cdot \mu(A_{f,y}\cap A_{g,x-\lambda y}) \\
&= \sum_{y\in X}\sum_{x\in X} \lambda y\cdot \mu(A_{f,y}\cap A_{g,x-\lambda y}) + \sum_{y\in X}\sum_{z\in X} z\cdot\mu(A_{f,y}\cap A_{g,z}),
\end{aligned}
$$

where we interchanged the finite sums. Now,

$$
\sum_{x\in X}\mu(A_{f,y}\cap A_{g,x-\lambda y}) = \mu\Big(A_{f,y}\cap \bigcup_{x\in X} A_{g,x-\lambda y}\Big) = \mu(A_{f,y})
$$

as well as

$$
\sum_{y\in X}\mu(A_{f,y}\cap A_{g,z}) = \mu\Big(\bigcup_{y\in X} A_{f,y}\cap A_{g,z}\Big) = \mu(A_{g,z}),
$$

and therefore we conclude

$$
\int_\Omega(\lambda f+g)\,\mathrm{d}\mu = \lambda\sum_{y\in X} y\cdot\mu(A_{f,y}) + \sum_{z\in X} z\cdot \mu(A_{g,z}) = \lambda\int_\Omega f\,\mathrm{d}\mu + \int_\Omega g\,\mathrm{d}\mu.
$$

In order to prove continuity, let $f \in S(\mu; X)$. We estimate

$$
\begin{aligned}
\left\|\int_\Omega f\,\mathrm{d}\mu\right\|_X &= \left\|\sum_{x\in f[\Omega]} x\cdot\mu(A_{f,x})\right\|_X \leqslant \sum_{x\in f[\Omega]} \|x\|_X\,\mu(A_{f,x}) \\
&= \int_\Omega \sum_{x\in f[\Omega]} \|x\|_X\,\mathbb{1}_{A_{f,x}}\,\mathrm{d}\mu \\
&= \int_\Omega \|f(\cdot)\|_X\,\mathrm{d}\mu = \|f\|_1\,.
\end{aligned}
$$

The remaining assertions now follow from Lemma 3.1.2 by continuous extension (see Corollary 2.1.5). □

The next proposition tells us how the Bochner-integral of a function behaves if we compose the function with a bounded or closed linear operator first. In what follows, let $X' := L(X, \mathbb{K})$ denote the dual space of X.

Proposition 3.1.6 *Let $f \in L_1(\mu; X)$, Y a Banach space.*

(a) *Let $B \in L(X, Y)$. Then $B \circ f \in L_1(\mu; Y)$ and*

$$\int_\Omega B \circ f \,\mathrm{d}\mu = B \int_\Omega f \,\mathrm{d}\mu.$$

(b) *If $X_0 \subseteq X$ is a closed subspace and $f(\omega) \in X_0$ for μ-a.e. $\omega \in \Omega$, then $\int_\Omega f \,\mathrm{d}\mu \in X_0$.*

(c) *(Theorem of Hille) Let $A\colon \operatorname{dom}(A) \subseteq X \to Y$ be a closed linear operator and assume that $f(\omega) \in \operatorname{dom}(A)$ for μ-a.e. $\omega \in \Omega$ and that $A \circ f \in L_1(\mu; Y)$. Then $\int_\Omega f \,\mathrm{d}\mu \in \operatorname{dom}(A)$ and*

$$A \int_\Omega f \,\mathrm{d}\mu = \int_\Omega A \circ f \,\mathrm{d}\mu.$$

Proof

(a) At first we observe that, if $f \in S(\mu; X)$, then

$$B \circ f = B \circ \sum_{x \in X \setminus \{0\}} x \cdot \mathbb{1}_{A_{f,x}} = \sum_{x \in X \setminus \{0\}} Bx \cdot \mathbb{1}_{A_{f,x}}.$$

Thus, $B \circ f \in S(\mu; Y)$ since $Bx \cdot \mathbb{1}_{A_{f,x}} \in S(\mu; Y)$, the sum is finite and $S(\mu; Y)$ is a vector space. Let now be $f \in L_1(\mu; X)$. Then there is $(f_n)_{n\in\mathbb{N}}$ a sequence in $S(\mu; X)$ such that $f_n \to f$ μ-a.e. Then $B \circ f_n \in S(\mu; Y)$ (see above) and due to the continuity of B we have that $B \circ f_n \to B \circ f$ μ-a.e., hence $B \circ f$ is Bochner-measurable. Moreover, $\|B \circ f(\cdot)\|_Y \leqslant \|B\| \, \|f(\cdot)\|_X$, which yields that $B \circ f \in L_1(\mu; Y)$. By continuity of both B and $\int_\Omega \mathrm{d}\mu$, it suffices to check the interchanging property for any $f \in S(\mu; X)$ alone. However, this is clear, since for a simple function f

$$B \circ f = B\left(\sum_{x \in X} x \cdot \mathbb{1}_{A_{f,x}}\right) = \sum_{x \in X} Bx \cdot \mathbb{1}_{A_{f,x}},$$

where the sum is actually finite and hence,

$$\begin{aligned}\int_\Omega B \circ f \,\mathrm{d}\mu &= \int_\Omega \sum_{x \in X} Bx \cdot \mathbb{1}_{A_{f,x}} \,\mathrm{d}\mu = \sum_{x \in X} \int_\Omega Bx \cdot \mathbb{1}_{A_{f,x}} \,\mathrm{d}\mu \\ &= \sum_{x \in X} Bx \cdot \mu(A_{f,x}) = B\left(\sum_{x \in X} x \cdot \mu(A_{f,x})\right) = B \int_\Omega f \,\mathrm{d}\mu,\end{aligned}$$

where in the third equality we have used that $Bx \cdot \mathbb{1}_{A_{f,x}}$ is a simple function.

(b) Let $x' \in X'$ with $x'|_{X_0} = 0$. It follows from (a) that

$$x'\left(\int_\Omega f \,\mathrm{d}\mu\right) = \int_\Omega x' \circ f \,\mathrm{d}\mu = 0,$$

and since x' was arbitrary, it follows that $\int_\Omega f \,\mathrm{d}\mu \in X_0$ from the Theorem of Hahn–Banach.

(c) Consider the space $L_1(\mu; X \times Y)$. By assumption, it follows that

$$(f, A \circ f) \in L_1(\mu; X \times Y).$$

However, $(f, A \circ f)(\omega) = (f(\omega), (A \circ f)(\omega)) \in A \subseteq X \times Y$ for μ-a.e. $\omega \in \Omega$, and since A is closed we can use (b) to derive that

$$\int_\Omega (f, A \circ f) \,\mathrm{d}\mu \in A. \tag{3.2}$$

Let π_1, π_2 be the projection from $X \times Y$ to X and Y, respectively. It then follows from part (a) that

$$\pi_1\left(\int_\Omega (f, A \circ f) \,\mathrm{d}\mu\right) = \int_\Omega \pi_1(f, A \circ f) \,\mathrm{d}\mu = \int_\Omega f \,\mathrm{d}\mu,$$

and analogously for π_2. Using these equalities we derive from (3.2) that $\int_\Omega f \,\mathrm{d}\mu \in \operatorname{dom}(A)$ and that $A \int_\Omega f \,\mathrm{d}\mu = \int_\Omega A \circ f \,\mathrm{d}\mu$. □

As a consequence of the latter proposition, we derive the fundamental theorem of calculus for Banach space-valued functions.

Corollary 3.1.7 (Fundamental Theorem of Calculus) *Let $a, b \in \mathbb{R}, a < b$ and consider the measure space $([a, b], \mathcal{B}([a, b]), \lambda)$, where $\mathcal{B}([a, b])$ denotes the Borel-σ-algebra of $[a, b]$ and λ is the Lebesgue measure. Let $f\colon [a, b] \to X$ be continuously differentiable.*[2] *Then*

$$f(b) - f(a) = \int_{[a,b]} f' \,\mathrm{d}\lambda.$$

Proof Note first of all that continuous functions are Bochner-measurable (which can be easily seen using Theorem 3.1.10 below). Thus, the integral on the right-hand side is well-defined. Let $\varphi \in X'$. Then $\varphi \circ f\colon [a, b] \to \mathbb{K}$ is continuously differentiable, and $(\varphi \circ f)'(t) = (\varphi \circ f')(t)$. Using Proposition 3.1.6 (a) together

[2] By this we mean that f is continuous on $[a, b]$, continuously differentiable on (a, b) and f' has a continuous extension to $[a, b]$.

with the fundamental theorem of calculus for the scalar-valued case we get

$$\varphi\left(\int_{[a,b]} f'\,\mathrm{d}\lambda\right) = \int_{[a,b]} (\varphi \circ f')\,\mathrm{d}\lambda = \varphi(f(b)) - \varphi(f(a)) = \varphi(f(b) - f(a)).$$

Since this holds for all $\varphi \in X'$, the assertion follows from the Theorem of Hahn–Banach. □

Next we state a density result, which will be useful throughout the course.

Lemma 3.1.8 *Let* $1 \leqslant p < \infty$, $\mathcal{D} \subseteq L_p(\mu)$ *be total in* $L_p(\mu)$ *and* X *a Banach space. Then the set* $\{\varphi(\cdot)x\,;\ x \in X,\ \varphi \in \mathcal{D}\}$ *is total in* $L_p(\mu; X)$.

Proof By Lemma 3.1.2, we know that $S(\mu; X)$ is dense in $L_p(\mu; X)$. Thus, it suffices to approximate $\mathbb{1}_A x$ for some $A \in \Sigma$ with $\mu(A) < \infty$ and $x \in X$. For this, however, take a sequence $(\phi_n)_n$ in the linear hull of $\mathcal{D}$ with $\phi_n \to \mathbb{1}_A$ in $L_p(\mu)$ as $n \to \infty$. Then

$$\|\mathbb{1}_A x - \phi_n x\|_{L_p(\mu;X)} = \|x\|_X\,\|\mathbb{1}_A - \phi_n\|_{L_p(\mu)} \to 0 \quad (n \to \infty).$$

Thus, the claim follows. □

The following application of Lemma 3.1.8 also deals with a dense subset of X.

Lemma 3.1.9 *Let* $1 \leqslant p < \infty$, $\mathcal{D} \subseteq L_p(\mu)$ *be total in* $L_p(\mu)$, X *a Banach space,* $D_0 \subseteq X$ *total in* X. *Then* $\{\varphi(\cdot)x\,;\ x \in D_0, \varphi \in \mathcal{D}\}$ *is total in* $L_p(\mu; X)$.

Proof The proof follows upon realising that the set $\{\varphi(\cdot)x\,;\ x \in D_0,\ \varphi \in \mathcal{D}\}$ is total in the set $\{\varphi(\cdot)x\,;\ x \in X,\ \varphi \in \mathcal{D}\}$. From here we just apply Lemma 3.1.8. □

We conclude this section by stating and proving the celebrated Theorem of Pettis, which characterises Bochner-measurability in terms of weak measurability.

Theorem 3.1.10 (Theorem of Pettis) *Let* $f : \Omega \to X$. *Then* f *is Bochner-measurable if and only if*

(a) f *is* weakly Bochner-measurable*; that is,* $x' \circ f : \Omega \to \mathbb{K}$ *is Bochner-measurable for each* $x' \in X'$, *and*
(b) f *is* almost separably-valued*; that is,* $\overline{\operatorname{lin} f[\Omega \setminus N_0]}$ *is separable for some* $N_0 \in \Sigma$ *with* $\mu(N_0) = 0$.

Proof If f is Bochner-measurable, then clearly it is weakly Bochner-measurable. Further, as f is the almost everywhere limit of simple functions, it is almost separably-valued, since each simple function attains values in a finite-dimensional subspace of X.

Assume now conversely that f satisfies (a) and (b). We define $Y := \overline{\operatorname{lin} f[\Omega \setminus N_0]}$, which is a separable Banach space by (b). Thus, there exists a sequence $(x_n')_{n\in\mathbb{N}}$ in X' such that

$$\|y\| = \sup_{n\in\mathbb{N}} |x_n'(y)| \quad (y \in Y).$$

Since for each $n \in \mathbb{N}$ the function $g_n := |x_n' \circ f|$ is Bochner-measurable by (a) and Remark 3.1.1(d), we find a μ-nullset N_n and a measurable function $\widetilde{g}_n : \Omega \to \mathbb{R}$ such that $g_n = \widetilde{g}_n$ on $\Omega \setminus N_n$ by Remark 3.1.1(b). Then $\sup_{n\in\mathbb{N}} \widetilde{g}_n(\cdot)$ is measurable and

$$\|f(\omega)\| = \sup_{n\in\mathbb{N}} \widetilde{g}_n(\omega) \quad (\omega \in \Omega \setminus N),$$

where $N := \bigcup_{n\in\mathbb{N}_0} N_n$, which shows that $\|f(\cdot)\|$ is Bochner-measurable. Let $\varepsilon > 0$, $(y_n)_{n\in\mathbb{N}}$ a dense sequence in Y. Applying the previous argument to the function $f_k(\cdot) := f(\cdot) - y_k$ for $k \in \mathbb{N}$ we infer that $\|f_k(\cdot)\|$ is Bochner-measurable and hence, there is a μ-nullset N_k' and a measurable funtion $\widetilde{f}_k : \Omega \to \mathbb{R}$ such that $\|f_k\| = \widetilde{f}_k$ on $\Omega \setminus N_k'$. Consequently, the sets

$$E_k := [\widetilde{f}_k \leqslant \varepsilon] = \{\omega \in \Omega\,;\ \widetilde{f}_k(\omega) \leqslant \varepsilon\} \quad (k \in \mathbb{N})$$

are measurable. Moreover, by the density of $\{y_n\,;\ n \in \mathbb{N}\}$ in Y, we get that $\Omega \setminus N' \subseteq \bigcup_{k\in\mathbb{N}} E_k$ with $N' := \bigcup_{k=1}^{\infty} N_k' \cup N_0$. Setting $F_1 := E_1$ and $F_{n+1} = E_{n+1} \setminus \bigcup_{k=1}^{n} F_k$ for $n \in \mathbb{N}$, we obtain a sequence of pairwise disjoint measurable sets $(F_n)_{n\in\mathbb{N}}$ with $\Omega \setminus N' \subseteq \bigcup_{n\in\mathbb{N}} F_n$. We set

$$g := \sum_{k=1}^{\infty} y_k \mathbb{1}_{F_k}$$

and obtain $\|f(\omega) - g(\omega)\| \leqslant \varepsilon$ for each $\omega \in \Omega \setminus N'$. Hence, if g is Bochner-measurable, then f is Bochner-measurable as well. Indeed, we find a sequence of such functions converging to f μ-almost everywhere and so Proposition 3.1.3 applies. For showing the Bochner-measurability of g, let $(\Omega_k)_{k\in\mathbb{N}}$ be a sequence of pairwise disjoint measurable sets such that $\bigcup_{k\in\mathbb{N}} \Omega_k = \Omega$ and $\mu(\Omega_k) < \infty$ for each $k \in \mathbb{N}$. For $n \in \mathbb{N}$ we set

$$g_n := \sum_{k,j=1}^{n} y_k \mathbb{1}_{F_k \cap \Omega_j}.$$

Then $(g_n)_{n\in\mathbb{N}}$ is a sequence of simple functions with $g_n \to g$ pointwise as $n \to \infty$ and thus, g is Bochner-measurable. □

3.2 The Time Derivative as a Normal Operator

Now let H be a Hilbert space over $\mathbb{K} \in \{\mathbb{R}, \mathbb{C}\}$. For $\nu \in \mathbb{R}$ and $p \in [1, \infty)$ we define the measure

$$\mu_{p,\nu}(A) := \int_A \mathrm{e}^{-p\nu t}\,\mathrm{d}\lambda(t)$$

for A in the Borel-σ-algebra, $\mathcal{B}(\mathbb{R})$, of $\mathbb{R}$. As our underlying Hilbert space for the time derivative we set

$$L_{2,\nu}(\mathbb{R}; H) := L_2(\mu_{2,\nu}; H).$$

In the same way we define

$$L_{p,\nu}(\mathbb{R}; H) := L_p(\mu_{p,\nu}; H)$$

for $p \in [1, \infty)$. If $H = \mathbb{K}$ we abbreviate $L_{p,\nu}(\mathbb{R}) := L_{p,\nu}(\mathbb{R}; \mathbb{K})$. Thus, $f \in L_{p,\nu}(\mathbb{R}; H)$ if and only if f is Bochner measurable and

$$\int_{\mathbb{R}} \|f(t)\|_H^p\,\mathrm{d}\mu_{p,\nu}(t) = \int_{\mathbb{R}} \|f(t)\|_H^p\,\mathrm{e}^{-p\nu t}\,\mathrm{d}t < \infty.$$

Our aim is to define the time derivative on $L_{2,\nu}(\mathbb{R}; H)$. For this, we define a suitable anti-derivative as an operator, which for $\nu \neq 0$ turns out to be one-to-one and bounded. Then we introduce the time derivative as the inverse of this anti-derivative. The reason for doing it that way is to easily get a formula for the adjoint for the time derivative using the boundedness of the anti-derivative.

We start our considerations with the definition of convolution operators in $L_{2,\nu}(\mathbb{R}; H)$.

Lemma 3.2.1 *Let $k \in L_{1,\nu}(\mathbb{R})$. We define the convolution operator*

$$k*\colon L_{2,\nu}(\mathbb{R}; H) \to L_{2,\nu}(\mathbb{R}; H)$$

by

$$(k * f)(t) := \int_{\mathbb{R}} k(s) f(t-s)\,\mathrm{d}s,$$

which exists for a.e. $t \in \mathbb{R}$. Then, $k*$ is linear and bounded with $\|k*\| \leqslant \|k\|_{L_{1,\nu}(\mathbb{R})}$.

Proof Let $f \in L_{2,\nu}(\mathbb{R}; H)$. We first prove that $s \mapsto k(s)f(t-s) \in L_1(\mathbb{R}; H)$ for a.e. $t \in \mathbb{R}$. The Bochner-measurability is clear since k and f are both Bochner-measurable. Moreover,

$$\begin{aligned}
&\int_{\mathbb{R}} \left(\int_{\mathbb{R}} \|k(s)f(t-s)\|_H \,\mathrm{d}s\right)^2 \mathrm{e}^{-2\nu t}\,\mathrm{d}t \\
&= \int_{\mathbb{R}} \left(\int_{\mathbb{R}} |k(s)|^{\frac{1}{2}}\,\mathrm{e}^{-\frac{\nu}{2}s}\,|k(s)|^{\frac{1}{2}}\,\mathrm{e}^{-\frac{\nu}{2}s}\,\|f(t-s)\|_H\,\mathrm{e}^{-\nu(t-s)}\,\mathrm{d}s\right)^2 \mathrm{d}t \\
&\leqslant \int_{\mathbb{R}} \left(\int_{\mathbb{R}} |k(s)|\,\mathrm{e}^{-\nu s}\,\mathrm{d}s\right)\left(\int_{\mathbb{R}} |k(s)|\,\mathrm{e}^{-\nu s}\,\|f(t-s)\|_H^2\,\mathrm{e}^{-2\nu(t-s)}\,\mathrm{d}s\right) \mathrm{d}t \\
&= \|k\|_{L_{1,\nu}(\mathbb{R})} \int_{\mathbb{R}} |k(s)| \int_{\mathbb{R}} \|f(t-s)\|^2\,\mathrm{e}^{-2\nu(t-s)}\,\mathrm{d}t\,\mathrm{e}^{-\nu s}\,\mathrm{d}s \\
&= \|k\|_{L_{1,\nu}(\mathbb{R})}^2\,\|f\|_{L_{2,\nu}(\mathbb{R};H)}^2,
\end{aligned}$$

which on the one hand proves that

$$\int_{\mathbb{R}} \|k(s)f(t-s)\|_H\,\mathrm{d}s < \infty$$

for a.e. $t \in \mathbb{R}$ and on the other hand shows the norm estimate, once we have shown the Bochner-measurability of $k*f$. For proving the latter, we apply Theorem 3.1.10. Since f is Bochner-measurable, we find a nullset N such that $H_0 := \overline{\operatorname{lin} f[\mathbb{R} \setminus N]}$ is separable. Hence, for almost every $t \in \mathbb{R}$ we have

$$(k*f)(t) = \int_{\mathbb{R}} k(s)f(t-s)\,\mathrm{d}s = \int_{\mathbb{R}\setminus N} k(t-s)f(s)\,\mathrm{d}s \in H_0$$

by Proposition 3.1.6(b). Thus, $k*f$ is almost separably-valued. Moreover, for $x' \in H'$ we have by Proposition 3.1.6(a)

$$x' \circ (k*f) = k*(x' \circ f)$$

almost everywhere and thus, the weak Bochner-measurability follows from the fact that the convolution of two measurable scalar-valued functions is measurable. Since the linearity of $k*$ is clear the proof is done. □

Definition For $\nu \neq 0$ we define the operator

$$I_\nu \colon L_{2,\nu}(\mathbb{R}; H) \to L_{2,\nu}(\mathbb{R}; H)$$

by

$$I_\nu := \begin{cases} \mathbb{1}_{[0,\infty)}*, & \text{if } \nu > 0, \\ -\mathbb{1}_{(-\infty,0]}*, & \text{if } \nu < 0. \end{cases}$$

Note that, by Lemma 3.2.1, I_ν is bounded with $\|I_\nu\| \leqslant \frac{1}{|\nu|}$.

Remark 3.2.2 For $\nu > 0$, $f \in L_{2,\nu}(\mathbb{R}; H)$ we have

$$I_\nu f(t) = \mathbb{1}_{[0,\infty)} * f(t) = \int_0^\infty f(t-s)\,\mathrm{d}s = \int_{-\infty}^t f(s)\,\mathrm{d}s \quad (\text{a.e. } t \in \mathbb{R}).$$

Analogously, for $\nu < 0$, $f \in L_{2,\nu}(\mathbb{R}; H)$ we have

$$I_\nu f(t) = -\int_t^\infty f(s)\,\mathrm{d}s \quad (\text{a.e. } t \in \mathbb{R}).$$

Proposition 3.2.3 *Let $\nu \neq 0$. Then I_ν is one-to-one and $C_\mathrm{c}^1(\mathbb{R}; H)$, the space of continuously differentiable, compactly supported functions on $\mathbb{R}$ with values in H, is in the range of I_ν.*

Proof We just prove the assertion for the case when $\nu > 0$. Let $f \in L_{2,\nu}(\mathbb{R}; H)$ satisfy $I_\nu f = 0$. In particular, we obtain for all $t \in \mathbb{R} \setminus N$ that $0 = I_\nu f(t) = \int_{-\infty}^t f(s)\,\mathrm{d}s$ for some Lebesgue nullset, $N \subseteq \mathbb{R}$. Then for $a, b \in \mathbb{R} \setminus N$ with $a < b$ and $x \in H$ we have that

$$\begin{aligned} \left\langle f, \mathrm{e}^{2\nu(\cdot)}\mathbb{1}_{[a,b]} \cdot x \right\rangle_{L_{2,\nu}(\mathbb{R}; H)} &= \int_\mathbb{R} \left\langle f(t), \mathrm{e}^{2\nu t}\mathbb{1}_{[a,b]}(t) \cdot x \right\rangle_H \mathrm{e}^{-2\nu t}\,\mathrm{d}t \\ &= \left\langle \int_a^b f(t)\,\mathrm{d}t, x \right\rangle_H \\ &= \langle (I_\nu f)(b) - (I_\nu f)(a), x \rangle_H = 0. \end{aligned}$$

Thus $f = 0$. Indeed, since $\mathbb{R} \setminus N$ is dense in $\mathbb{R}$, $\left\{\mathrm{e}^{2\nu(\cdot)}\mathbb{1}_{[a,b]}\,;\, a, b \in \mathbb{R} \setminus N\right\}$ is total in $L_{2,\nu}(\mathbb{R})$. Hence, $\left\{\mathrm{e}^{2\nu(\cdot)}\mathbb{1}_{[a,b]} \cdot x\,;\, a, b \in \mathbb{R} \setminus N,\ x \in H\right\}$ is total in $L_{2,\nu}(\mathbb{R}; H)$ by Lemma 3.1.8. This proves the injectivity of I_ν. Moreover, if $\varphi \in C_\mathrm{c}^1(\mathbb{R}; H)$ then by Corollary 3.1.7 we have

$$\varphi(t) = \int_{-\infty}^t \varphi'(s)\,\mathrm{d}s = \left(I_\nu \varphi'\right)(t) \quad (\text{a.e. } t \in \mathbb{R}). \qquad \square$$

Definition For $\nu \neq 0$ we define the *time derivative*, $\partial_{t,\nu}$, on $L_{2,\nu}(\mathbb{R}; H)$ by

$$\partial_{t,\nu} := I_\nu^{-1}.$$

Note that by Lemma 3.2.1 and Proposition 3.2.3, $\partial_{t,\nu}$ is a closed linear operator for which $C_c^1(\mathbb{R}; H) \subseteq \operatorname{dom}(\partial_{t,\nu})$. Since

$$C_c^1(\mathbb{R}; H) \supseteq \operatorname{lin}\left\{\varphi \cdot x\,;\ \varphi \in C_c^1(\mathbb{R}),\ x \in H\right\}$$

we infer that $\partial_{t,\nu}$ is densely defined by Lemma 3.1.8 and Exercise 3.2. Moreover, since $I_\nu \varphi' = \varphi$ for $\varphi \in C_c^1(\mathbb{R}; H)$ we get that

$$\partial_{t,\nu}\varphi = \varphi';$$

that is, $\partial_{t,\nu}$ extends the classical derivative of continuously differentiable functions. We shall discuss the actual domain of $\partial_{t,\nu}$ in the next chapter.

Proposition 3.2.4 *Let $\nu \neq 0$. Then $\mathcal{D}_H := \operatorname{lin}\left\{\varphi \cdot x\,;\ \varphi \in C_c^\infty(\mathbb{R}),\ x \in H\right\}$ is a core for $\partial_{t,\nu}$. Here, $C_c^\infty(\mathbb{R})$ denotes the space of smooth functions on $\mathbb{R}$ with compact support.*

Proof We first prove that

$$\left\{\varphi'\,;\ \varphi \in C_c^\infty(\mathbb{R})\right\} \tag{3.3}$$

is dense in $L_{2,\nu}(\mathbb{R})$. As $C_c^\infty(\mathbb{R})$ is dense in $L_{2,\nu}(\mathbb{R})$ (see Exercise 3.2), it suffices to approximate functions in $C_c^\infty(\mathbb{R})$. For this, let $f \in C_c^\infty(\mathbb{R})$. We now define

$$\varphi_n(t) := \begin{cases} \int_{-\infty}^t f(s) - f(s-n)\,\mathrm{d}s & \text{if } \nu > 0, \\ \int_{-\infty}^t f(s) - f(s+n)\,\mathrm{d}s & \text{if } \nu < 0 \end{cases} \quad (t \in \mathbb{R}, n \in \mathbb{N}).$$

Then $\varphi_n \in C_c^\infty(\mathbb{R})$ for each $n \in \mathbb{N}$ and

$$\varphi_n'(t) = \begin{cases} f(t) - f(t-n) & \text{if } \nu > 0, \\ f(t) - f(t+n) & \text{if } \nu < 0 \end{cases} \quad (t \in \mathbb{R}, n \in \mathbb{N}).$$

Consequently,

$$\begin{aligned} \|\varphi_n' - f\|_{L_{2,\nu}(\mathbb{R})}^2 &= \begin{cases} \int_{\mathbb{R}} |f(t-n)|^2 \mathrm{e}^{-2\nu t}\,\mathrm{d}t & \text{if } \nu > 0, \\ \int_{\mathbb{R}} |f(t+n)|^2 \mathrm{e}^{-2\nu t}\,\mathrm{d}t & \text{if } \nu < 0 \end{cases} \\ &= \|f\|_{L_{2,\nu}(\mathbb{R})}^2 \mathrm{e}^{-2|\nu| n} \to 0 \quad (n \to \infty), \end{aligned}$$

which shows the density of (3.3) in $L_{2,\nu}(\mathbb{R})$. By Lemma 3.1.8 we have that

$$\left\{\varphi' \cdot x\,;\ \varphi \in C_c^\infty(\mathbb{R}),\ x \in H\right\}$$

is total in $L_{2,\nu}(\mathbb{R}; H)$ and so $\partial_{t,\nu}[\mathcal{D}_H]$ is dense in $L_{2,\nu}(\mathbb{R}; H)$. Now let $f \in \operatorname{dom}(\partial_{t,\nu})$ and $\varepsilon > 0$. By what we have shown above there exists some $\varphi \in \mathcal{D}_H$ such that

$$\|\partial_{t,\nu}\varphi - \partial_{t,\nu} f\|_{L_{2,\nu}(\mathbb{R}; H)} \leqslant \varepsilon.$$

Since $\partial_{t,\nu}^{-1} = I_\nu$ is bounded with $\left\|\partial_{t,\nu}^{-1}\right\| \leqslant \frac{1}{|\nu|}$, the latter implies that

$$\|\varphi - f\|_{L_{2,\nu}(\mathbb{R}; H)} \leqslant \frac{\varepsilon}{|\nu|},$$

and hence, $\mathcal{D}_H$ is indeed a core for $\partial_{t,\nu}$. □

Corollary 3.2.5 *For $\nu \in \mathbb{R}$ the mapping*

$$\begin{aligned} \exp(-\nu \mathrm{m}) : L_{2,\nu}(\mathbb{R}; H) &\to L_2(\mathbb{R}; H) \\ f &\mapsto (t \mapsto \mathrm{e}^{-\nu t} f(t)) \end{aligned}$$

is unitary, and for $\nu, \mu \neq 0$ one has

$$\exp(-\nu \mathrm{m})(\partial_{t,\nu} - \nu) \exp(-\nu \mathrm{m})^{-1} = \exp(-\mu \mathrm{m})(\partial_{t,\mu} - \mu) \exp(-\mu \mathrm{m})^{-1}.$$

Proof The proof is left as Exercise 3.5. For this we recall that the equality to be proven is an equality of relations and particularly includes the equality of the (natural) domains of the operators involved. Furthermore, note that it suffices to show equality on $C_c^\infty(\mathbb{R}; H)$ and then to use an appropriate density result. □

By Corollary 3.2.5 we can now define $\partial_{t,0}$. Let $\nu \neq 0$. Then

$$\partial_{t,0} := \exp(-\nu \mathrm{m})(\partial_{t,\nu} - \nu) \exp(-\nu \mathrm{m})^{-1}.$$

Note that in view of Corollary 3.2.5, the assertion of Proposition 3.2.4 now also holds for $\nu = 0$.

Finally, we want to compute the adjoint of $\partial_{t,\nu}$.

Corollary 3.2.6 *Let $\nu \in \mathbb{R}$. The adjoint of $\partial_{t,\nu}$ is given by*

$$\partial_{t,\nu}^* = -\partial_{t,\nu} + 2\nu.$$

In particular, $\partial_{t,\nu}$ is a normal operator with $\operatorname{Re} \partial_{t,\nu} := \frac{1}{2}\left(\overline{\partial_{t,\nu} + \partial_{t,\nu}^*}\right) = \nu$, *and $\partial_{t,0}$ is skew-selfadjoint.*

Proof Let $\nu \neq 0$ first. Integrating by parts, one obtains

$$\begin{aligned}\int_{\mathbb{R}} \langle \partial_{t,\nu}\varphi(t), \psi(t)\rangle \mathrm{e}^{-2\nu t}\,\mathrm{d}t &= \int_{\mathbb{R}} \langle \varphi'(t), \psi(t)\rangle \mathrm{e}^{-2\nu t}\,\mathrm{d}t \\ &= \int_{\mathbb{R}} \langle \varphi(t), -\psi'(t) + 2\nu\psi(t)\rangle \mathrm{e}^{-2\nu t}\,\mathrm{d}t\end{aligned}$$

for $\varphi, \psi \in C_c^\infty(\mathbb{R}; H)$. Since $C_c^\infty(\mathbb{R}; H)$ is a core for $\partial_{t,\nu}$ by Proposition 3.2.4, the latter shows

$$\partial_{t,\nu} \subseteq -\partial_{t,\nu}^* + 2\nu.$$

Since we know that $\partial_{t,\nu}$ is onto, it suffices to prove that $-\partial_{t,\nu}^* + 2\nu$ is one-to-one, since this would imply equality in the latter operator inclusion. For doing so, we apply Theorem 2.2.5 to compute

$$\ker(-\partial_{t,\nu}^* + 2\nu) = \operatorname{ran}(-\partial_{t,\nu} + 2\nu)^\perp.$$

Moreover, we have that $-\partial_{t,\nu} + 2\nu$ is unitarily equivalent to $-\partial_{t,-\nu}$ by Corollary 3.2.5 and since $\partial_{t,-\nu}$ is onto, so is $-\partial_{t,\nu} + 2\nu$ and thus $\ker(-\partial_{t,\nu}^* + 2\nu) = L_{2,\nu}(\mathbb{R}; H)^\perp = \{0\}$, which yields the assertion.

The case $\nu = 0$ follows directly from the definition of $\partial_{t,0}$. □

3.3 Comments

Standard references for Bochner integration and related results are [6, 31].

Considering the derivative operator in an exponentially weighted space goes back (at least) to Morgenstern [67], where ordinary differential equations were considered in a classical setting. In fact, we shall return to this observation in the next chapter when we devote our study to some implications of the already developed concepts on ordinary and delay differential equations.

A first occurrence of the derivative operator in exponentially weighted L_2-spaces can be found in [83], where a corresponding spectral theorem has been focussed on. We will prove in a later chapter that the spectral representation of the time derivative as a multiplication operator can be realised by a shifted variant of the Fourier transformation—the so-called Fourier–Laplace transformation.

In an applied context, the time derivative operator discussed here has been introduced in [82].

Exercises

Exercise 3.1 A sequence $(\varphi_n)_n$ in $C_c^\infty(\mathbb{R}^d)$ is called a *δ-sequence* if

(a) $\varphi_n \geqslant 0$ for $n \in \mathbb{N}$,

(b) $\operatorname{spt} \varphi_n \subseteq \left[-\frac{1}{n}, \frac{1}{n}\right]^d$ for $n \in \mathbb{N}$,

(c) $\int_{\mathbb{R}^d} \varphi_n = 1$ for $n \in \mathbb{N}$.

Let $\varphi \in C_c^\infty(\mathbb{R}^d)$ with $\operatorname{spt} \varphi \subseteq [-1, 1]^d$, $\varphi \geqslant 0$ and $\int_{\mathbb{R}^d} \varphi = 1$. Prove that $(\varphi_n)_n$ given by $\varphi_n(x) := n^d \varphi(nx)$ for $x \in \mathbb{R}^d$, $n \in \mathbb{N}$ defines a δ-sequence. Moreover, give an example for such a function φ.

Exercise 3.2 It is well-known that $\{\mathbb{1}_I \,;\, I \text{ } d\text{-dimensional bounded interval}\}$ is total in $L_2(\mathbb{R}^d)$.

(a) Let $\varphi \in C_c^\infty(\mathbb{R}^d)$, $f \in L_2(\mathbb{R}^d)$. Define as usual

$$f * \varphi := \Big(x \mapsto \int_{\mathbb{R}^d} f(x-y)\varphi(y)\,\mathrm{d}y\Big).$$

Prove that $f * \varphi \in C^\infty(\mathbb{R}^d)$ with $\partial^\alpha (f * \varphi) = f * \partial^\alpha \varphi$ for all $\alpha \in \mathbb{N}_0^d$, where $\partial^\alpha \varphi = \partial_1^{\alpha_1} \ldots \partial_d^{\alpha_d} \varphi$. Moreover, prove that $\operatorname{spt} f * \varphi \subseteq \operatorname{spt} f + \operatorname{spt} \varphi$.

(b) Let $(\varphi_n)_n$ be a δ-sequence and $f \in L_2(\mathbb{R}^d)$. Show that $f * \varphi_n \to f$ in $L_2(\mathbb{R}^d)$ as $n \to \infty$.
Hint: Prove that $\mathbb{1}_I * \varphi_n \to \mathbb{1}_I$ in $L_2(\mathbb{R}^d)$ for all d-dimensional bounded intervals and use that $\|f * \varphi_n\|_2 \leqslant \|f\|_2$ (see also Lemma 3.2.1).

(c) Prove that $C_c^\infty(\mathbb{R}^d)$ is dense in $L_2(\mathbb{R}^d)$.

Exercise 3.3 Let $a < b$, X_0, X_1, X_2 be Banach spaces, $f\colon (a,b) \to X_0$ and $g\colon (a,b) \to X_1$ both continuously differentiable, $\ell\colon X_0 \times X_1 \to X_2$ bilinear and continuous. Prove that $h\colon (a,b) \to X_2$ given by

$$h(t) := \ell(f(t), g(t)) \quad (t \in (a,b))$$

is continuously differentiable with

$$h'(t) = \ell(f'(t), g(t)) + \ell(f(t), g'(t)) \quad (t \in (a,b)).$$

If f, f', g, g' have continuous extensions to $[a,b]$, prove the integration by parts formula:

$$\int_a^b \ell(f'(t), g(t))\,\mathrm{d}t = \ell(f(b), g(b)) - \ell(f(a), g(a)) - \int_a^b \ell(f(t), g'(t))\,\mathrm{d}t.$$

Exercise 3.4 For $\nu \neq 0$, show that $\|I_\nu\| = \frac{1}{|\nu|}$.

Exercise 3.5 Prove Corollary 3.2.5.

Exercise 3.6 Let $\nu \in \mathbb{R}$ and H be a complex Hilbert space. Prove that $\sigma(\partial_{t,\nu}) \subseteq \{it + \nu\,;\, t \in \mathbb{R}\}$, where $\partial_{t,0}$ is defined in Corollary 3.2.6.
Hint: For $f \in \operatorname{dom}(\partial_{t,\nu}), z \in \mathbb{C}$ compute $\operatorname{Re}\langle (z - \partial_{t,\nu})f, f\rangle_{L_{2,\nu}(\mathbb{R};H)}$ by using Corollary 3.2.6. For proving the surjectivity of $z - \partial_{t,\nu}$ for a suitable z, use the formula

$$\overline{\operatorname{ran}}(z - \partial_{t,\nu}) = \ker(z^* - \partial_{t,\nu}^*)^\perp.$$

Remark: Later we will see that, actually, $\sigma(\partial_{t,\nu}) = \{it + \nu\,;\, t \in \mathbb{R}\}$.

Exercise 3.7 Consider the differential equation

$$\left(\partial_{t,\nu}^2 - 1\right) u = \mathbb{1}_{[-1,1]}.$$

Since $\partial_{t,\nu}^2 - 1 = \left(\partial_{t,\nu} - 1\right)\left(\partial_{t,\nu} + 1\right)$, it follows by Exercise 3.6 that there is a unique $u \in L_{2,\nu}(\mathbb{R})$ solving this equation if $\nu \notin \{-1, 1\}$. Compute these solutions.
Hint: For $u \in \operatorname{dom}(\partial_{t,\nu})$ use the fact that u is necessarily continuous (which we shall establish in the next chapter).

References

6. W. Arendt et al., *Vector-Valued Laplace Transforms and Cauchy Problems. 2nd edn.* (Birkhäuser, Basel, 2011)
31. J. Diestel, J.J. Uhl Jr., *Vector Measures*. With a foreword by B. J. Pettis, Mathematical Surveys, No. 15 (American Mathematical Society, Providence, RI, 1977)
67. D. Morgenstern, Beträge zur nichtlinearen Funktionalanalysis. PhD thesis. TU Berlin, 1952
82. R. Picard, A structural observation for linear material laws in classical mathematical physics. Math. Methods Appl. Sci. **32**, 1768–1803 (2009)
83. R. Picard, *Hilbert Space Approach to Some Classical Transforms* (Wiley, New York, 1989)

Chapter 4
Ordinary Differential Equations

In this chapter, we discuss a first application of the time derivative operator constructed in the previous chapter. More precisely, we analyse well-posedness of ordinary differential equations and will at the same time provide a Hilbert space proof of the classical Picard–Lindelöf theorem.[1] We shall furthermore see that the abstract theory developed here also allows for more general differential equations to be considered. In particular, we will have a look at so-called delay differential equations with finite or infinite delay; neutral differential equations are considered in the exercises section.

We start with some information on the time derivative and its domain.

4.1 The Domain of $\partial_{t,\nu}$ and the Sobolev Embedding Theorem

Let H be a Hilbert space. Readers familiar with the notion of Sobolev spaces might have already realised that the domain of $\partial_{t,\nu}$ can be described as $L_{2,\nu}(\mathbb{R}; H)$-functions with distributional derivative lying in $L_{2,\nu}(\mathbb{R}; H)$. We shall also use

$$H^1_\nu(\mathbb{R}; H) := \operatorname{dom}(\partial_{t,\nu}) \subseteq L_{2,\nu}(\mathbb{R}; H),$$

if we want to emphasise the target Hilbert space of the $\operatorname{dom}(\partial_{t,\nu})$-functions. In order to stress the distributional character of the derivative introduced, we include the following result. Later on, we have the opportunity to have a more detailed look at Sobolev spaces in more general contexts.

[1] There are different notions for this theorem. It is also called existence and uniqueness theorem for initial value problems for ordinary differential equations as well as Cauchy–Lipschitz theorem.

C. Seifert et al., *Evolutionary Equations*, Operator Theory: Advances and Applications 287, https://doi.org/10.1007/978-3-030-89397-2_4

Proposition 4.1.1 *Let $\nu \in \mathbb{R}$ and $f, g \in L_{2,\nu}(\mathbb{R}; H)$. Then the following conditions are equivalent:*

(i) *$f \in \operatorname{dom}(\partial_{t,\nu})$ and $\partial_{t,\nu} f = g$.*
(ii) *For all $\phi \in C_c^\infty(\mathbb{R})$ we have*

$$-\int_{\mathbb{R}} \phi' f = \int_{\mathbb{R}} \phi g,$$

where these integrals are Bochner integrals of the H-valued functions $t \mapsto \phi'(t) f(t)$ and $t \mapsto \phi(t) g(t)$, respectively.

Proof Assume that $f \in \operatorname{dom}(\partial_{t,\nu})$. By Proposition 3.2.4 and Corollary 3.2.6, we have that $\mathcal{D}_H = \operatorname{lin}\left\{\varphi \cdot x\,;\, \varphi \in C_c^\infty(\mathbb{R}),\ x \in H\right\} \subseteq \operatorname{dom}(\partial_{t,\nu}^*)$ (which also holds for $\nu = 0$) and

$$\left\langle \partial_{t,\nu} f, \psi \cdot x \right\rangle_{L_{2,\nu}} = \left\langle f, \left(-\psi' + 2\nu\psi\right) \cdot x \right\rangle_{L_{2,\nu}}$$

for all $x \in H$ and $\psi \in C_c^\infty(\mathbb{R})$. Hence, we obtain for all $\psi \in C_c^\infty(\mathbb{R})$

$$\int_{\mathbb{R}} \left(-\psi' + 2\nu\psi\right) f \mathrm{e}^{-2\nu\cdot} = \int_{\mathbb{R}} \psi \partial_{t,\nu} f \mathrm{e}^{-2\nu\cdot};$$

putting $\phi := \mathrm{e}^{-2\nu\cdot}\psi$ and using that multiplication by $\mathrm{e}^{-2\nu\cdot}$ is a bijection on $C_c^\infty(\mathbb{R})$, we deduce the claimed formula with $g = \partial_{t,\nu} f$.

On the other hand, the equation involving g applied to $\phi = \mathrm{e}^{-2\nu\cdot}\psi$ for $\psi \in C_c^\infty(\mathbb{R})$ implies that

$$\int_{\mathbb{R}} \left(-\psi' + 2\nu\psi\right) f \mathrm{e}^{-2\nu\cdot} = \int_{\mathbb{R}} \psi g \mathrm{e}^{-2\nu\cdot}.$$

Testing this equation with $x \in H$ yields

$$\langle g, \psi \cdot x \rangle_{L_{2,\nu}} = \left\langle f, \left(-\psi' + 2\nu\psi\right) \cdot x \right\rangle_{L_{2,\nu}} = \left\langle f, \left(-\partial_{t,\nu}\psi \cdot x + 2\nu\psi \cdot x\right) \right\rangle_{L_{2,\nu}}.$$

Since $\mathcal{D}_H$ is dense in $\operatorname{dom}(\partial_{t,\nu})$ by Proposition 3.2.4, we infer that

$$\langle g, h \rangle_{L_{2,\nu}} = \left\langle f, \left(-\partial_{t,\nu} h + 2\nu h\right) \right\rangle_{L_{2,\nu}}$$

for all $h \in \operatorname{dom}(\partial_{t,\nu})$. Now, Corollary 3.2.6, yields

$$\langle g, h \rangle_{L_{2,\nu}} = \left\langle f, \partial_{t,\nu}^* h \right\rangle_{L_{2,\nu}} \quad (h \in \operatorname{dom}(\partial_{t,\nu}^*)).$$

Thus, $f \in \operatorname{dom}(\partial_{t,\nu}^{**}) = \operatorname{dom}(\partial_{t,\nu})$ and $\partial_{t,\nu} f = g$. □

The next result is a version of the Sobolev embedding theorem. It particularly confirms that functions in the domain of $\partial_{t,\nu}$ are continuous. This result was announced in Exercise 3.7. Here, we make use of the explicit form of the domain of $\partial_{t,\nu}$ as being the range space of the integral operator I_ν. We define

$$C_\nu(\mathbb{R}; H) := \left\{ f : \mathbb{R} \to H\,;\ f \text{ continuous},\ \|f\|_{\nu,\infty} := \sup_{t\in\mathbb{R}} \left\| \mathrm{e}^{-\nu t} f(t) \right\|_H < \infty \right\}$$

and regard it as being endowed with the obvious norm.

Theorem 4.1.2 (Sobolev Embedding Theorem) *Let $\nu \in \mathbb{R}$. Then every $f \in \operatorname{dom}(\partial_{t,\nu})$ has a continuous representative, and the mapping*

$$\operatorname{dom}(\partial_{t,\nu}) \ni f \mapsto f \in C_\nu(\mathbb{R}; H)$$

is continuous.

Proof We restrict ourselves to the case when $\nu > 0$; the remaining cases can be proved by invoking Corollary 3.2.5. Let $f \in \operatorname{dom}(\partial_{t,\nu})$. By definition, we find $g \in L_{2,\nu}(\mathbb{R}; H)$ such that $f = \partial_{t,\nu}^{-1} g = I_\nu g$. Then for all $t \in \mathbb{R}$ we compute

$$\begin{aligned}
\int_{-\infty}^{t} \|g(\tau)\|\, \mathrm{d}\tau &= \int_{-\infty}^{t} \|g(\tau)\|\, \mathrm{e}^{-\nu\tau}\mathrm{e}^{\nu\tau}\, \mathrm{d}\tau \leqslant \sqrt{\int_{-\infty}^{t} \|g(\tau)\|^2\, \mathrm{e}^{-2\nu\tau}\, \mathrm{d}\tau}\sqrt{\int_{-\infty}^{t} \mathrm{e}^{2\nu\tau}\, \mathrm{d}\tau} \\
&\leqslant \left\| \partial_{t,\nu} f \right\|_{L_{2,\nu}} \sqrt{\frac{1}{2\nu}}\mathrm{e}^{\nu t}.
\end{aligned}$$

Thus, g is integrable on $(-\infty, t]$ for all $t \in \mathbb{R}$ and dominated convergence implies that

$$f = \left(t \mapsto \int_{-\infty}^{t} g(s)\, \mathrm{d}s \right)$$

is continuous. Moreover, for $t \in \mathbb{R}$ we obtain

$$\|f(t)\| \leqslant \int_{-\infty}^{t} \|g(\tau)\|\, \mathrm{d}\tau \leqslant \left\| \partial_{t,\nu} f \right\|_{L_{2,\nu}} \sqrt{\frac{1}{2\nu}}\mathrm{e}^{\nu t}$$

which yields the claimed continuity. □

Corollary 4.1.3 *For all $f \in \operatorname{dom}(\partial_{t,\nu})$, we have that $\left\| \mathrm{e}^{-\nu t} f(t) \right\|_H \to 0$ as $t \to \pm\infty$.*

The proof is left as Exercise 4.2.

4.2 The Picard–Lindelöf Theorem

The prototype of the Picard–Lindelöf theorem will be formulated for so-called uniformly Lipschitz continuous functions. We first need a preparation.

Definition Let X be a Banach space. Then we define

$$S_{\mathrm{c}}(\mathbb{R}; X) := \{f : \mathbb{R} \to X\,;\ f \text{ simple},\ \operatorname{spt} f \text{ compact}\}$$

to be the set of *simple functions from* $\mathbb{R}$ *to* X *with compact support*.

Lemma 4.2.1 *Let X be a Banach space and $\nu, \eta \in \mathbb{R}$. Then $S_{\mathrm{c}}(\mathbb{R}; X)$ is dense in $L_{2,\nu}(\mathbb{R}; X) \cap L_{2,\eta}(\mathbb{R}; X)$; that is, for all $f \in L_{2,\nu}(\mathbb{R}; X) \cap L_{2,\eta}(\mathbb{R}; X)$ there exists $(f_n)_n$ in $S_{\mathrm{c}}(\mathbb{R}; X)$ such that $f_n \to f$ in both $L_{2,\nu}(\mathbb{R}; X)$ and $L_{2,\eta}(\mathbb{R}; X)$. In particular, $S_{\mathrm{c}}(\mathbb{R}; X)$ is dense in $L_{2,\nu}(\mathbb{R}; X)$.*

Proof Let $f \in L_{2,\nu}(\mathbb{R}; X) \cap L_{2,\eta}(\mathbb{R}; X)$. Then for all $n \in \mathbb{N}$ we have that $\mathbb{1}_{[-n,n]} f \in L_{2,\nu}(\mathbb{R}; X) \cap L_{2,\eta}(\mathbb{R}; X)$ and $\mathbb{1}_{[-n,n]} f \to f$ in $L_{2,\nu}(\mathbb{R}; X)$ and in $L_{2,\eta}(\mathbb{R}; X)$ as $n \to \infty$. For $n \in \mathbb{N}$ let $(\widetilde{f}_{n,k})_k$ be in $S(\mu_{2,\nu}; X)$ such that $\widetilde{f}_{n,k} \to \mathbb{1}_{[-n,n]} f$ in $L_{2,\nu}(\mathbb{R}; X)$ as $k \to \infty$. We put $f_{n,k} := \mathbb{1}_{[-n,n]} \widetilde{f}_{n,k} \in S_{\mathrm{c}}(\mathbb{R}; X)$. Then $f_{n,k} \to \mathbb{1}_{[-n,n]} f$ in $L_{2,\nu}(\mathbb{R}; X)$ and in $L_{2,\eta}(\mathbb{R}; X)$ as $k \to \infty$. □

In order to define the notion of uniformly Lipschitz continuous functions, we first need the Lipschitz semi-norm.

Definition Let X_0, X_1 be normed spaces, and $F : X_0 \to X_1$ Lipschitz continuous. Then

$$\|F\|_{\mathrm{Lip}} := \sup_{\substack{x,y \in X_0 \\ x \neq y}} \frac{\|F(x) - F(y)\|}{\|x - y\|}$$

is the *Lipschitz semi-norm* of F.

Definition Let H_0, H_1 be Hilbert spaces, $\mu \in \mathbb{R}$. Then a function $F : S_{\mathrm{c}}(\mathbb{R}; H_0) \to \bigcap_{\nu \geqslant \mu} L_{2,\nu}(\mathbb{R}; H_1)$ is called *uniformly Lipschitz continuous* if for all $\nu \geqslant \mu$ we have that F considered in $L_{2,\nu}(\mathbb{R}; H_0) \times L_{2,\nu}(\mathbb{R}; H_1)$ is Lipschitz continuous, and for the unique Lipschitz continuous extensions F^{ν}, $\nu \geqslant \mu$, we have that

$$\sup_{\nu \geqslant \mu} \left\| F^{\nu} \right\|_{\mathrm{Lip}} < \infty.$$

Remark 4.2.2 Another way to introduce uniformly Lipschitz continuous mappings is the following. Let H_0, H_1 be Hilbert spaces, $\mu \in \mathbb{R}$. Let $(F^{\nu})_{\nu \geqslant \mu}$ be a family of Lipschitz continuous mappings $F^{\nu} : L_{2,\nu}(\mathbb{R}; H_0) \to L_{2,\nu}(\mathbb{R}; H_1)$ such that

$$\sup_{\nu \geqslant \mu} \left\| F^{\nu} \right\|_{\mathrm{Lip}} < \infty$$

and the mappings are consistent in the sense that for all $\nu, \eta \geqslant \mu$ and $f \in L_{2,\nu}(\mathbb{R}; H_0) \cap L_{2,\eta}(\mathbb{R}; H_0)$ we have

$$F^{\nu}(f) = F^{\eta}(f).$$

Then, for $\nu \geqslant \mu$ and $f \in S_c(\mathbb{R}; H_0)$ we have $F^{\nu}(f) \in \bigcap_{\eta \geqslant \mu} L_{2,\eta}(\mathbb{R}; H_1)$ and $F^{\nu}|_{S_c(\mathbb{R}; H_0)}$ is uniformly Lipschitz continuous.

Theorem 4.2.3 (Picard–Lindelöf—Hilbert Space Version) *Let H be a Hilbert space, $\mu \in \mathbb{R}$ and $F\colon S_c(\mathbb{R}; H) \to \bigcap_{\nu \geqslant \mu} L_{2,\nu}(\mathbb{R}; H)$ uniformly Lipschitz continuous with $L := \sup_{\nu \geqslant \mu} \|F^{\nu}\|_{\mathrm{Lip}}$. Then for all $\nu > \max\{L, \mu\}$ the equation*

$$\partial_{t,\nu} u_{\nu} = F^{\nu}(u_{\nu})$$

admits a unique solution $u_{\nu} \in \operatorname{dom}(\partial_{t,\nu})$. Furthermore, for all $\nu > \max\{L, \mu\}$ the following properties hold:

(a) *If $F^{\nu}(u_{\nu})$ is continuous in a neighbourhood of $a \in \mathbb{R}$, then u_{ν} is continuously differentiable in a neighbourhood of a.*
(b) *For all $a \in \mathbb{R}$, $\mathbb{1}_{(-\infty,a]} u_{\nu}$ is the unique fixed point $v \in L_{2,\nu}(\mathbb{R}; H)$ of $\mathbb{1}_{(-\infty,a]} \partial_{t,\nu}^{-1} F^{\nu}$, that is, v uniquely solves*

$$v = \mathbb{1}_{(-\infty,a]} \partial_{t,\nu}^{-1} F^{\nu}(v).$$

(c) *For all $\eta \geqslant \nu$ we have that $u_{\nu} = u_{\eta}$.*
(d) *For all $f \in L_{2,\nu}(\mathbb{R}; H)$ the equation*

$$\partial_{t,\nu} v = F^{\nu}(v) + f$$

admits a unique solution $v_{\nu,f} \in \operatorname{dom}(\partial_{t,\nu})$, and if $f, g \in L_{2,\nu}(\mathbb{R}; H)$ satisfy $f = g$ on $(-\infty, a]$ for some $a \in \mathbb{R}$, then $v_{\nu,f} = v_{\nu,g}$ on $(-\infty, a]$.

Proof of Theorem 4.2.3—First Part Define $\Phi\colon L_{2,\nu}(\mathbb{R}; H) \to L_{2,\nu}(\mathbb{R}; H)$ by

$$\Phi(u) = \partial_{t,\nu}^{-1} F^{\nu}(u).$$

Since $\left\|\partial_{t,\nu}^{-1}\right\| \leqslant \frac{1}{\nu}$ and $\nu > L$ it follows that Φ is a contraction and thus admits a unique fixed point, which by definition solves the equation in question. Moreover, we have that $u_{\nu} = \Phi(u_{\nu}) = \partial_{t,\nu}^{-1} F^{\nu}(u_{\nu}) \in \operatorname{dom}(\partial_{t,\nu})$.

Differentiability of u_{ν} as in (a) follows from Exercise 4.1 and the continuity of $F^{\nu}(u_{\nu})$.

For the unique existence asserted in (d), note that the unique existence of $v_{\nu,f}$ follows from the above considerations after realising that $\Psi(v) := \partial_{t,\nu}^{-1} F^{\nu}(v) + \partial_{t,\nu}^{-1} f$ defines a contraction in $L_{2,\nu}(\mathbb{R}; H)$. For the remaining statements in (d) and the statements in (b) and (c), we need some prerequisites. □

Definition Let H_0, H_1 be Hilbert spaces, $\nu \in \mathbb{R}$ and $F\colon L_{2,\nu}(\mathbb{R}; H_0) \to L_{2,\nu}(\mathbb{R}; H_1)$. Then, F is called *causal* if for all $a \in \mathbb{R}$ and all $f, g \in L_{2,\nu}(\mathbb{R}; H_0)$ with $f = g$ on $(-\infty, a]$, we have that $F(f) = F(g)$ on $(-\infty, a]$.

Remark 4.2.4 Let $\nu \in \mathbb{R}$, $a \in \mathbb{R}$. If $f \in L(L_{2,\nu}(\mathbb{R}; H))$ with $\operatorname{spt} f \subseteq (-\infty, a]$ then $f \in \bigcap_{\eta \leqslant \nu} L_{2,\eta}(\mathbb{R}; H)$ and

$$\|f\|_{L_{2,\eta}(\mathbb{R};H)} \leqslant \mathrm{e}^{(\nu-\eta)a} \|f\|_{L_{2,\nu}(\mathbb{R};H)} \quad (\eta \leqslant \nu).$$

Likewise, if $\operatorname{spt} f \subseteq [a, \infty)$, we get $f \in \bigcap_{\rho \geqslant \nu} L_{2,\rho}(\mathbb{R}; H)$ with

$$\|f\|_{L_{2,\rho}(\mathbb{R};H)} \leqslant \mathrm{e}^{(\nu-\rho)a} \|f\|_{L_{2,\nu}(\mathbb{R};H)} \quad (\rho \geqslant \nu).$$

Lemma 4.2.5 *Let H_0, H_1 be Hilbert spaces, $\mu \in \mathbb{R}$, $F\colon S_{\mathrm{c}}(\mathbb{R}; H_0) \to \bigcap_{\nu \geqslant \mu} L_{2,\nu}(\mathbb{R}; H_1)$ uniformly Lipschitz continuous. Then the following statements hold:*

(a) *F^ν is causal for all $\nu \geqslant \mu$.*
(b) *The mapping $\partial_{t,\nu}^{-1} F^\nu$ is causal if $\nu \geqslant \max\{\mu, 0\}$ and $\nu \neq 0$.*
(c) *For all $\nu \geqslant \eta \geqslant \mu$, we have that $F^\nu = F^\eta$ on $L_{2,\nu}(\mathbb{R}; H_0) \cap L_{2,\eta}(\mathbb{R}; H_0)$.*

Proof (a) We divide the proof into three steps.

(i) Let $\nu \geqslant \mu$. In order to show causality of F^ν, we first note that it suffices to have $F^\nu(f) = F^\nu(g)$ on $(-\infty, a]$ for all $f, g \in S_{\mathrm{c}}(\mathbb{R}; H_0)$ with $f = g$ on $(-\infty, a]$. Indeed, let $f, g \in L_{2,\nu}(\mathbb{R}; H)$ with $f = g$ on $(-\infty, a]$ for some $a \in \mathbb{R}$. By Lemma 4.2.1 we find $(f_n)_n$ and $(\widetilde{g}_n)_n$ in $S_{\mathrm{c}}(\mathbb{R}; H_0)$ such that $f_n \to f$ and $\widetilde{g}_n \to g$ in $L_{2,\nu}(\mathbb{R}; H_0)$. Next, $\mathbb{1}_{(-\infty,a]} f_n \to \mathbb{1}_{(-\infty,a]} f = \mathbb{1}_{(-\infty,a]} g$ as $n \to \infty$ in $L_{2,\nu}(\mathbb{R}; H_0)$. Thus, putting $g_n := \mathbb{1}_{(-\infty,a]} f_n + \mathbb{1}_{(a,\infty)} \widetilde{g}_n$ for all $n \in \mathbb{N}$ we obtain that $g_n \to g$ in $L_{2,\nu}(\mathbb{R}; H_0)$. Since $F^\nu(f_n) = F^\nu(g_n)$ on $(-\infty, a]$ for all $n \in \mathbb{N}$ and $F^\nu\colon L_{2,\nu}(\mathbb{R}; H_0) \to L_{2,\nu}(\mathbb{R}; H_1)$ is continuous, taking the limit $n \to \infty$ yields $F^\nu(f) = F^\nu(g)$ on $(-\infty, a]$.

(ii) Let $a \in \mathbb{R}$, $c \geqslant 0$ and $f \in S_{\mathrm{c}}(\mathbb{R}; H_0)$ such that $f = 0$ on $(-\infty, a]$, $g \in \bigcap_{\nu \geqslant \mu} L_{2,\nu}(\mathbb{R}; H_1)$ such that $\|g\|_{L_{2,\nu}(\mathbb{R};H_1)} \leqslant c \|f\|_{L_{2,\nu}(\mathbb{R};H_0)}$ for all $\nu \geqslant \mu$. Then

$$\begin{aligned}\int_{-\infty}^a \|g(t)\|_{H_1}^2 \mathrm{e}^{2\nu(a-t)} \,\mathrm{d}t &\leqslant \int_{\mathbb{R}} \|g(t)\|_{H_1}^2 \mathrm{e}^{2\nu(a-t)} \,\mathrm{d}t \\ &\leqslant c^2 \int_a^\infty \|f(t)\|_{H_0}^2 \mathrm{e}^{2\nu(a-t)} \,\mathrm{d}t \to 0\end{aligned}$$

as $\nu \to \infty$. Since $\mathrm{e}^{2\nu(a-t)} \to \infty$ as $\nu \to \infty$ for all $t < a$, the monotone convergence theorem implies $g = 0$ on $(-\infty, a]$.

(iii) Let $f, g \in S_{\mathrm{c}}(\mathbb{R}; H_0)$ such that $f = g$ on $(-\infty, a]$ for some $a \in \mathbb{R}$. Then $f - g = 0$ on $(-\infty, a]$. Since F is uniformly Lipschitz continuous, with $L := \sup_{\nu \geqslant \mu} \|F^\nu\|_{\mathrm{Lip}}$ we obtain $\|F^\nu(f) - F^\nu(g)\|_{L_{2,\nu}(\mathbb{R};H_1)} \leqslant L \|f - g\|_{L_{2,\nu}(\mathbb{R};H_0)}$ for all $\nu \geqslant \mu$. By (ii) we conclude $F^\nu(f) = F^\nu(g)$ on $(-\infty, a]$ for all $\nu \geqslant \mu$, which by (i) yields the assertion.

The statement in (b) directly follows from (a). Note that $\partial_{t,\nu}^{-1} F^{\nu}$ is uniformly Lipschitz continuous only for $\nu > 0$.

Let us prove (c). Since $F^{\nu}(f) = F(f) = F^{\eta}(f)$ for $f \in S_{\mathrm{c}}(\mathbb{R}; H_0)$, the set $S_{\mathrm{c}}(\mathbb{R}; H_0)$ is dense in $L_{2,\nu}(\mathbb{R}; H_0) \cap L_{2,\mu}(\mathbb{R}; H_0)$ by Lemma 4.2.1, and F^{ν} and F^{η} are Lipschitz-continuous, we obtain the assertion. □

Proof of Theorem 4.2.3—Second Part The remaining part in (d): Let $f, g \in L_{2,\nu}(\mathbb{R}; H)$ with $f = g$ on $(-\infty, a]$. Since $\nu > L \geqslant 0$, we compute using Lemma 4.2.5(b) and causality of $\partial_{t,\nu}^{-1}$ that

$$\begin{aligned}
\mathbb{1}_{(-\infty,a]} v_{\nu,f} &= \mathbb{1}_{(-\infty,a]} \partial_{t,\nu}^{-1} F^{\nu}\left(v_{\nu,f}\right) + \mathbb{1}_{(-\infty,a]} \partial_{t,\nu}^{-1} f \\
&= \mathbb{1}_{(-\infty,a]} \partial_{t,\nu}^{-1} F^{\nu}\left(\mathbb{1}_{(-\infty,a]} v_{\nu,f}\right) + \mathbb{1}_{(-\infty,a]} \partial_{t,\nu}^{-1} \mathbb{1}_{(-\infty,a]} f \\
&= \mathbb{1}_{(-\infty,a]} \partial_{t,\nu}^{-1} F^{\nu}\left(\mathbb{1}_{(-\infty,a]} v_{\nu,f}\right) + \mathbb{1}_{(-\infty,a]} \partial_{t,\nu}^{-1} \mathbb{1}_{(-\infty,a]} g.
\end{aligned}$$

The same computation also yields that

$$\mathbb{1}_{(-\infty,a]} v_{\nu,g} = \mathbb{1}_{(-\infty,a]} \partial_{t,\nu}^{-1} F^{\nu}\left(\mathbb{1}_{(-\infty,a]} v_{\nu,g}\right) + \mathbb{1}_{(-\infty,a]} \partial_{t,\nu}^{-1} \mathbb{1}_{(-\infty,a]} g.$$

It is easy to see that $u \mapsto \mathbb{1}_{(-\infty,a]} \partial_{t,\nu}^{-1} F^{\nu}(u) + \mathbb{1}_{(-\infty,a]} \partial_{t,\nu}^{-1} \mathbb{1}_{(-\infty,a]} g$ defines a contraction in $L_{2,\nu}(\mathbb{R}; H)$. Hence, the contraction mapping principle implies that $\mathbb{1}_{(-\infty,a]} v_{\nu,f} = \mathbb{1}_{(-\infty,a]} v_{\nu,g}$.

The statement in (b) follows from the fact that $u \mapsto \mathbb{1}_{(-\infty,a]} \partial_{t,\nu}^{-1} F^{\nu}(u)$ defines a contraction and Lemma 4.2.5(b).

For the proof of (c), we observe that for all $n \in \mathbb{N}$, we have $\mathbb{1}_{(-\infty,n]} u_{\eta} \in L_{2,\nu}(\mathbb{R}; H) \cap L_{2,\eta}(\mathbb{R}; H)$. Hence, by (b) and Lemma 4.2.5(c), it follows that

$$\mathbb{1}_{(-\infty,n]} u_{\eta} = \mathbb{1}_{(-\infty,n]} \partial_{t,\eta}^{-1} F^{\eta}\left(\mathbb{1}_{(-\infty,n]} u_{\eta}\right) = \mathbb{1}_{(-\infty,n]} \partial_{t,\nu}^{-1} F^{\nu}\left(\mathbb{1}_{(-\infty,n]} u_{\eta}\right).$$

As $\mathbb{1}_{(-\infty,n]} u_{\nu}$ satisfies the same fixed point equation, we deduce $\mathbb{1}_{(-\infty,n]} u_{\eta} = \mathbb{1}_{(-\infty,n]} u_{\nu}$ for all $n \in \mathbb{N}$, which yields the assertion. □

As a first application of Theorem 4.2.3 we state and prove the classical version of the Theorem of Picard–Lindelöf.

Theorem 4.2.6 (Picard–Lindelöf—Classical Version) *Let H be a Hilbert space, $\Omega \subseteq \mathbb{R} \times H$ be open, $f : \Omega \to H$ continuous, $(t_0, x_0) \in \Omega$. Assume there exists $L \geqslant 0$ such that for all $(t, x), (t, y) \in \Omega$ we have*

$$\|f(t, x) - f(t, y)\| \leqslant L \|x - y\|.$$

Then, there exists $\delta > 0$ such that the initial value problem

$$\begin{cases} u'(t) = f(t, u(t)) & (t \in (t_0, t_0 + \delta)), \\ u(t_0) = x_0, \end{cases} \tag{4.1}$$

admits a unique continuously differentiable solution, $u\colon [t_0, t_0 + \delta] \to H$, which satisfies $(t, u(t)) \in \Omega$ for all $t \in [t_0, t_0 + \delta]$.

Proof First of all we observe that we may assume, without loss of generality, that $x_0 = 0$. Indeed, to solve the initial value problem

$$\begin{cases} v'(t) = f(t, v(t) + x_0) & (t \in (t_0, t_0 + \delta)), \\ v(t_0) = 0, \end{cases}$$

for a continuously differentiable $v\colon [t_0, t_0 + \delta] \to H$ is equivalent to solving the problem in Theorem 4.2.6 for u by setting $u = v + \mathbb{1}_{[t_0, t_0+\delta]} x_0$. Appropriately shifting the time coordinate, we may also assume that $t_0 = 0$.

Thus, let $(0, 0) \in \Omega$. Then $[0, \delta'] \times B[0, \varepsilon] \subseteq \Omega$ for some $\delta', \varepsilon > 0$. Denote by $P\colon H \to H$ the projection onto $B[0, \varepsilon]$; that is, for $x \in H$, $Px \in B[0, \varepsilon]$ is the unique element satisfying

$$\|x - Px\|_H = \inf_{y \in B[0,\varepsilon]} \|x - y\|_H .$$

By Exercise 4.4, P is Lipschitz continuous with Lipschitz semi-norm bounded by 1. We then define

$$\begin{aligned} F\colon S_c(\mathbb{R}; H) &\to \bigcap_{\nu \geqslant 0} L_{2,\nu}(\mathbb{R}; H) \\ g &\mapsto \big(t \mapsto \mathbb{1}_{[0,\delta')}(t) f(t, P(g(t)))\big) \end{aligned}$$

and will prove that F is well-defined and uniformly Lipschitz continuous. Since the mapping $t \mapsto \mathbb{1}_{[0,\delta')}(t) f(t, 0)$ is supported on $\big[0, \delta'\big]$, we obtain for $\nu \geqslant 0$ that $F(0) \in L_{2,\nu}(\mathbb{R}; H)$. Moreover, for $\nu \geqslant 0$ and $g, h \in S_c(\mathbb{R}; H)$ we estimate

$$\begin{aligned} &\|F(g) - F(h)\|^2_{L_{2,\nu}(\mathbb{R};H)} \\ &= \int_{\mathbb{R}} \|F(g)(t) - F(h)(t)\|^2 \, \mathrm{e}^{-2\nu t} \, \mathrm{d}t = \int_0^{\delta'} \|f(t, P(g(t))) - f(t, P(h(t)))\|^2 \, \mathrm{e}^{-2\nu t} \, \mathrm{d}t \\ &\leqslant L^2 \int_0^{\delta'} \|P(g(t)) - P(h(t))\|^2 \, \mathrm{e}^{-2\nu t} \, \mathrm{d}t \leqslant L^2 \int_0^{\delta'} \|g(t) - h(t)\|^2 \, \mathrm{e}^{-2\nu t} \, \mathrm{d}t \\ &\leqslant L^2 \, \|g - h\|^2_{L_{2,\nu}(\mathbb{R};H)} , \end{aligned}$$

which shows that F is well-defined and uniformly Lipschitz continuous.

By Theorem 4.2.3, there exists $v \in \mathrm{dom}(\partial_{t,\nu})$ with $\nu > L$ such that

$$\partial_{t,\nu} v = F^{\nu}(v).$$

We read off from $v = \partial_{t,\nu}^{-1} F^{\nu}(v)$ that $v = 0$ on $(-\infty, 0]$, and that v is continuous by Theorem 4.1.2. Moreover, we obtain that

$$v(t) = \int_{-\infty}^{t} \mathbb{1}_{[0,\delta')}(\tau) f(\tau, P(v(\tau)))\,\mathrm{d}\tau = \int_{0}^{\min\{t,\delta'\}} f(\tau, P(v(\tau)))\,\mathrm{d}\tau,$$

from which we read off that v is continuously differentiable on $\left(0, \delta'\right)$ since f and P are also continuous. The same equality implies for $0 < t \leqslant \delta := \min\{\frac{\varepsilon}{M}, \delta'\}$, where $M := \sup_{(t,x)\in[0,\delta']\times B[0,\varepsilon]} \|f(t,x)\|$, that

$$\|v(t)\| \leqslant \int_{0}^{t} \|f(\tau, P(v(\tau)))\|\,\mathrm{d}\tau \leqslant M\delta \leqslant \varepsilon.$$

Thus, $(t, v(t)) \in \left[0, \delta'\right] \times B\left[0, \varepsilon\right] \subseteq \Omega$ for all $0 \leqslant t \leqslant \delta$ and so $Pv(t) = v(t)$ for $0 \leqslant t \leqslant \delta$. Thus, $u := v|_{[0,\delta]}$ satisfies (4.1).

Finally, concerning uniqueness, let $\widetilde{u}\colon [0, \delta] \to H$ be a continuously differentiable solution of (4.1). Let $\widetilde{v}$ be the extension of $\widetilde{u}$ by 0 to the whole of $\mathbb{R}$. Then we get that

$$\begin{aligned}
\mathbb{1}_{(-\infty,\delta]}\widetilde{v} &= \mathbb{1}_{(-\infty,\delta]} \int_{0}^{\cdot} \mathbb{1}_{[0,\delta')}(\tau) f(\tau, \widetilde{v}(\tau))\,\mathrm{d}\tau \\
&= \mathbb{1}_{(-\infty,\delta]} \int_{-\infty}^{\cdot} \mathbb{1}_{[0,\delta')}(\tau) f(\tau, P(\widetilde{v}(\tau)))\,\mathrm{d}\tau \\
&= \mathbb{1}_{(-\infty,\delta]} \partial_{t,\nu}^{-1} F^{\nu}(\mathbb{1}_{(-\infty,\delta]}\widetilde{v}).
\end{aligned}$$

Since $\mathbb{1}_{(-\infty,\delta]}v$ is the unique solution of the equation $w = \mathbb{1}_{(-\infty,\delta]}\partial_{t,\nu}^{-1} F^{\nu}(w)$, we obtain that $\mathbb{1}_{(-\infty,\delta]}\widetilde{v} = \mathbb{1}_{(-\infty,\delta]}v$, which yields $u = \widetilde{u}$. □

Remark 4.2.7 The reason for the proof of the classical Picard–Lindelöf theorem being seemingly complicated is two-fold. First of all, the Hilbert space solution theory is for L_2-functions rather than continuous (or continuously differentiable) functions. The second, maybe more important point is that the Hilbert space Picard–Lindelöf asserts a solution theory, which provides *global* existence in the time variable. The main body of the proof of the classical Picard–Lindelöf theorem presented here is therefore devoted to 'localisation' of the abstract theorem. Furthermore, note that the method of proof for obtaining uniqueness and the admittance of the initial value rests on causality. This effect will resurface when we discuss partial differential equations.

4.3 Delay Differential Equations

In this section, our study will not be as in depth as done for the local Picard–Lindelöf theorem. Of course, the solution theory would not be a very good one if it was only applicable to, arguably, the easiest case of ordinary differential equations. We shall see next that the developed theory applies to more elaborate examples.

In what follows, let H be a Hilbert space over $\mathbb{K}$. We start out with a delay differential equation with so-called 'discrete delay'. For this, we introduce, for $h \in \mathbb{R}$, the *time-shift operator*

$$\tau_h \colon S_c(\mathbb{R}; H) \to \bigcap_{\nu \in \mathbb{R}} L_{2,\nu}(\mathbb{R}; H),$$
$$f \mapsto f(\cdot + h).$$

Lemma 4.3.1 *Let $\mu \in \mathbb{R}$. The mapping $\tau_h \colon S_c(\mathbb{R}; H) \to \bigcap_{\nu \geqslant \mu} L_{2,\nu}(\mathbb{R}; H)$ is uniformly Lipschitz continuous if and only if $h \leqslant 0$. More precisely, for $\nu \in \mathbb{R}$ we have*

$$\|\tau_h\|_{L(L_{2,\nu}(\mathbb{R}; H))} = e^{h\nu}.$$

Proof Let $f \in S_c(\mathbb{R}; H)$. Then for $\nu \in \mathbb{R}$ we compute

$$\begin{aligned}\|\tau_h f\|^2_{L_{2,\nu}(\mathbb{R}; H)} &= \int_{\mathbb{R}} \|f(t+h)\|^2 e^{-2\nu t} \, dt = \int_{\mathbb{R}} \|f(t)\|^2 e^{-2\nu(t-h)} \, dt \\ &= \|f\|^2_{L_{2,\nu}(\mathbb{R}; H)} e^{2\nu h}.\end{aligned}$$

Since $\sup_{\nu \geqslant \mu} e^{2\nu h} < \infty$ if and only if $h \leqslant 0$ we obtain the equivalence. Moreover, the above equality also yields the norm of τ_h on $L_{2,\nu}(\mathbb{R}; H)$. □

We will reuse τ_h for the Lipschitz continuous extensions to $L_{2,\nu}(\mathbb{R}; H)$. The well-posedness theorem for delay equations with discrete delay is contained in the next theorem. We note here that we only formulate the respective result for right-hand sides that are globally Lipschitz continuous. With a localisation technique, as has already been carried out for the classical Picard–Lindelöf theorem, it is also possible to obtain local results.

Theorem 4.3.2 *Let H be a Hilbert space, $\mu \in \mathbb{R}$, $N \in \mathbb{N}$, $h_1, \ldots, h_N \in (-\infty, 0]$, and*

$$G \colon S_c(\mathbb{R}; H^N) \to \bigcap_{\nu \geqslant \mu} L_{2,\nu}(\mathbb{R}; H)$$

uniformly Lipschitz. Then there exists an $\eta \in \mathbb{R}$ *such that for all* $\nu \geqslant \eta$ *the equation*

$$\partial_{t,\nu} u = G^{\nu}\left(\tau_{h_1} u, \ldots, \tau_{h_N} u\right)$$

admits a solution $u \in \operatorname{dom}(\partial_{t,\nu})$ *which is unique in* $\bigcup_{\nu \geqslant \eta} L_{2,\nu}(\mathbb{R}; H)$*. Moreover, for all* $a \in \mathbb{R}$ *the function* $u_a := \mathbb{1}_{(-\infty,a]} u$ *satisfies*

$$u_a = \mathbb{1}_{(-\infty,a]} \partial_{t,\nu}^{-1} G^{\nu}\left(\tau_{h_1} u_a, \ldots, \tau_{h_N} u_a\right).$$

Proof The assertion follows from Theorem 4.2.3 applied to $F := G \circ \left(\tau_{h_1}, \ldots, \tau_{h_N}\right)$ in conjunction with Lemma 4.3.1. □

Next, we formulate an initial value problem for a subclass of the latter type of equations.

Theorem 4.3.3 *Let* $h > 0$*,* $f \colon \mathbb{R}_{\geq 0} \times H \times H \to H$ *continuous, and* $f(\cdot, 0, 0) \in L_{2,\mu}(\mathbb{R}; H)$ *for some* $\mu > 0$*. Assume that there exists* $L \geqslant 0$ *with*

$$\|f(t,x,y) - f(t,u,v)\| \leqslant L \,\|(x,y) - (u,v)\| \quad \big((t,x,y),(t,u,v) \in \mathbb{R}_{\geq 0} \times H \times H\big).$$

Let $u_0 \in C\left([-h,0]; H\right)$*. Then the initial value problem*

$$\begin{cases} u'(t) = f(t, u(t), u(t-h)) & (t > 0), \\ u(\tau) = u_0(\tau) & (\tau \in [-h, 0]) \end{cases} \tag{4.2}$$

admits a unique continuous solution $u \colon [-h, \infty) \to H$*, continuously differentiable on* $(0, \infty)$*.*

Proof For $t < 0$ let $f(t, \cdot, \cdot) := 0$. We define $F \colon S_{\mathrm{c}}(\mathbb{R}; H) \to \bigcap_{\nu \geqslant \mu} L_{2,\nu}(\mathbb{R}; H)$ by

$$\begin{aligned} &F(\phi)(t) \\ &:= f\big(t, \phi(t) + \mathbb{1}_{[0,\infty)}(t) u_0(0), \phi(t-h) + \mathbb{1}_{[0,\infty)}(t-h) u_0(0) + \mathbb{1}_{[0,h)}(t) u_0(t-h)\big) \end{aligned}$$

for all $t \in \mathbb{R}$. It is easy to see that F is uniformly Lipschitz continuous. Thus, by Theorem 4.2.3, we find $\eta \geqslant \mu$ such that for all $\nu \geqslant \eta$ the equation

$$\partial_{t,\nu} v = F^{\nu}(v)$$

admits a solution $v \in \bigcap_{\nu \geqslant \eta} \operatorname{dom}(\partial_{t,\nu})$ which is unique in $\bigcup_{\nu \geqslant \eta} L_{2,\nu}(\mathbb{R}; H)$. Note that $\operatorname{spt} F^{\nu}(v) \subseteq [0, \infty)$. Hence, $v = 0$ on $(-\infty, 0]$. By Theorem 4.1.2, we obtain that $v(0) = 0$. We claim that $u := v + \mathbb{1}_{[0,\infty)}(\cdot) u_0(0) + \mathbb{1}_{[-h,0)} u_0$ is a solution of (4.2). First of all note that u is continuous on $[-h, \infty)$. Next, for $0 < t < h$ we have that $t - h < 0$ and thus $v(t-h) = 0$ and so we see that

$$\begin{aligned}
&F^{\nu}(v)(t) \\
&\quad = f(t, v(t) + \mathbb{1}_{[0,\infty)}(t)u_0(0), v(t-h) + \mathbb{1}_{[0,\infty)}(t-h)u_0(0) + \mathbb{1}_{[0,h)}(t)u_0(t-h)) \\
&\quad = f(t, u(t), u_0(t-h)).
\end{aligned}$$

Similarly, for $t \geqslant h$ we obtain

$$F^{\nu}(v)(t) = f(t, u(t), u(t-h))$$

and thus, by continuity of f, u_0 and u, it follows that v is continuously differentiable on $(0, \infty)$ and

$$u'(t) = v'(t) = \partial_{t,\nu} v(t) = f(t, u(t), u(t-h)).$$

It remains to show uniqueness. For this, let $w\colon [-h, \infty) \to H$ be a solution of (4.2). Then

$$w(t) = u_0(0) + \int_0^t f(s, w(s), w(s-h))\,\mathrm{d}s \quad (t \geqslant 0)$$

and $w(t) = u_0(t)$ if $t \in [-h, 0]$. Extend w by 0 on $(-\infty, -h)$ and set $\widetilde{v} := w - \mathbb{1}_{[0,\infty)}(\cdot)u_0(0) - \mathbb{1}_{[-h,0)}u_0$. We infer

$$\begin{aligned}
\widetilde{v}(t) &= \int_0^t f(s, w(s), w(s-h))\,\mathrm{d}s \\
&= \int_{-\infty}^t f\big(s, \widetilde{v}(s) + \mathbb{1}_{[0,\infty)}(s)u_0(0), \\
&\qquad\qquad \widetilde{v}(s-h) + \mathbb{1}_{[0,\infty)}(s-h)u_0(0) + \mathbb{1}_{[0,h)}(s)u_0(s-h)\big)\,\mathrm{d}s
\end{aligned}$$

for all $t \in \mathbb{R}$. For $a \in \mathbb{R}$ we set $\widetilde{v}_a := \mathbb{1}_{(-\infty,a]}\widetilde{v} \in \bigcap_{\nu\in\mathbb{R}} L_{2,\nu}(\mathbb{R}; H)$ and obtain, using the above formula for $\widetilde{v}$,

$$\widetilde{v}_a = \mathbb{1}_{(-\infty,a]}\partial_{t,\nu}^{-1} F^{\nu}(\widetilde{v}_a).$$

By uniqueness of the solution of

$$\mathbb{1}_{(-\infty,a]}v = \mathbb{1}_{(-\infty,a]}\partial_{t,\nu}^{-1} F^{\nu}\left(\mathbb{1}_{(-\infty,a]}v\right)$$

it follows that $\widetilde{v}_a = \mathbb{1}_{(-\infty,a]}v$ for all $a \in \mathbb{R}$ and, thus, $u = w$. □

The equation to come involves the whole history of the unknown; that is, the unknown evaluated at $(-\infty, 0]$. For a mapping $u\colon \mathbb{R} \to H$ and $t \in \mathbb{R}$ we define the 'history' of u up to time t as $u_t\colon \mathbb{R}_{\leq 0} \to H$, $u_t(\theta) := u(t+\theta)$ for all $\theta \in \mathbb{R}_{\leq 0}$. Moreover, we define the mapping

$$u_{(\cdot)}\colon \mathbb{R} \ni t \mapsto u_t,$$

which maps each $t \in \mathbb{R}$ to the history of u up to time t.

Lemma 4.3.4 *Let $\mu > 0$. Then*

$$\begin{aligned}\Theta\colon S_{\mathrm{c}}(\mathbb{R}; H) &\to \bigcap_{\nu \geqslant \mu} L_{2,\nu}\big(\mathbb{R}; L_2(\mathbb{R}_{\leq 0}; H)\big)\\ u &\mapsto u_{(\cdot)}\end{aligned}$$

is uniformly Lipschitz continuous. More precisely, for all $\nu > 0$ we have

$$\|\Theta^\nu\| = \frac{1}{\sqrt{2\nu}}.$$

Proof Let $u \in S_{\mathrm{c}}(\mathbb{R}; H)$. Then $\Theta u(t) = u_t \in L_2(\mathbb{R}_{\leq 0}; H)$ for all $t \in \mathbb{R}$ and we compute

$$\begin{aligned}\|\Theta u\|^2_{L_{2,\nu}\left(\mathbb{R}; L_2(\mathbb{R}_{\leq 0}; H)\right)} &= \int_{\mathbb{R}} \int_{\mathbb{R}_{\leq 0}} \|u(t+\theta)\|^2 \,\mathrm{d}\theta\, \mathrm{e}^{-2\nu t}\,\mathrm{d}t\\ &= \int_{\mathbb{R}} \int_{\mathbb{R}_{\leq 0}} \|u(t)\|^2\, \mathrm{e}^{-2\nu(t-\theta)}\,\mathrm{d}\theta\,\mathrm{d}t\\ &= \frac{1}{2\nu}\int_{\mathbb{R}} \|u(t)\|^2\, \mathrm{e}^{-2\nu t}\,\mathrm{d}t.\end{aligned}$$

□

Theorem 4.3.5 *Let H be a Hilbert space, $\mu \in \mathbb{R}$ and let $\Phi\colon S_{\mathrm{c}}\big(\mathbb{R}; L_2(\mathbb{R}_{\leq 0}; H)\big) \to \bigcap_{\nu \geqslant \mu} L_{2,\nu}(\mathbb{R}; H)$ be uniformly Lipschitz. Then, there exists $\eta > 0$ such that for all $\nu \geqslant \eta$ the equation*

$$\partial_{t,\nu} u = \Phi^\nu(u_{(\cdot)})$$

admits a solution $u \in \bigcap_{\nu \geqslant \eta} \operatorname{dom}(\partial_{t,\nu})$ unique in $\bigcup_{\nu \geqslant \eta} L_{2,\nu}(\mathbb{R}; H)$.

Proof This is another application of Theorem 4.2.3. □

4.4 Comments

In a way, the proof of Theorem 4.2.6 is standard PDE-theory in a nutshell; a solution theory for L_p-spaces is used to deduce existence and uniqueness of solutions and a posteriori regularity theory provides more information on the properties of the solution.

Note that—of course—other proofs are available for the Picard–Lindelöf theorem. We chose, however, to present this proof here in order to provide a perspective on classical results. Furthermore, we mention that in order to obtain unique existence for the solution, it suffices to assume that f satisfies a uniform Lipschitz condition with respect to the second variable and that f is measurable. Continuity of f is needed in order to obtain C^1-solutions.

A more detailed exposition and more examples of the theory applied to delay differential equations can be found in [52] and—in a Banach space setting—[85].

There is also a way of dealing with delay differential equations by expanding the state space the problem is formulated in. In this case, it is possible to make use of the rich theory of C_0-semigroups. We refer to [10] for this.

Causality is one of the main concepts for evolutionary equations. We have provided this notion for mappings defined on $L_{2,\nu}$-type spaces only. The situation becomes different if one considers merely densely defined mappings. Then it is a priori unclear, whether for a Lipschitz continuous mapping the continuous extension is also causal. For this we refer to Exercise 4.7 below and to [51, 131], and [138, Chapter 2] as well as to references mentioned there.

Exercises

Exercise 4.1

(a) Let X be a Banach space, $u\colon [a,b] \to X$ continuous. Show that $v\colon (a,b) \to X$ given by

$$v(t) = \int_a^t u(\tau)\,\mathrm{d}\tau$$

is continuously differentiable with $v'(t) = u(t)$ for all $t \in (a,b)$.

(b) Let H be a Hilbert space, and $\nu \in \mathbb{R}$. Let $u \in \operatorname{dom}(\partial_{t,\nu})$ with $\partial_{t,\nu}u$ continuous. Show that u is continuously differentiable and $u' = \partial_{t,\nu}u$.

Exercise 4.2 Prove Corollary 4.1.3.

Exercise 4.3 Let H be a Hilbert space. Show that

$$\operatorname{dom}(\partial_{t,\nu}) \hookrightarrow C_\nu^{1/2}(\mathbb{R}; H) := \left\{ f \in C_\nu(\mathbb{R}; H)\,;\; \mathrm{e}^{-\nu\cdot} f \text{ is } \tfrac{1}{2}\text{-Hölder continuous} \right\},$$

where a function $g: \mathbb{R} \to H$ is said to be $\frac{1}{2}$-*Hölder continuous* if

$$\sup_{\substack{s,t\in\mathbb{R}\\ t\neq s}} \frac{\|g(t)-g(s)\|}{|t-s|^{1/2}} < \infty.$$

Exercise 4.4 Let H be a Hilbert space, $C \subseteq H$ non-empty, closed and convex. Show that the projection, P, of H onto C defines a Lipschitz continuous mapping with Lipschitz semi-norm bounded by 1, where for $x \in H$, $Px \in C$ is the unique element satisfying

$$\|x - Px\|_H = \inf_{y\in C} \|x-y\|_H .$$

Exercise 4.5 Let $h: \mathbb{R} \times \mathbb{R}_{\leq 0} \times \mathbb{R}^n \to \mathbb{R}^n$ be continuous satisfying

$$\|h(t,s,x) - h(t,s,y)\| \leqslant L\,\|x-y\|$$

with $h(\cdot,\cdot,0) = 0$. Let $R > 0$ and $u_0 \in C(\mathbb{R}_{\leq 0}; \mathbb{R}^n)$ have compact support. Show that the initial value problem

$$\begin{cases} u'(t) = \int_{-R}^{0} h(t,s,u_{(t)}(s))\,\mathrm{d}s & (t>0),\\ u(t) = u_0(t) & (t \leqslant 0)\end{cases}$$

admits a unique continuous solution $u: \mathbb{R} \to \mathbb{R}^n$, which is continuously differentiable on $\mathbb{R}_{>0}$.
Hint: Modify Θ from Lemma 4.3.4.

Exercise 4.6 Let H be a Hilbert space. Show that for a uniformly Lipschitz continuous $\Phi: S_c\big(\mathbb{R}; L_2(\mathbb{R}_{<0}; H)^2\big) \to \bigcap_{\nu\geqslant\mu} L_{2,\nu}(\mathbb{R}; H)$ the equation

$$\partial_{t,\nu} u = \Phi^{\nu}\left(u_{(\cdot)}, \left(\partial_{t,\nu} u\right)_{(\cdot)}\right)$$

admits a unique solution $u \in \operatorname{dom}(\partial_{t,\nu})$ for ν large enough.

Exercise 4.7 Let $D \subseteq L_2(\mathbb{R})$ be dense and suppose that $F: D \subseteq L_2(\mathbb{R}) \to L_2(\mathbb{R})$ admits a Lipschitz continuous extension F^0.

(a) Show that F^0 is causal if and only if for all $\phi \in S_c(\mathbb{R}; \mathbb{R})$ and all $a \in \mathbb{R}$ there exists $L \geqslant 0$ such that

$$\left|\left\langle \mathbb{1}_{(-\infty,a]} \cdot (F(f) - F(g)), \phi\right\rangle_{L_2(\mathbb{R})}\right| \leqslant L\,\left\|\mathbb{1}_{(-\infty,a]} \cdot (f-g)\right\|_{L_2(\mathbb{R})}$$

for all $f, g \in D$; that is, the mapping

$$\left(D, \left\|\mathbb{1}_{(-\infty,a]} \cdot (\cdot - \cdot)\right\|_{L_2(\mathbb{R})}\right) \ni f \mapsto F(f) \in \left(L_2(\mathbb{R}), \left|\langle \mathbb{1}_{(-\infty,a]} \cdot (\cdot - \cdot), \phi\rangle_{L_2(\mathbb{R})}\right|\right)$$

is Lipschitz continuous.

(b) For $a \in \mathbb{R}$ let $\operatorname{dom}(F) \cap \operatorname{dom}(F\mathbb{1}_{(-\infty,a]})$ be dense in $L_2(\mathbb{R})$ and if $f, g \in D = \operatorname{dom}(F)$ and $f = g$ on $(-\infty, a]$ then also $F(f) = F(g)$ on $(-\infty, a]$. Show that F^0 is causal.

(c) Assume for all $f, g \in D$ and $a \in \mathbb{R}$ that $f = g$ on $(-\infty, a]$ implies that $F(f) = F(g)$ on $(-\infty, a]$. Show that this is not sufficient for F^0 to be causal. *Hint:* Find a dense subspace $D = \operatorname{dom}(F)$ so that the first condition in (b) is not satisfied.

References

10. A. Bátkai, S. Piazzera, *Semigroups for Delay Equations*, vol. 10. Research Notes in Mathematics. (A. K. Peters, Ltd., Wellesley, MA, 2005)
51. B. Jacob, J.R. Partington, Graphs, closability, and causality of linear time-invariant discrete-time systems. Int. J. Control **73**(11), 1051–1060 (2000)
52. A. Kalauch et al., A Hilbert space perspective on ordinary differential equations with memory term. J. Dyn. Differ. Equ. **26**(2), 369–399 (2014)
85. R. Picard, S. Trostorff, M. Waurick, A functional analytic perspective to delay differential equations. Oper. Matrices **8**(1), 217–236 (2014)
131. M. Waurick, A note on causality in Banach spaces. Indagationes Mathematicae **26**(2), 404–412 (2015)
138. M. Waurick, On the continuous dependence on the coefficients of evolutionary equations. Habilitation. Technische Universität Dresden, 2016. http://arxiv.org/abs/1606.07731

Chapter 5
The Fourier–Laplace Transformation and Material Law Operators

In this chapter we introduce the Fourier–Laplace transformation and use it to define operator-valued functions of $\partial_{t,\nu}$; the so-called material law operators. These operators will play a crucial role when we deal with partial differential equations. In the equations of classical mathematical physics, like the heat equation, wave equation or Maxwell's equation, the involved material parameters, such as heat conductivity or permeability of the underlying medium, are incorporated within these operators. Hence, these operators are called "material law operators". We start our chapter by defining the Fourier transformation and proving Plancherel's theorem in the Hilbert space-valued case, which states that the Fourier transformation defines a unitary operator on $L_2(\mathbb{R}; H)$.

Throughout, let H be a complex Hilbert space.

5.1 The Fourier Transformation

We start by defining the Fourier transformation on $L_1(\mathbb{R}; H)$.

Definition For $f \in L_1(\mathbb{R}; H)$ we define the *Fourier transform* $\hat{f}$ *of* f by

$$\hat{f}(s) := \frac{1}{\sqrt{2\pi}} \int_{\mathbb{R}} \mathrm{e}^{-\mathrm{i}st} f(t)\, \mathrm{d}t \quad (s \in \mathbb{R}).$$

We also introduce

$$C_{\mathrm{b}}(\mathbb{R}; H) := \{f : \mathbb{R} \to H\,;\; f \text{ continuous, bounded}\}$$

endowed with the sup-norm, $\|\cdot\|_\infty$.

C. Seifert et al., *Evolutionary Equations*, Operator Theory: Advances and Applications 287, https://doi.org/10.1007/978-3-030-89397-2_5

Lemma 5.1.1 (Riemann–Lebesgue) *Let* $f \in L_1(\mathbb{R}; H)$. *Then* $\widehat{f} \in C_{\mathrm{b}}(\mathbb{R}; H)$ *and* $\lim_{|t|\to\infty} \left\| \widehat{f}(t) \right\| = 0$. *Moreover,*

$$\left\| \widehat{f} \right\|_\infty \leqslant \frac{1}{\sqrt{2\pi}} \left\| f \right\|_1 .$$

Proof First, note that $\widehat{f}$ is continuous by dominated convergence and bounded with

$$\left\| \widehat{f} \right\|_\infty \leqslant \frac{1}{\sqrt{2\pi}} \left\| f \right\|_1 .$$

This shows that the mapping

$$L_1(\mathbb{R}; H) \to C_{\mathrm{b}}(\mathbb{R}; H), \quad f \mapsto \widehat{f} \tag{5.1}$$

defines a bounded linear operator. Moreover, for $\varphi \in C_{\mathrm{c}}^1(\mathbb{R}; H)$ we compute

$$\widehat{\varphi}(s) = \frac{1}{\sqrt{2\pi}} \int_{\mathbb{R}} \mathrm{e}^{-\mathrm{i}st} \varphi(t) \, \mathrm{d}t = \frac{1}{\sqrt{2\pi}} \frac{1}{\mathrm{i}s} \int_{\mathbb{R}} \mathrm{e}^{-\mathrm{i}st} \varphi'(t) \, \mathrm{d}t$$

for $s \neq 0$ and thus,

$$\limsup_{|s|\to\infty} \|\widehat{\varphi}(s)\| \leqslant \limsup_{|s|\to\infty} \frac{1}{|s|} \frac{1}{\sqrt{2\pi}} \left\| \varphi' \right\|_1 = 0,$$

which shows that $\lim_{|s|\to\infty} \|\widehat{\varphi}(s)\| = 0$. By the facts that $C_{\mathrm{c}}^1(\mathbb{R}; H)$ is dense in $L_1(\mathbb{R}; H)$ (see Lemma 3.1.8), $\left\{ f \in C_{\mathrm{b}}(\mathbb{R}; H) \, ; \, \lim_{|t|\to\infty} \|f(t)\| = 0 \right\}$ is a closed subspace of $C_{\mathrm{b}}(\mathbb{R}; H)$ and the operator in (5.1) is bounded, the assertion follows. □

It is our main goal to extend the definition of the Fourier transformation to functions in $L_2(\mathbb{R}; H)$. For doing so, we make use of the Schwartz space of rapidly decreasing functions.

Definition We define

$$\mathcal{S}(\mathbb{R}; H) := \left\{ f \in C^\infty(\mathbb{R}; H) \, ; \, \forall n, k \in \mathbb{N}_0 : \left(t \mapsto t^k f^{(n)}(t) \right) \in C_{\mathrm{b}}(\mathbb{R}; H) \right\}$$

to be the *Schwartz space* of rapidly decreasing functions on $\mathbb{R}$ with values in H.

As usual we abbreviate $\mathcal{S}(\mathbb{R}) := \mathcal{S}(\mathbb{R}; \mathbb{K})$.

Remark 5.1.2 $\mathcal{S}(\mathbb{R}; H)$ is a Fréchet space with respect to the seminorms

$$\mathcal{S}(\mathbb{R}; H) \ni f \mapsto \sup_{t\in\mathbb{R}} \left\| t^k f^{(n)}(t) \right\| \quad (n, k \in \mathbb{N}_0).$$

Moreover, $\mathcal{S}(\mathbb{R}; H) \subseteq \bigcap_{p\in[1,\infty]} L_p(\mathbb{R}; H)$. Indeed, $\mathcal{S}(\mathbb{R}; H) \subseteq L_\infty(\mathbb{R}; H)$ by definition, and for $f \in \mathcal{S}(\mathbb{R}; H)$ and $1 \leqslant p < \infty$ we have that

$$\int_{\mathbb{R}} \|f(t)\|^p \,\mathrm{d}t = \int_{\mathbb{R}} \frac{1}{(1+|t|)^{2p}} \left\|(1+|t|)^2 f(t)\right\|^p \,\mathrm{d}t$$
$$\leqslant \sup_{t\in\mathbb{R}} \left\|(1+|t|)^2 f(t)\right\|^p \int_{\mathbb{R}} \frac{1}{(1+|t|)^{2p}} \,\mathrm{d}t < \infty.$$

Proposition 5.1.3 *For $f \in \mathcal{S}(\mathbb{R}; H)$ we have $\widehat{f} \in \mathcal{S}(\mathbb{R}; H)$ and the mapping*

$$\mathcal{S}(\mathbb{R}; H) \to \mathcal{S}(\mathbb{R}; H), \quad f \mapsto \widehat{f}$$

is bijective. Moreover, for $f, g \in L_1(\mathbb{R}; H)$ we have that

$$\int_{\mathbb{R}} \left\langle \widehat{f}(t), g(t) \right\rangle \mathrm{d}t = \int_{\mathbb{R}} \langle f(t), \widehat{g}(-t) \rangle \,\mathrm{d}t. \tag{5.2}$$

Additionally, if $f, \widehat{f} \in L_1(\mathbb{R}; H)$ then

$$f(t) = \widehat{\widehat{f}}(-t) \quad (t \in \mathbb{R}). \tag{5.3}$$

Proof Let $f \in \mathcal{S}(\mathbb{R}; H)$. By Exercise 5.1 we have

$$\widehat{f}'(s) = \frac{1}{\sqrt{2\pi}} \int_{\mathbb{R}} (-\mathrm{i}t)\mathrm{e}^{-\mathrm{i}st} f(t) \,\mathrm{d}t = -\mathrm{i}\big(\widehat{t \mapsto tf(t)}\big)(s) \quad (s \in \mathbb{R}) \tag{5.4}$$

and

$$s\widehat{f}(s) = \frac{\mathrm{i}}{\sqrt{2\pi}} \int_{\mathbb{R}} (-\mathrm{i}s)\,\mathrm{e}^{-\mathrm{i}st} f(t) \,\mathrm{d}t = -\mathrm{i}\widehat{f'}(s) \quad (s \in \mathbb{R}). \tag{5.5}$$

Using these formulas, one can show that $\widehat{f} \in \mathcal{S}(\mathbb{R}; H)$. Since the bijectivity of the Fourier transformation on $\mathcal{S}(\mathbb{R}; H)$ would follow from (5.3), it suffices to prove the formulas (5.2) and (5.3). Let $f, g \in L_1(\mathbb{R}; H)$. Then we compute using Proposition 3.1.6 and Fubini's theorem

$$\int_{\mathbb{R}} \left\langle \widehat{f}(t), g(t) \right\rangle \mathrm{d}t = \int_{\mathbb{R}} \frac{1}{\sqrt{2\pi}} \left\langle \int_{\mathbb{R}} \mathrm{e}^{-\mathrm{i}st} f(s) \,\mathrm{d}s, g(t) \right\rangle \mathrm{d}t$$
$$= \int_{\mathbb{R}} \int_{\mathbb{R}} \frac{1}{\sqrt{2\pi}} \mathrm{e}^{\mathrm{i}st} \langle f(s), g(t) \rangle \,\mathrm{d}s \,\mathrm{d}t$$

$$= \int_{\mathbb{R}} \left\langle f(s), \frac{1}{\sqrt{2\pi}} \int_{\mathbb{R}} e^{ist} g(t)\, dt \right\rangle ds$$

$$= \int_{\mathbb{R}} \langle f(s), \widehat{g}(-s) \rangle \, ds,$$

which yields (5.2). For proving formula (5.3), we consider the function γ defined by $\gamma(t) := e^{-\frac{t^2}{2}}$ for $t \in \mathbb{R}$. Clearly, $\gamma \in \mathcal{S}(\mathbb{R})$. We claim that $\widehat{\gamma} = \gamma$. Indeed, we observe that γ solves the initial value problem $y' + ty = 0$ subject to $y(0) = 1$; if we can show that $\widehat{\gamma}$ solves the same initial value problem, then their equality would follow from the uniqueness of the solution. First, we observe that $\widehat{\gamma}(0) = \frac{1}{\sqrt{2\pi}} \int_{\mathbb{R}} e^{-\frac{t^2}{2}} \, dt = 1$. Second, we compute using the formulas (5.4) and (5.5) that

$$\widehat{\gamma}'(s) = -i\big(\widehat{t \mapsto t\gamma(t)}\big)(s) = i\widehat{\gamma'}(s) = -s\widehat{\gamma}(s) \quad (s \in \mathbb{R}).$$

Altogether, we have shown that $\widehat{\gamma}$ solves the same initial value problem as γ and hence, $\widehat{\gamma} = \gamma$. Let now $f \in L_1(\mathbb{R}; H)$ with $\widehat{f} \in L_1(\mathbb{R}; H)$, $a > 0$ and $x \in H$. Then we compute using (5.2)

$$\begin{aligned}
\left\langle \int_{\mathbb{R}} \widehat{f}(t)\gamma(at)e^{ist}\, dt, x \right\rangle &= \int_{\mathbb{R}} \left\langle \widehat{f}(t), \gamma(at)xe^{-ist} \right\rangle dt = \int_{\mathbb{R}} \left\langle f(t), \big(\widehat{\gamma(a\cdot)xe^{-is(\cdot)}}\big)(-t) \right\rangle dt \\
&= \int_{\mathbb{R}} \left\langle f(t), \frac{1}{\sqrt{2\pi}} \int_{\mathbb{R}} \gamma(ar)xe^{-isr}e^{itr}\, dr \right\rangle dt \\
&= \frac{1}{a} \int_{\mathbb{R}} \left\langle f(t), \widehat{\gamma}\left(\frac{s-t}{a}\right)x \right\rangle dt = \frac{1}{a} \int_{\mathbb{R}} \left\langle f(t), \gamma\left(\frac{s-t}{a}\right)x \right\rangle dt \\
&= \int_{\mathbb{R}} \langle f(s-at), \gamma(t)x \rangle \, dt = \left\langle \int_{\mathbb{R}} f(s-at)\gamma(t)\, dt, x \right\rangle
\end{aligned}$$

for each $s \in \mathbb{R}$. Since this holds for all $x \in H$ we get

$$\int_{\mathbb{R}} \widehat{f}(t)\gamma(at)e^{ist}\, dt = \int_{\mathbb{R}} f(s-at)\gamma(t)\, dt \quad (s \in \mathbb{R}).$$

Letting $a \to 0$ in the latter equality, we obtain

$$\int_{\mathbb{R}} \widehat{f}(t)e^{ist}\, dt = \lim_{a\to 0} \int_{\mathbb{R}} f(s-at)\gamma(t)\, dt \quad (s \in \mathbb{R}), \tag{5.6}$$

where we have used dominated convergence for the term on the left-hand side. In order to compute the limit on the right-hand side, we first observe that

$$\int_{\mathbb{R}} \left\| \int_{\mathbb{R}} f(s-at)\gamma(t)\, dt \right\| ds \leqslant \int_{\mathbb{R}} \int_{\mathbb{R}} \|f(s-at)\|\, ds\, \gamma(t)\, dt = \|f\|_1 \, \|\gamma\|_1 ,$$

and hence, for each $a > 0$ the operator

$$S_a \colon L_1(\mathbb{R}; H) \to L_1(\mathbb{R}; H),$$
$$f \mapsto \left(s \mapsto \int_{\mathbb{R}} f(s - at)\gamma(t) \, \mathrm{d}t \right)$$

is bounded by $\|\gamma\|_1$. Moreover, since $S_a\psi \to \psi(\cdot) \|\gamma\|_1$ as $a \to 0$ for $\psi \in C_c(\mathbb{R}; H)$, we infer that

$$S_a f \to f(\cdot) \|\gamma\|_1 \quad (a \to 0)$$

for each $f \in L_1(\mathbb{R}; H)$. Hence, passing to a suitable sequence $(a_n)_n$ in $\mathbb{R}_{>0}$ tending to 0, we get

$$\lim_{n\to\infty} \left(S_{a_n} f\right)(s) \to f(s) \|\gamma\|_1 \quad (\text{a.e. } s \in \mathbb{R}).$$

Using this identity for the right-hand side of (5.6), we get

$$\int_{\mathbb{R}} \widehat{f}(t)\mathrm{e}^{\mathrm{i}st} \, \mathrm{d}t = f(s) \|\gamma\|_1 \quad (\text{a.e. } s \in \mathbb{R}),$$

and since $\|\gamma\|_1 = \sqrt{2\pi}$, we derive (5.3). □

With these preparations at hand, we are now able to prove the main theorem of this section.

Theorem 5.1.4 (Plancherel) *The mapping*

$$\mathcal{F} \colon \mathcal{S}(\mathbb{R}; H) \subseteq L_2(\mathbb{R}; H) \to L_2(\mathbb{R}; H), \ f \mapsto \widehat{f}$$

extends to a unitary operator on $L_2(\mathbb{R}; H)$, *again denoted by* $\mathcal{F}$, *the* Fourier transformation. *Moreover,* $\mathcal{F}^* = \mathcal{F}^{-1}$ *is given by* $f \mapsto \widehat{f}(-\cdot)$.

Proof Using (5.2) and (5.3) we obtain that

$$\langle \widehat{f}, \widehat{g} \rangle_2 = \int_{\mathbb{R}} \langle \widehat{f}(t), \widehat{g}(t) \rangle \, \mathrm{d}t = \int_{\mathbb{R}} \langle f(t), \widehat{\widehat{g}}(-t) \rangle \, \mathrm{d}t = \int_{\mathbb{R}} \langle f(t), g(t) \rangle \, \mathrm{d}t = \langle f, g \rangle_2$$

for all $f, g \in \mathcal{S}(\mathbb{R}; H)$ and thus, in particular,

$$\|f\|_2 = \|\mathcal{F}f\|_2 . \tag{5.7}$$

Moreover, $\mathrm{dom}(\mathcal{F}) = \mathrm{ran}(\mathcal{F}) = \mathcal{S}(\mathbb{R}; H)$ is dense in $L_2(\mathbb{R}; H)$ and hence, the first assertion follows by Exercise 5.2. As $\mathcal{F}$ is unitary, we have $\mathcal{F}^* = \mathcal{F}^{-1}$, thus, by (5.2) applied to $f, g \in \mathcal{S}(\mathbb{R}; H)$, we read off (using Proposition 2.3.8) that $\mathcal{F}^{-1} = (f \mapsto \widehat{f}(-\cdot))$, which yields all the claims of the theorem at hand. □

Remark 5.1.5 We emphasise that for $f \in L_2(\mathbb{R}; H)$ the Fourier transform $\mathcal{F}f$ is not given by the integral expression for L_1-functions, simply because the integral does not need to exist. However, by dominated convergence

$$\mathcal{F}f = \lim_{R\to\infty} \frac{1}{\sqrt{2\pi}} \int_{-R}^{R} \mathrm{e}^{-\mathrm{i}t(\cdot)} f(t)\,\mathrm{d}t,$$

where the limit is taken in $L_2(\mathbb{R}; H)$.

5.2 The Fourier–Laplace Transformation and Its Relation to the Time Derivative

We now use the Fourier transformation to define an analogous transformation on our exponentially weighted L_2-type spaces; the so-called Fourier–Laplace transformation. We recall from Corollary 3.2.5 that for $\nu \in \mathbb{R}$ the mapping

$$\exp(-\nu\mathrm{m})\colon L_{2,\nu}(\mathbb{R}; H) \to L_2(\mathbb{R}; H),\ f \mapsto \big(t \mapsto \mathrm{e}^{-\nu t} f(t)\big)$$

is unitary. In a similar fashion, we obtain that

$$\exp(-\nu\mathrm{m})\colon L_{1,\nu}(\mathbb{R}; H) \to L_1(\mathbb{R}; H),\ f \mapsto \big(t \mapsto \mathrm{e}^{-\nu t} f(t)\big)$$

defines an isometry.

Definition Let $\nu \in \mathbb{R}$. We define the *Fourier–Laplace transformation* as

$$\mathcal{L}_\nu\colon L_{2,\nu}(\mathbb{R}; H) \to L_2(\mathbb{R}; H),\ f \mapsto \mathcal{F}\exp(-\nu\mathrm{m})f.$$

We can also consider the Fourier–Laplace transformation as a mapping from $L_{1,\nu}(\mathbb{R}; H)$ to $C_\mathrm{b}(\mathbb{R}; H)$; that is,

$$\mathcal{L}_\nu\colon L_{1,\nu}(\mathbb{R}; H) \to C_\mathrm{b}(\mathbb{R}; H),\ f \mapsto \mathcal{F}\exp(-\nu\mathrm{m})f.$$

Remark 5.2.1 Note that $\mathcal{L}_\nu = \mathcal{F}\exp(-\nu\mathrm{m})$ is unitary as an operator from $L_{2,\nu}(\mathbb{R}; H)$ to $L_2(\mathbb{R}; H)$ since it is the composition of two unitary operators. For $\varphi \in C_\mathrm{c}^\infty(\mathbb{R}; H)$, we have the expression

$$(\mathcal{L}_\nu\varphi)(t) = \frac{1}{\sqrt{2\pi}} \int_{\mathbb{R}} \mathrm{e}^{-(\mathrm{i}t+\nu)s} \varphi(s)\,\mathrm{d}s \quad (t \in \mathbb{R}),$$

which shows that $\mathcal{L}_\nu$ can be interpreted as a shifted variant of the Fourier transformation, where the real part in the exponent equals ν instead of zero.

Our next goal is to show that the Fourier–Laplace transformation provides a spectral representation of our time derivative, $\partial_{t,\nu}$.

Definition Let $V\colon \mathbb{R} \to \mathbb{K}$ be measurable. We define the *multiplication-by-V operator* as

$$V(\mathrm{m})\colon \operatorname{dom}(V(\mathrm{m})) \subseteq L_2(\mathbb{R}; H) \to L_2(\mathbb{R}; H),\ f \mapsto \big(t \mapsto V(t)f(t)\big)$$

with

$$\operatorname{dom}(V(\mathrm{m})) := \big\{f \in L_2(\mathbb{R}; H)\,;\ \big(t \mapsto V(t)f(t)\big) \in L_2(\mathbb{R}; H)\big\}.$$

In particular, if V is the identity on $\mathbb{R}$ we will just write m instead of $\mathrm{id}(\mathrm{m})$ and call it the *multiplication-by-the-argument operator.*

Remark 5.2.2 Note that the multiplication-by-V operator is a vector-valued analogue of the multiplication operator seen in Theorems 2.4.3 and 2.4.7. The statements in these theorems generalise (easily) to the vector-valued situation at hand. Thus, as in Theorem 2.4.3, one shows that m is selfadjoint. Moreover, when $H \neq \{0\}$, in a similar fashion to the arguments carried out in Theorem 2.4.7 one shows that

$$\sigma(\mathrm{m}) = \mathbb{R}.$$

In order to avoid trivial cases, we shall assume throughout that $H \neq \{0\}$.

Theorem 5.2.3 *Let $\nu \in \mathbb{R}$. Then*

$$\partial_{t,\nu} = \mathcal{L}_\nu^*(\mathrm{im} + \nu)\mathcal{L}_\nu.$$

In particular,

$$\sigma(\partial_{t,\nu}) = \{\mathrm{i}t + \nu\,;\ t \in \mathbb{R}\}.$$

Proof We first prove the assertion for $\nu \neq 0$ and show that

$$I_\nu = \mathcal{L}_\nu^*\left(\frac{1}{\mathrm{im} + \nu}\right)\mathcal{L}_\nu.$$

The assertion will then follow by Theorem 2.4.3(d). Note that $\frac{1}{\mathrm{im}+\nu} \in L(L_2(\mathbb{R}; H))$ by Proposition 2.4.6, and hence, both operators I_ν and $\mathcal{L}_\nu^*(\frac{1}{\mathrm{im}+\nu})\mathcal{L}_\nu$ are bounded and defined on the whole of $L_{2,\nu}(\mathbb{R}; H)$. Thus, it suffices to prove the equality on a dense subset of $L_{2,\nu}(\mathbb{R}; H)$, like $C_\mathrm{c}(\mathbb{R}; H)$. We will just do the computation for the case when $\nu > 0$. So, let $\varphi \in C_c(\mathbb{R}; H)$ and compute

$$\begin{aligned}(\mathcal{L}_\nu I_\nu \varphi)(t) &= \frac{1}{\sqrt{2\pi}} \int_{\mathbb{R}} \mathrm{e}^{-(\mathrm{i}t+\nu)s} \int_{-\infty}^{s} \varphi(r)\,\mathrm{d}r\,\mathrm{d}s = \frac{1}{\sqrt{2\pi}} \int_{\mathbb{R}} \int_{r}^{\infty} \mathrm{e}^{-(\mathrm{i}t+\nu)s}\,\mathrm{d}s\,\varphi(r)\,\mathrm{d}r \\ &= \frac{1}{\sqrt{2\pi}} \frac{1}{\mathrm{i}t+\nu} \int_{\mathbb{R}} \mathrm{e}^{-(\mathrm{i}t+\nu)r} \varphi(r)\,\mathrm{d}r = \frac{1}{\mathrm{i}t+\nu} (\mathcal{L}_\nu \varphi)(t)\end{aligned}$$

for $t \in \mathbb{R}$. For $\nu < 0$ the computation is analogous. In the case when $\nu = 0$ we observe that

$$\begin{aligned}\partial_{t,0} &= \exp(-\nu\mathrm{m})(\partial_{t,\nu} - \nu)\exp(-\nu\mathrm{m})^{-1} = \exp(-\nu\mathrm{m})\mathcal{L}_\nu^*(\mathrm{im} + \nu - \nu)\mathcal{L}_\nu \exp(-\nu\mathrm{m})^{-1} \\ &= \mathcal{L}_0^*(\mathrm{im})\mathcal{L}_0. \qquad \square\end{aligned}$$

5.3 Material Law Operators

Using the multiplication operator representation of $\partial_{t,\nu}$ via the Fourier–Laplace transformation, we can assign a functional calculus to this operator. We will do this in the following and define operator-valued functions of $\partial_{t,\nu}$. The class of functions used for this calculus are the so-called material laws. We begin by defining this function class.

Definition A mapping $M\colon \operatorname{dom}(M) \subseteq \mathbb{C} \to L(H)$ is called a *material law* if

(a) $\operatorname{dom}(M)$ is open and M is holomorphic (i.e., complex differentiable; see also Exercise 5.3),
(b) there exists some $\nu \in \mathbb{R}$ such that $\mathbb{C}_{\mathrm{Re}>\nu} \subseteq \operatorname{dom}(M)$ and

$$\|M\|_{\infty,\mathbb{C}_{\mathrm{Re}>\nu}} := \sup_{z\in\mathbb{C}_{\mathrm{Re}>\nu}} \|M(z)\| < \infty.$$

Moreover, we set

$$\mathrm{s_b}(M) := \inf\left\{\nu \in \mathbb{R}\,;\ \mathbb{C}_{\mathrm{Re}>\nu} \subseteq \operatorname{dom}(M) \text{ and } \|M\|_{\infty,\mathbb{C}_{\mathrm{Re}>\nu}} < \infty\right\}$$

to be the *abscissa of boundedness* of M.

Example 5.3.1 Let us state various examples of material laws.

(a) Polynomials in z^{-1}: Let $n \in \mathbb{N}_0$, $M_0, \ldots, M_n \in L(H)$. Then

$$M(z) := \sum_{k=0}^{n} z^{-k} M_k \quad (z \in \mathbb{C} \setminus \{0\})$$

defines a material law with

$$\mathrm{s_b}(M) = \begin{cases} -\infty & \text{if } M_1 = \ldots = M_n = 0, \\ 0 & \text{otherwise.} \end{cases}$$

(b) Series in z^{-1}: Let $(M_k)_{k\in\mathbb{N}}$ in $L(H)$ such that $\sum_{k=0}^{\infty} \|M_k\| r^{-k} < \infty$ for some $r > 0$. Then

$$M(z) := \sum_{k=0}^{\infty} z^{-k} M_k \quad (z \in \mathbb{C} \setminus \{0\})$$

defines a material law with $\mathrm{s_b}(M) \leqslant r$.

(c) Exponentials: Let $h \in \mathbb{R}$, $M_0 \in L(H)$ where $M_0 \neq 0$ and set

$$M(z) := M_0 \mathrm{e}^{zh} \quad (z \in \mathbb{C}).$$

Then M is a material law if and only if $h \leqslant 0$. In this case, $\mathrm{s_b}(M) = -\infty$.

(d) Laplace transforms: Let $\nu \in \mathbb{R}$ and $k \in L_{1,\nu}(\mathbb{R})$ with $\operatorname{spt} k \subseteq \mathbb{R}_{\geq 0}$. Then

$$M(z) := \sqrt{2\pi}(\mathcal{L}k)(z) := \int_0^{\infty} \mathrm{e}^{-zt} k(t)\,\mathrm{d}t \quad (z \in \mathbb{C}_{\mathrm{Re}>\nu})$$

defines a material law with $\mathrm{s_b}(M) \leqslant \nu$.

(e) Fractional powers: Let $M_0 \in L(H)$, $M_0 \neq 0$, $\alpha \in \mathbb{R}$ and set

$$M(z) := M_0 z^{-\alpha} \quad (z \in \mathbb{C} \setminus \mathbb{R}_{\leq 0}),$$

where we set

$$\left(r\mathrm{e}^{\mathrm{i}\theta}\right)^{-\alpha} := r^{-\alpha}\mathrm{e}^{-\mathrm{i}\alpha\theta} \quad (r > 0, \theta \in (-\pi, \pi)).$$

Then M is a material law if and only if $\alpha \geqslant 0$ and

$$\mathrm{s_b}(M) = \begin{cases} -\infty & \text{if } \alpha = 0, \\ 0 & \text{otherwise.} \end{cases}$$

For material laws M we now define the corresponding material law operators in terms of the functional calculus induced by the spectral representation of $\partial_{t,\nu}$.

Proposition 5.3.2 *Let* $M\colon \operatorname{dom}(M) \subseteq \mathbb{C} \to L(H)$ *be a material law. Then, for* $\nu > \mathrm{s_b}(M)$, *the operator*

$$M(\mathrm{im} + \nu)\colon L_2(\mathbb{R}; H) \to L_2(\mathbb{R}; H),\ f \mapsto \left(t \mapsto M(\mathrm{i}t + \nu) f(t)\right)$$

is bounded. Moreover, we define the material law operator

$$M(\partial_{t,\nu}) := \mathcal{L}_\nu^* M(\mathrm{im} + \nu)\mathcal{L}_\nu \in L(L_{2,\nu}(\mathbb{R}; H))$$

and obtain

$$\left\|M(\partial_{t,\nu})\right\| \leqslant \|M\|_{\infty,\mathbb{C}_{\mathrm{Re}>\nu}}.$$

Proof The proof is clear. □

Remark 5.3.3 The set of material laws is an algebra and the mapping of assigning a material law to its corresponding material law operator is an algebra homomorphism in the following sense. For $j \in \{1, 2\}$ let $M_j : \mathrm{dom}(M_j) \subseteq \mathbb{C} \to L(H)$ be material laws, $\lambda \in \mathbb{C}$. Then $M_1 + M_2$ (with domain $\mathrm{dom}(M_1) \cap \mathrm{dom}(M_2)$), λM_1 and $M_1 \cdot M_2$ (with domain $\mathrm{dom}(M_1) \cap \mathrm{dom}(M_2)$) are material laws as well. Moreover, $\mathrm{s_b}(M_1 + M_2), \mathrm{s_b}(M_1 \cdot M_2) \leqslant \max\{\mathrm{s_b}(M_1), \mathrm{s_b}(M_2)\}$. Furthermore, if $M_2(z)$ is a scalar for all $z \in \mathrm{dom}(M_2)$, then for $\nu > \max\{\mathrm{s_b}(M_1), \mathrm{s_b}(M_2)\}$ we have $(M_1M_2)(\partial_{t,\nu}) = M_1(\partial_{t,\nu})M_2(\partial_{t,\nu}) = M_2(\partial_{t,\nu})M_1(\partial_{t,\nu}) = (M_2M_1)(\partial_{t,\nu})$.

Example 5.3.4 We now revisit the material laws presented in Example 5.3.1 and compute their corresponding operators, $M(\partial_{t,\nu})$.

(a) Let $n \in \mathbb{N}_0$, $M_0, \ldots, M_n \in L(H)$ and

$$M(z) := \sum_{k=0}^{n} z^{-k} M_k \quad (z \in \mathbb{C} \setminus \{0\}).$$

Then, for $\nu > 0$, one obviously has

$$M(\partial_{t,\nu}) = \sum_{k=0}^{n} \partial_{t,\nu}^{-k} M_k,$$

due to Theorem 5.2.3.

(b) Let $(M_k)_{k\in\mathbb{N}}$ in $L(H)$ such that $\sum_{k=0}^{\infty} \|M_k\| r^{-k} < \infty$ for some $r > 0$ and

$$M(z) := \sum_{k=0}^{\infty} z^{-k} M_k \quad (z \in \mathbb{C} \setminus \{0\}).$$

Then, for $\nu > r$, one has

$$M(\partial_{t,\nu}) = \sum_{k=0}^{\infty} \partial_{t,\nu}^{-k} M_k$$

again on account of Theorem 5.2.3.

(c) Let $h \leqslant 0$, $M_0 \in L(H)$ and

$$M(z) := M_0 \mathrm{e}^{zh} \quad (z \in \mathbb{C}).$$

Then, for $\nu \in \mathbb{R}$, we have

$$M(\partial_{t,\nu}) = M_0 \tau_h,$$

where

$$\tau_h \colon L_{2,\nu}(\mathbb{R}; H) \to L_{2,\nu}(\mathbb{R}; H), \; f \mapsto \big(t \mapsto f(t+h)\big).$$

Indeed, for $\varphi \in C_c(\mathbb{R}; H)$ we compute

$$\begin{aligned} (\mathcal{L}_\nu M_0 \tau_h \varphi)(t) &= \frac{1}{\sqrt{2\pi}} \int_{\mathbb{R}} \mathrm{e}^{-(\mathrm{i}t+\nu)s} M_0 \varphi(s+h)\,\mathrm{d}s \\ &= M_0 \frac{1}{\sqrt{2\pi}} \int_{\mathbb{R}} \mathrm{e}^{-(\mathrm{i}t+\nu)(s-h)} \varphi(s)\,\mathrm{d}s = M(\mathrm{i}t+\nu)\,(\mathcal{L}_\nu \varphi)(t) \end{aligned}$$

for all $t \in \mathbb{R}$, where we have used Proposition 3.1.6 in the second line. Hence,

$$M_0 \tau_h \varphi = \mathcal{L}_\nu^* M(\mathrm{i}\mathrm{m} + \nu) \mathcal{L}_\nu \varphi = M(\partial_{t,\nu})\varphi$$

and since $C_c(\mathbb{R}; H)$ is dense in $L_{2,\nu}(\mathbb{R}; H)$ the assertion follows.

(d) Let $\nu \in \mathbb{R}$ and $k \in L_{1,\nu}(\mathbb{R})$ with $\operatorname{spt} k \subseteq \mathbb{R}_{\geq 0}$ and

$$M(z) := \sqrt{2\pi}\,(\mathcal{L}k)(z) \quad (z \in \mathbb{C}_{\mathrm{Re} > \nu}).$$

Then, by Exercise 5.4,

$$M(\partial_{t,\mu}) = k *$$

for each $\mu > \nu$.

(e) Let $M_0 \in L(H)$, $\alpha > 0$ and

$$M(z) := M_0 z^{-\alpha} \quad (z \in \mathbb{C} \setminus \mathbb{R}_{\leq 0}).$$

Then for $\nu > 0$ we have

$$\big(M(\partial_{t,\nu}) f\big)(t) = M_0 \int_{-\infty}^{t} \frac{1}{\Gamma(\alpha)} (t-s)^{\alpha-1} f(s)\,\mathrm{d}s \quad (\text{a.e. } t \in \mathbb{R}) \tag{5.8}$$

for each $f \in L_{2,\nu}(\mathbb{R}; H)$; see Exercise 5.5. This formula gives rise to the definition

$$\left(\partial_{t,\nu}^{-\alpha} f\right)(t) := \int_{-\infty}^{t} \frac{1}{\Gamma(\alpha)}(t-s)^{\alpha-1} f(s)\,\mathrm{d}s \quad (t \in \mathbb{R}),$$

which is known as the *(Riemann–Liouville) fractional integral of order* α.

Throughout the previous examples, the operator $M(\partial_{t,\nu})$ did not depend on the actual value of ν. Indeed, this is true for all material laws. In order to see this, we need the following lemma.

Lemma 5.3.5 *Let $\mu, \nu \in \mathbb{R}$ with $\mu < \nu$, and set $U := \{z \in \mathbb{C};\ \operatorname{Re} z \in (\mu, \nu)\}$. Let $g \colon \overline{U} \to H$ be continuous and holomorphic on U such that $g(\mathrm{i}\cdot + \nu), g(\mathrm{i}\cdot + \mu) \in L_2(\mathbb{R}; H)$ and there exists a sequence $(R_n)_{n\in\mathbb{N}}$ in $\mathbb{R}_{\geqslant 0}$ such that $R_n \to \infty$ and*

$$\int_{\mu}^{\nu} \|g(\pm \mathrm{i}R_n + \rho)\|\,\mathrm{d}\rho \to 0 \quad (n \to \infty). \tag{5.9}$$

Then

$$\mathcal{L}_\mu^* g(\mathrm{i}\cdot + \mu) = \mathcal{L}_\nu^* g(\mathrm{i}\cdot + \nu).$$

Proof Let $t \in \mathbb{R}$. By Cauchy's integral theorem, we have that

$$\int_{\gamma_{R_n}} g(z)\mathrm{e}^{zt}\,\mathrm{d}z = 0,$$

where γ_{R_n} is the rectangular closed path with corners $\pm\mathrm{i}R_n + \mu, \pm\mathrm{i}R_n + \nu$ (see Fig. 5.1). Thus, we have that

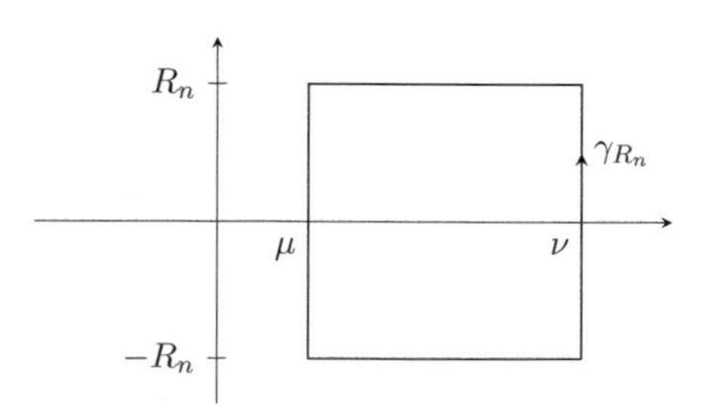

Fig. 5.1 Curve γ_{R_n}

$$\begin{aligned} &\mathrm{i}\int_{-R_n}^{R_n} g(\mathrm{i}s+\nu)\mathrm{e}^{(\mathrm{i}s+\nu)t}\,\mathrm{d}s - \mathrm{i}\int_{-R_n}^{R_n} g(\mathrm{i}s+\mu)\mathrm{e}^{(\mathrm{i}s+\mu)t}\,\mathrm{d}s \\ &= -\int_{\mu}^{\nu} g(-\mathrm{i}R_n+\rho)\mathrm{e}^{(-\mathrm{i}R_n+\rho)t}\,\mathrm{d}\rho + \int_{\mu}^{\nu} g(\mathrm{i}R_n+\rho)\mathrm{e}^{(\mathrm{i}R_n+\rho)t}\,\mathrm{d}\rho. \end{aligned} \tag{5.10}$$

Note that with the help of the formula for the inverse Fourier transformation (see Theorem 5.1.4) and $\mathcal{L}_\nu^* = (\mathcal{F}\exp(-\nu \mathrm{m}))^* = \exp(-\nu \mathrm{m})^{-1}\mathcal{F}^*$ the left-hand side of (5.10) is nothing but

$$\sqrt{2\pi}\mathrm{i}\left(\left(\mathcal{L}_\nu^*\mathbb{1}_{[-R_n,R_n]}g(\mathrm{i}\cdot+\nu)\right)(t) - \left(\mathcal{L}_\mu^*\mathbb{1}_{[-R_n,R_n]}g(\mathrm{i}\cdot+\mu)\right)(t)\right),$$

and hence, there is a subsequence of $(R_n)_n$ (which we do not relabel) such that the left-hand side of (5.10) tends to

$$\sqrt{2\pi}\mathrm{i}\left(\left(\mathcal{L}_\nu^* g(\mathrm{i}\cdot+\nu)\right)(t) - \left(\mathcal{L}_\mu^* g(\mathrm{i}\cdot+\mu)\right)(t)\right)$$

for almost every $t \in \mathbb{R}$ as $n \to \infty$. As such, all we need to show is that the right-hand side of (5.10) tends to 0 as $n \to \infty$, which obviously follows by (5.9). □

Theorem 5.3.6 *Let $M\colon \operatorname{dom}(M) \subseteq \mathbb{C} \to L(H)$ be a material law. Then, for $\mu, \nu > \mathrm{s_b}(M)$ and $f \in L_{2,\nu}(\mathbb{R}; H) \cap L_{2,\mu}(\mathbb{R}; H)$, we have*

$$M(\partial_{t,\nu})f = M(\partial_{t,\mu})f.$$

Moreover, $M(\partial_{t,\nu})$ is causal for all $\nu > \mathrm{s_b}(M)$.

Proof Let $\mu < \nu$. We prove the assertion for $f = \mathbb{1}_{[a,b]} \cdot x$ with $a < b$ and $x \in H$ first. For $\rho \in \mathbb{R}$ we compute

$$\left(\mathcal{L}_\rho f\right)(t) = \frac{1}{\sqrt{2\pi}}\int_a^b x\mathrm{e}^{-(\mathrm{i}t+\rho)s}\,\mathrm{d}s = \frac{1}{\sqrt{2\pi}}\frac{1}{\mathrm{i}t+\rho}\left(\mathrm{e}^{-(\mathrm{i}t+\rho)a} - \mathrm{e}^{-(\mathrm{i}t+\rho)b}\right)x.$$

for all $t \in \mathbb{R} \setminus \{0\}$. Moreover, we define

$$g(z) := \frac{1}{\sqrt{2\pi}}M(z)x\frac{1}{z}\left(\mathrm{e}^{-za} - \mathrm{e}^{-zb}\right) \quad (z \in \mathbb{C}_{\mathrm{Re}\geqslant\mu} \setminus \{0\})$$

and prove that g satisfies the assumptions of Lemma 5.3.5. First, we note that g is bounded on $\{z \in \mathbb{C}\,;\ \mu \leqslant \operatorname{Re} z \leqslant \nu\} \setminus \{0\}$. Indeed, we only need to prove that it is bounded near 0 provided that $\mu \leqslant 0$. To that end, we observe

$$\frac{1}{z}(\mathrm{e}^{-za} - \mathrm{e}^{-zb}) = \mathrm{e}^{-za}\frac{1 - \mathrm{e}^{-z(b-a)}}{z} \to b - a \quad (z \to 0).$$

Thus, g is bounded near 0. In particular, $z = 0$ is a removable singularity and, hence, g can be extended holomorphically to $\mathbb{C}_{\mathrm{Re}\geqslant\mu}$. Moreover, for $\rho \geqslant \mu$ we have that

$$\int_{\mathbb{R}} \|g(\mathrm{i}t+\rho)\|^2 \,\mathrm{d}t = \int_{-1}^{1} \|g(\mathrm{i}t+\rho)\|^2 \,\mathrm{d}t + \int_{|t|>1} \|g(\mathrm{i}t+\rho)\|^2 \,\mathrm{d}t.$$

The first term on the right-hand side is finite since g is bounded, while the second term can be estimated by

$$\int_{|t|>1} \|g(\mathrm{i}t+\rho)\|^2 \,\mathrm{d}t \leqslant \|M\|_{\infty,\mathbb{C}_{\mathrm{Re}>\mu}}^2 \|x\|^2 \frac{(\mathrm{e}^{-\rho a}+\mathrm{e}^{-\rho b})^2}{2\pi} \int_{|t|>1} \frac{1}{t^2+\rho^2} \,\mathrm{d}t < \infty.$$

This proves that $g(\mathrm{i}\cdot+\rho) \in L_2(\mathbb{R}; H)$ for each $\rho \geqslant \mu$ and hence, particularly for $\rho = \mu$ and $\rho = \nu$. Finally, for $\rho \geqslant \mu$ we have that

$$\|g(\mathrm{i}t+\rho)\| \leqslant \frac{1}{\sqrt{2\pi}} \|M\|_{\infty,\mathbb{C}_{\mathrm{Re}>\mu}} \|x\| \frac{1}{\sqrt{t^2+\rho^2}} \left(\mathrm{e}^{-\rho a}+\mathrm{e}^{-\rho b}\right) \to 0 \quad (|t| \to \infty),$$

which together with the boundedness of g yields (5.9) by dominated convergence. This shows that g satisfies the assumptions of Lemma 5.3.5 and thus

$$M(\partial_{t,\nu})f = \mathcal{L}_\nu^* g(\mathrm{i}\cdot+\nu) = \mathcal{L}_\mu^* g(\mathrm{i}\cdot+\mu) = M(\partial_{t,\mu})f.$$

By linearity, this equality extends to $S_{\mathrm{c}}(\mathbb{R}; H)$ and so,

$$F\colon S_{\mathrm{c}}(\mathbb{R}; H) \to \bigcap_{\nu\geqslant\mu} L_{2,\nu}(\mathbb{R}; H),\ f \mapsto M(\partial_{t,\nu})f$$

is well-defined. Moreover, F is uniformly Lipschitz continuous (observe that $\sup_{\nu\geqslant\mu} \|F^\nu\| \leq \|M\|_{\infty,\mathbb{C}_{\mathrm{Re}>\mu}}$) and hence, the assertions follow from Lemma 4.2.5. □

5.4 Comments

The Fourier and the Fourier–Laplace transformation introduced in this chapter are used to define an operator-valued functional calculus for the time derivative, $\partial_{t,\nu}$. This functional calculus can be defined since the Fourier–Laplace transformation provides the unitary transformation yielding the spectral representation of the time derivative as multiplication operator. This fact was already noticed in [83], which eventually led to evolutionary equations in [82].

We emphasise that we have used the fundamental property that both $\mathcal{F}$ and $\mathcal{L}_\nu$ are unitary. It is noteworthy that the Fourier transformation is an isometric isomorphism on $L_2(\mathbb{R}; X)$ if and only if X is a Hilbert space, see [58]. In the Banach space-valued case one has to further restrict the class of functions used to define a functional calculus. For the topic of functional calculus we refer to the 21st Internet Seminar [46] by Markus Haase and to his monograph, [47].

Material laws and the corresponding material law operators were also considered in [82, Section 3], including a physical motivation. Note that the definition in [82] is slightly different compared to the one presented here.

Exercises

Exercise 5.1 Let (Ω, Σ, μ) be a σ-finite measure space, X a Banach space and $I \subseteq \mathbb{R}$ an open interval. Let $g\colon I \times \Omega \to X$ such that $g(t, \cdot) \in L_1(\mu; X)$ for each $t \in I$, and define

$$h\colon I \to X,\ t \mapsto \int_\Omega g(t, \omega)\, \mathrm{d}\mu(\omega).$$

(a) Assume that $g(\cdot, \omega)$ is continuous for μ-almost every $\omega \in \Omega$ and let $f \in L_1(\mu)$ such that

$$\|g(t, \omega)\| \leqslant f(\omega) \quad (t \in I, \omega \in \Omega).$$

Prove that h is continuous.

(b) Assume that $g(\cdot, \omega)$ is differentiable for μ-almost every $\omega \in \Omega$ and let $f \in L_1(\mu)$ such that

$$\|\partial_t g(t, \omega)\| \leqslant f(\omega) \quad (t \in I, \mu - a.a.\ \omega \in \Omega).$$

Prove that h is differentiable with

$$h'(t) = \int_\Omega \partial_t g(t, \omega)\, \mathrm{d}\mu(\omega).$$

Exercise 5.2 Let H_0, H_1 be two Hilbert spaces and $U\colon \operatorname{dom}(U) \subseteq H_0 \to H_1$ linear such that

- $\operatorname{dom}(U)$ is dense in H_0 and $\operatorname{ran}(U)$ is dense in H_1.
- $\forall x \in \operatorname{dom}(U): \|Ux\|_{H_1} = \|x\|_{H_0}$.

Show that U can be uniquely extended to a unitary operator between H_0 and H_1.

Exercise 5.3 Let $\Omega \subseteq \mathbb{C}$ be open, X a complex Banach space and $f\colon \Omega \to X$. Prove that the following statements are equivalent:

(i) f is holomorphic.
(ii) For all $x' \in X'$ the mapping $x' \circ f\colon \Omega \to \mathbb{C}$ is holomorphic.
(iii) f is locally bounded and $x' \circ f\colon \Omega \to \mathbb{C}$ is holomorphic for all $x' \in D$, where $D \subseteq X'$ is a norming set[1] for X.

[1] $D \subseteq X'$ is called a norming set for X if $\|x\| = \sup_{x' \in D\setminus\{0\}} \frac{1}{\|x'\|} \left|x'(x)\right|$ for each $x \in X$. Note that X' is norming for X by the Hahn–Banach theorem.

(iv) f is analytic, i.e. for each $z_0 \in \Omega$ there is $r > 0$ and $(a_n)_n$ in X with $B(z_0, r) \subseteq \Omega$ and

$$f(z) = \sum_{n=0}^{\infty} a_n (z - z_0)^n \quad (z \in B(z_0, r)).$$

Assume now that $X = L(X_1, X_2)$ for two complex Banach spaces X_1, X_2, let $D_1 \subseteq X_1$ be dense and $D_2 \subseteq X_2'$ norming for X_2. Prove that the statements (i) to (iv) are equivalent to

(v) f is locally bounded and $\Omega \ni z \mapsto x_2'(f(z)(x_1)) \in \mathbb{C}$ is holomorphic for all $x_1 \in D_1$ and $x_2' \in D_2$.

Hint: For the difficult implications one might also consult [6, Appendix A]. In the same source one can find that in part (iii) it is enough for D to be separating.

Exercise 5.4 Let $\nu \in \mathbb{R}$ and $k \in L_{1,\nu}(\mathbb{R})$. Prove that

$$\mathcal{L}_\nu (k * f) = \sqrt{2\pi}\, (\mathcal{L}_\nu k) \cdot (\mathcal{L}_\nu f)$$

for $f \in L_{2,\nu}(\mathbb{R}; H)$.

Exercise 5.5 Let $\alpha > 0$ and define $g_\alpha(t) := \mathbb{1}_{[0,\infty)}(t) t^{\alpha-1}$ for $t \in \mathbb{R}$. Show that $g_\alpha \in L_{1,\nu}(\mathbb{R})$ for each $\nu > 0$ and that

$$(\mathcal{L}_\nu g_\alpha)(t) = \frac{1}{\sqrt{2\pi}} \Gamma(\alpha)(it + \nu)^{-\alpha}.$$

Use this formula and Exercise 5.4 to derive (5.8).
Hint: To compute the Fourier–Laplace transform of g_α, derive that $\mathcal{L}_\nu g_\alpha$ solves a first order ordinary differential equation and use separation of variables to solve this equation.

Exercise 5.6 Let $\mu, \nu \in \mathbb{R}$ with $\mu < \nu$ and $f \in L_{2,\nu}(\mathbb{R}; H) \cap L_{2,\mu}(\mathbb{R}; H)$. Moreover, set $U := \{z \in \mathbb{C}\,;\ \mu < \operatorname{Re} z < \nu\}$. Show that $f \in \bigcap_{\mu<\rho<\nu} L_{2,\rho}(\mathbb{R}; H) \cap L_{1,\rho}(\mathbb{R}; H)$ and that

$$U \ni z \mapsto (\mathcal{L}_{\operatorname{Re} z} f)(\operatorname{Im} z)$$

is holomorphic.

Exercise 5.7 Let H_0, H_1 be Hilbert spaces and $T\colon L_{2,\nu}(\mathbb{R}; H_0) \to L_{2,\nu}(\mathbb{R}; H_1)$ linear and bounded. We call T *autonomous* if $T\tau_h = \tau_h T$ for each $h \in \mathbb{R}$ (τ_h denotes the translation operator defined in Example 5.3.4). Prove that for autonomous T, the following statements are equivalent:

(i) T is causal.
(ii) For all $f \in L_{2,\nu}(\mathbb{R}; H_0)$ with $\operatorname{spt} f \subseteq [0, \infty)$ one has $\operatorname{spt} Tf \subseteq [0, \infty)$.

Moreover, prove that for a material law M, the operator $M(\partial_{t,\nu})$ is autonomous for each $\nu > \mathrm{s_b}(M)$.

References

6. W. Arendt et al., *Vector-Valued Laplace Transforms and Cauchy Problems*, 2nd edn. (Birkhäuser, Basel, 2011)
46. M. Haase, *Functional Calculus*. 21st Internet Seminar. 2017/2018.
47. M. Haase, *The Functional Calculus for Sectorial Operators*, vol. 169 (Birkhäuser, Basel, 2006)
58. S. Kwapien, Isomorphic characterizations of inner product spaces by orthogonal series with vector valued coefficients. Stud. Math. **44**(6), 583–595 (1972)
82. R. Picard, A structural observation for linear material laws in classical mathematical physics. Math. Method Appl. Sci. **32**, 1768–1803 (2009)
83. R. Picard, *Hilbert Space Approach to Some Classical Transforms* (Wiley, New York, 1989)

Chapter 6
Solution Theory for Evolutionary Equations

In this chapter, we shall discuss and present the first major result of the manuscript: Picard's theorem on the solution theory for evolutionary equations which is the main result of [82]. In order to stress the applicability of this theorem, we shall deal with applications first and provide a proof of the actual result afterwards. With an initial interest in applications in mind, we start off with the introduction of some operators related to vector calculus.

6.1 First Order Sobolev Spaces

Throughout this section let $\Omega \subseteq \mathbb{R}^d$ be an open set.

Definition We define

$$\begin{aligned} \operatorname{grad}_{\mathrm{c}} \colon C_{\mathrm{c}}^{\infty}(\Omega) \subseteq L_2(\Omega) &\to L_2(\Omega)^d \\ \phi &\mapsto \left(\partial_j \phi\right)_{j\in\{1,\ldots,d\}}, \\ \operatorname{div}_{\mathrm{c}} \colon C_{\mathrm{c}}^{\infty}(\Omega)^d \subseteq L_2(\Omega)^d &\to L_2(\Omega) \\ \left(\phi_j\right)_{j\in\{1,\ldots,d\}} &\mapsto \sum_{j\in\{1,\ldots,d\}} \partial_j \phi_j, \end{aligned}$$

C. Seifert et al., *Evolutionary Equations*, Operator Theory: Advances and Applications 287, https://doi.org/10.1007/978-3-030-89397-2_6

and if $d = 3$,

$$\operatorname{curl}_{\mathrm{c}} \colon C_{\mathrm{c}}^{\infty}(\Omega)^3 \subseteq L_2(\Omega)^3 \to L_2(\Omega)^3$$
$$\left(\phi_j\right)_{j\in\{1,2,3\}} \mapsto \begin{pmatrix} \partial_2\phi_3 - \partial_3\phi_2 \\ \partial_3\phi_1 - \partial_1\phi_3 \\ \partial_1\phi_2 - \partial_2\phi_1 \end{pmatrix}.$$

Furthermore, we put

$$\operatorname{div} := -\operatorname{grad}_{\mathrm{c}}^*, \quad \operatorname{grad} := -\operatorname{div}_{\mathrm{c}}^*, \quad \operatorname{curl} := \operatorname{curl}_{\mathrm{c}}^*$$

and

$$\operatorname{div}_0 := -\operatorname{grad}^*, \quad \operatorname{grad}_0 := -\operatorname{div}^*, \quad \operatorname{curl}_0 := \operatorname{curl}^*.$$

Proposition 6.1.1 *The relations* $\operatorname{div}, \operatorname{div}_0, \operatorname{grad}, \operatorname{grad}_0, \operatorname{curl}$ *and* curl_0 *are all densely defined, closed linear operators.*

Proof The operators $\operatorname{grad}_{\mathrm{c}}$, $\operatorname{div}_{\mathrm{c}}$ and $\operatorname{curl}_{\mathrm{c}}$ are densely defined by Exercise 6.3. Thus, div, grad and curl are closed linear operators by Lemma 2.2.7. Moreover, it follows from integration by parts that $\operatorname{grad}_{\mathrm{c}} \subseteq \operatorname{grad}$, $\operatorname{div}_{\mathrm{c}} \subseteq \operatorname{div}$ and $\operatorname{curl}_{\mathrm{c}} \subseteq \operatorname{curl}$. Thus, div, grad and curl are also densely defined. This, in turn, implies that $\operatorname{grad}_{\mathrm{c}}$, $\operatorname{div}_{\mathrm{c}}$ and $\operatorname{curl}_{\mathrm{c}}$ are closable by Lemma 2.2.7 with respective closures grad_0, div_0 and curl_0 by Lemma 2.2.4. □

We shall describe the domains of these operators in more detail in the next theorem.

Theorem 6.1.2 *If* $f \in L_2(\Omega)$ *and* $g = (g_j)_{j\in\{1,\ldots,d\}} \in L_2(\Omega)^d$ *then the following statements hold:*

(a) $f \in \operatorname{dom}(\operatorname{grad})$ *and* $g = \operatorname{grad} f$ *if and only if*

$$\forall \phi \in C_{\mathrm{c}}^{\infty}(\Omega), j \in \{1, \ldots, d\} \colon -\int_{\Omega} f \partial_j \phi = \int_{\Omega} g_j \phi.$$

(b) $f \in \operatorname{dom}(\operatorname{grad}_0)$ *and* $g = \operatorname{grad}_0 f$ *if and only if there exists* $(f_k)_k$ *in* $C_{\mathrm{c}}^{\infty}(\Omega)$ *such that* $f_k \to f$ *in* $L_2(\Omega)$ *and* $\operatorname{grad} f_k \to g$ *in* $L_2(\Omega)^d$ *as* $k \to \infty$.

(c) $g \in \operatorname{dom}(\operatorname{div})$ *and* $f = \operatorname{div} g$ *if and only if*

$$\forall \phi \in C_{\mathrm{c}}^{\infty}(\Omega) \colon -\int_{\Omega} g \cdot \operatorname{grad} \phi = \int_{\Omega} f \phi.$$

(d) $g \in \operatorname{dom}(\operatorname{div}_0)$ *and* $f = \operatorname{div}_0 g$ *if and only if there exists* $(g_k)_k$ *in* $C_{\mathrm{c}}^{\infty}(\Omega)^d$ *such that* $g_k \to g$ *in* $L_2(\Omega)^d$ *and* $\operatorname{div} g_k \to f$ *in* $L_2(\Omega)$ *as* $k \to \infty$.

If $d = 3$ and $f, g \in L_2(\Omega)^3$ then the following statements hold:

(e) $f \in \operatorname{dom}(\operatorname{curl})$ *and* $g = \operatorname{curl} f$ *if and only if*

$$\forall \phi \in C_c^\infty(\Omega)^3 \colon \int_\Omega f \cdot \operatorname{curl} \phi = \int_\Omega g \cdot \phi.$$

(f) $f \in \operatorname{dom}(\operatorname{curl}_0)$ *and* $g = \operatorname{curl}_0 f$ *if and only if there exists* $(f_k)_k$ *in* $C_c^\infty(\Omega)^3$ *such that* $f_k \to f$ *in* $L_2(\Omega)^3$ *and* $\operatorname{curl} f_k \to g$ *in* $L_2(\Omega)^3$ *as* $k \to \infty$.

All the statements in Theorem 6.1.2 are elementary consequences of the integration by parts formula, the definitions of the adjoint and Lemma 2.2.4. We ask the reader to prove these statements in Exercise 6.4.

We introduce the following notation:

$$\begin{aligned} H^1(\Omega) &:= \operatorname{dom}(\operatorname{grad}), \\ H_0^1(\Omega) &:= \operatorname{dom}(\operatorname{grad}_0), \\ H(\operatorname{div}, \Omega) &:= \operatorname{dom}(\operatorname{div}), \\ H(\operatorname{curl}, \Omega) &:= \operatorname{dom}(\operatorname{curl}). \end{aligned}$$

Following the rationale of appending zero as an index for $H_0^1(\Omega)$, we shall also use

$$\begin{aligned} H_0(\operatorname{div}, \Omega) &:= \operatorname{dom}(\operatorname{div}_0), \\ H_0(\operatorname{curl}, \Omega) &:= \operatorname{dom}(\operatorname{curl}_0). \end{aligned}$$

We caution the reader that other authors also use $H_0(\operatorname{div}, \Omega)$ and $H_0(\operatorname{curl}, \Omega)$ to denote the kernel of div and curl.

All the spaces just defined are so-called Sobolev spaces. We note that for $d = 3$ we clearly have $H^1(\Omega)^3 \subseteq H(\operatorname{div}, \Omega) \cap H(\operatorname{curl}, \Omega)$. On the other hand, note that $H(\operatorname{div}, \Omega)$ is neither a sub- nor a superset of $H(\operatorname{curl}, \Omega)$.

Remark 6.1.3 We emphasise that $H_0^1(\Omega) = \overline{C_c^\infty(\Omega)}^{H^1(\Omega)} \subseteq H^1(\Omega)$ is a proper inclusion for many open Ω. The '0' in the index is a reminder of '0'-boundary conditions. In fact, the only difference between these two spaces lies in the behaviour of their elements at the boundary of Ω. The space H_0^1 signifies all H^1-functions vanishing at $\partial\Omega$ in a generalised sense. The corresponding statements are true for the inclusions $H_0(\operatorname{div}, \Omega) \subseteq H(\operatorname{div}, \Omega)$ and $H_0(\operatorname{curl}, \Omega) \subseteq H(\operatorname{curl}, \Omega)$. The space $H_0(\operatorname{div}, \Omega)$ describes $H(\operatorname{div}, \Omega)$-vector fields with vanishing normal component and to lie in $H_0(\operatorname{curl}, \Omega)$ provides a handy generalisation of vanishing tangential component. We will anticipate these abstractions when we apply the solution theory of evolutionary equations for particular cases. In a later chapter we will come back to this issue when we discuss inhomogeneous boundary value problems.

For later use, we record the following relationships between the vector-analytical operators introduced above.

Proposition 6.1.4 *Let $d = 3$. We have the following inclusions:*

$$\begin{aligned}\overline{\operatorname{ran}}(\operatorname{curl}_0) &\subseteq \ker(\operatorname{div}_0),\\ \overline{\operatorname{ran}}(\operatorname{grad}_0) &\subseteq \ker(\operatorname{curl}_0),\\ \overline{\operatorname{ran}}(\operatorname{curl}) &\subseteq \ker(\operatorname{div}),\\ \overline{\operatorname{ran}}(\operatorname{grad}) &\subseteq \ker(\operatorname{curl}).\end{aligned}$$

Proof It is elementary to show that for given $\psi \in C_c^\infty(\Omega)^3$ and $\phi \in C_c^\infty(\Omega)$ we have $\operatorname{div}_0 \operatorname{curl}_0 \psi = 0$ as well as $\operatorname{curl}_0 \operatorname{grad}_0 \phi = 0$. Thus, we obtain $\operatorname{ran}(\operatorname{curl}_c) \subseteq \ker(\operatorname{div}_0)$ and $\operatorname{ran}(\operatorname{grad}_c) \subseteq \ker(\operatorname{curl}_0)$. Since $\ker(\operatorname{div}_0)$ and $\ker(\operatorname{curl}_0)$ are closed, and $C_c^\infty(\Omega)^3$ and $C_c^\infty(\Omega)$ are cores for curl_0 and grad_0 respectively, we obtain the first two inclusions. The last two inclusions follow from the first two by taking into account the orthogonal decompositions

$$L_2(\Omega)^3 = \overline{\operatorname{ran}}(\operatorname{grad}) \oplus \ker(\operatorname{div}_0) = \ker(\operatorname{curl}) \oplus \overline{\operatorname{ran}}(\operatorname{curl}_0)$$

and

$$L_2(\Omega)^3 = \overline{\operatorname{ran}}(\operatorname{grad}_0) \oplus \ker(\operatorname{div}) = \ker(\operatorname{curl}_0) \oplus \overline{\operatorname{ran}}(\operatorname{curl})$$

which follow from Corollary 2.2.6. □

6.2 Well-Posedness of Evolutionary Equations and Applications

The solution theory of evolutionary equations is contained in the next result, Picard's theorem. This result is central for all the derivations to come. In fact, with the notation of Theorem 6.2.1, we shall prove that for all (well-behaved) F there is a unique solution of

$$\big(\partial_{t,\nu} M(\partial_{t,\nu}) + A\big) U = F.$$

The solution U depends continuously and causally on the choice of F.

In order to formulate the result, for a Hilbert space H, $\nu \in \mathbb{R}$ and a given closed operator $A\colon \operatorname{dom}(A) \subseteq H \to H$ we define its extended operator in $L_{2,\nu}(\mathbb{R}; H)$, again denoted by A, by

$$\begin{aligned}L_{2,\nu}(\mathbb{R}; \operatorname{dom}(A)) \subseteq L_{2,\nu}(\mathbb{R}; H) &\to L_{2,\nu}(\mathbb{R}; H)\\ f &\mapsto \big(t \mapsto Af(t)\big).\end{aligned}$$

We have collected some properties of extended operators in Exercises 6.1 and 6.2.

Theorem 6.2.1 (Picard) *Let $\nu_0 \in \mathbb{R}$ and H be a Hilbert space. Let $M\colon \operatorname{dom}(M) \subseteq \mathbb{C} \to L(H)$ be a material law with $\mathrm{s_b}(M) < \nu_0$ and let $A\colon \operatorname{dom}(A) \subseteq H \to H$ be skew-selfadjoint. Assume that*

$$\operatorname{Re} \langle \phi, zM(z)\phi \rangle_H \geqslant c \, \|\phi\|_H^2 \quad (\phi \in H, z \in \mathbb{C}_{\operatorname{Re} \geqslant \nu_0})$$

for some $c > 0$. Then for all $\nu \geqslant \nu_0$ the operator $\partial_{t,\nu} M(\partial_{t,\nu}) + A$ is closable and

$$S_\nu := \left(\overline{\partial_{t,\nu} M(\partial_{t,\nu}) + A}\right)^{-1} \in L(L_{2,\nu}(\mathbb{R}; H)).$$

Furthermore, S_ν is causal and satisfies $\|S_\nu\|_{L(L_{2,\nu})} \leqslant 1/c$, and for all $F \in \operatorname{dom}(\partial_{t,\nu})$ we have

$$S_\nu F \in \operatorname{dom}(\partial_{t,\nu}) \cap \operatorname{dom}(A).$$

Furthermore, for $\eta, \nu \geqslant \nu_0$ and $F \in L_{2,\nu}(\mathbb{R}; H) \cap L_{2,\eta}(\mathbb{R}; H)$ we have that $S_\nu F = S_\eta F$.

The property that $S_\nu F = S_\eta F$ for all $F \in L_{2,\nu}(\mathbb{R}; H) \cap L_{2,\eta}(\mathbb{R}; H)$ where $\eta, \nu \geqslant \nu_0$, for some $\nu_0 \in \mathbb{R}$, will be referred to as S_ν being *eventually independent of* ν in what follows.

Remark 6.2.2 If $F \in \operatorname{dom}(\partial_{t,\nu})$, then $U := S_\nu F \in \operatorname{dom}(\partial_{t,\nu}) \cap \operatorname{dom}(A)$ by Theorem 6.2.1. Since $M(\partial_{t,\nu})$ leaves the space $\operatorname{dom}(\partial_{t,\nu})$ invariant, this gives that $M(\partial_{t,\nu})U \in \operatorname{dom}(\partial_{t,\nu})$ and thus, U solves the evolutionary equation literally; that is,

$$(\partial_{t,\nu} M(\partial_{t,\nu}) + A)U = F,$$

while for $F \in L_{2,\nu}(\mathbb{R}; H)$, in general, we just have

$$\left(\overline{\partial_{t,\nu} M(\partial_{t,\nu}) + A}\right)U = F.$$

Definition Let H be a Hilbert space and $T \in L(H)$. If T is selfadjoint, we write $T \geqslant c$ for some $c \in \mathbb{R}$ if

$$\forall x \in H : \langle x, Tx \rangle_H \geqslant c \, \|x\|_H^2 .$$

Moreover, we define the *real part of* T by $\operatorname{Re} T := \frac{1}{2}(T + T^*)$.

Note that if H is a Hilbert space and $T \in L(H)$ then $\operatorname{Re} T$ is selfadjoint. Moreover,

$$\langle x, (\operatorname{Re} T)x \rangle_H = \operatorname{Re} \langle x, Tx \rangle_H \quad (x \in H).$$

Hence, in Theorem 6.2.1 the assumption on the material law can be rephrased as

$$\operatorname{Re} zM(z) \geqslant c \quad (z \in \mathbb{C}_{\operatorname{Re}\geqslant \nu_0}).$$

The following operators will be prototypical examples needed for the applications of the previous theorem.

Proposition 6.2.3 *Let H_0, H_1 be Hilbert spaces.*

(a) *Let $B\colon \operatorname{dom}(B) \subseteq H_0 \to H_1$, $C\colon \operatorname{dom}(C) \subseteq H_1 \to H_0$ be densely defined linear operators. Then*

$$\begin{aligned}\begin{pmatrix} 0 & C \\ B & 0 \end{pmatrix} : \operatorname{dom}(B) \times \operatorname{dom}(C) \subseteq H_0 \times H_1 &\to H_0 \times H_1 \\ (\phi, \psi) &\mapsto (C\psi, B\phi)\end{aligned}$$

is densely defined, and we have

$$\begin{pmatrix} 0 & C \\ B & 0 \end{pmatrix}^* = \begin{pmatrix} 0 & B^* \\ C^* & 0 \end{pmatrix}.$$

(b) *Let $a \in L(H_0)$, and $c > 0$. Assume $\operatorname{Re} a \geqslant c$. Then $a^{-1} \in L(H_0)$ with $\|a^{-1}\| \leqslant \frac{1}{c}$ and $\operatorname{Re} a^{-1} \geqslant c \|a\|^{-2}$.*

Proof The proof of the first statement can be done in two steps. First, notice that the inclusion $\begin{pmatrix} 0 & B^* \\ C^* & 0 \end{pmatrix} \subseteq \begin{pmatrix} 0 & C \\ B & 0 \end{pmatrix}^*$ follows immediately. If, on the other hand, $\begin{pmatrix} \phi \\ \psi \end{pmatrix} \in \operatorname{dom}\left(\begin{pmatrix} 0 & C \\ B & 0 \end{pmatrix}^*\right)$ with $\begin{pmatrix} 0 & C \\ B & 0 \end{pmatrix}^* \begin{pmatrix} \phi \\ \psi \end{pmatrix} = \begin{pmatrix} \xi \\ \zeta \end{pmatrix}$ we get for all $x \in \operatorname{dom}(B)$ that

$$\begin{aligned}\langle Bx, \psi\rangle_{H_1} &= \left\langle \begin{pmatrix} 0 & C \\ B & 0 \end{pmatrix}\begin{pmatrix} x \\ 0 \end{pmatrix}, \begin{pmatrix} \phi \\ \psi \end{pmatrix}\right\rangle_{H_0\times H_1} = \left\langle \begin{pmatrix} x \\ 0 \end{pmatrix}, \begin{pmatrix} 0 & C \\ B & 0 \end{pmatrix}^*\begin{pmatrix} \phi \\ \psi \end{pmatrix}\right\rangle_{H_0\times H_1} \\ &= \left\langle \begin{pmatrix} x \\ 0 \end{pmatrix}, \begin{pmatrix} \xi \\ \zeta \end{pmatrix}\right\rangle_{H_0\times H_1} = \langle x, \xi\rangle_{H_0}.\end{aligned}$$

Hence, $\psi \in \operatorname{dom}(B^*)$ and $B^*\psi = \xi$. Similarly, we obtain $\phi \in \operatorname{dom}(C^*)$ and $C^*\phi = \zeta$.

For the second statement, we compute for all $\phi \in H_0$ using the Cauchy–Schwarz inequality

$$\|\phi\|_{H_0} \|a\phi\|_{H_0} \geqslant \left|\langle \phi, a\phi\rangle_{H_0}\right| \geqslant \operatorname{Re}\langle \phi, a\phi\rangle_{H_0} \geqslant c\,\langle \phi, \phi\rangle_{H_0} = c\,\|\phi\|_{H_0}^2.$$

Thus, a is one-to-one. Since $\operatorname{Re} a = \operatorname{Re} a^*$ it follows that a^* is one-to-one, as well. Thus, we get that a has dense range by Theorem 2.2.5. The inequality

$$\|a\phi\|_{H_0} \geqslant c\,\|\phi\|_{H_0}$$

implies that a^{-1} is bounded with $\|a^{-1}\| \leqslant \frac{1}{c}$. Hence, as a^{-1} is closed, $\operatorname{dom}(a^{-1}) = \operatorname{ran}(a)$ is closed by Lemma 2.1.3 and hence, $\operatorname{dom}(a^{-1}) = H_0$; that is, $a^{-1} \in L(H_0)$. To conclude, let $\psi \in H_0$ and put $\phi := a^{-1}\psi$. Then $\|\psi\|_{H_0} = \left\|aa^{-1}\psi\right\|_{H_0} \leqslant \|a\|\,\left\|a^{-1}\psi\right\|_{H_0}$ and so

$$\begin{aligned}\operatorname{Re}\left\langle \psi, a^{-1}\psi\right\rangle_{H_0} &= \operatorname{Re}\langle a\phi, \phi\rangle_{H_0} = \operatorname{Re}\langle \phi, a\phi\rangle_{H_0} \geqslant c\,\langle \phi, \phi\rangle_{H_0} = c\left\langle a^{-1}\psi, a^{-1}\psi\right\rangle_{H_0} \\ &\geqslant c\frac{1}{\|a\|^2}\,\|\psi\|_{H_0}^2 .\end{aligned}$$

□

The Heat Equation

The first example we will consider is the heat equation in an open subset $\Omega \subseteq \mathbb{R}^d$. Under a heat source, $Q\colon \mathbb{R}\times\Omega \to \mathbb{R}$, the heat distribution, $\theta\colon \mathbb{R}\times\Omega \to \mathbb{R}$, satisfies the so-called heat flux balance

$$\partial_t\theta + \operatorname{div} q = Q.$$

Here, $q\colon \mathbb{R}\times\Omega \to \mathbb{R}^d$ is the heat flux which is connected to θ via Fourier's law

$$q = -a \operatorname{grad}\theta,$$

where $a\colon \Omega \to \mathbb{R}^{d\times d}$ is the heat conductivity, which is measurable, bounded and uniformly strictly positive in the sense that

$$\operatorname{Re} a(x) \geqslant c$$

for all $x \in \Omega$ and some $c > 0$ in the sense of positive definiteness. Moreover, we assume that Ω is thermally isolated, which is modelled by requiring that the normal component of q vanishes at $\partial\Omega$; that is, $q \in \operatorname{dom}(\operatorname{div}_0)$ (see Remark 6.1.3). Written as a block matrix and incorporating the boundary condition, we obtain

$$\left(\partial_t \begin{pmatrix} 1 & 0 \\ 0 & 0 \end{pmatrix} + \begin{pmatrix} 0 & 0 \\ 0 & a^{-1} \end{pmatrix} + \begin{pmatrix} 0 & \operatorname{div}_0 \\ \operatorname{grad} & 0 \end{pmatrix}\right)\begin{pmatrix} \theta \\ q \end{pmatrix} = \begin{pmatrix} Q \\ 0 \end{pmatrix}.$$

Theorem 6.2.4 *For all $\nu > 0$, the operator*

$$\partial_{t,\nu} \begin{pmatrix} 1 & 0 \\ 0 & 0 \end{pmatrix} + \begin{pmatrix} 0 & 0 \\ 0 & a^{-1} \end{pmatrix} + \begin{pmatrix} 0 & \operatorname{div}_0 \\ \operatorname{grad} & 0 \end{pmatrix}$$

is densely defined and closable in $L_{2,\nu}\left(\mathbb{R}; L_2(\Omega) \times L_2(\Omega)^d\right)$. The respective closure is continuously invertible with causal inverse being eventually independent of ν.

Proof The assertion follows from Theorem 6.2.1 applied to

$$M(z) = \begin{pmatrix} 1 & 0 \\ 0 & 0 \end{pmatrix} + z^{-1} \begin{pmatrix} 0 & 0 \\ 0 & a^{-1} \end{pmatrix} \quad \text{and} \quad A = \begin{pmatrix} 0 & \operatorname{div}_0 \\ \operatorname{grad} & 0 \end{pmatrix}.$$

Note that M is a material law with $s_b(M) = 0$ by Example 5.3.1. Moreover, for $(x, y) \in L_2(\Omega) \times L_2(\Omega)^d$ and $z \in \mathbb{C}_{\mathrm{Re} \geqslant \nu}$ with $\nu > 0$ we estimate

$$\begin{aligned} \operatorname{Re} \langle (x, y), zM(z)(x, y) \rangle_{L_2(\Omega) \times L_2(\Omega)^d} &\geqslant \operatorname{Re} z \, \|x\|^2_{L_2(\Omega)} + c \, \|a\|^{-2} \, \|y\|^2_{L_2(\Omega)^d} \\ &\geqslant \min\{\nu, c \, \|a\|^{-2}\} \, \|(x, y)\|^2_{L_2(\Omega) \times L_2(\Omega)^d}, \end{aligned}$$

where we have used Proposition 6.2.3(b) in the first inequality. Moreover, A is skew-selfadjoint by Proposition 6.2.3(a). □

Remark 6.2.5 Assume that $Q \in \operatorname{dom}(\partial_{t,\nu})$. It then follows from Theorem 6.2.1 that

$$\begin{aligned} \begin{pmatrix} \theta \\ q \end{pmatrix} &:= \left(\overline{\partial_{t,\nu} \begin{pmatrix} 1 & 0 \\ 0 & 0 \end{pmatrix} + \begin{pmatrix} 0 & 0 \\ 0 & a^{-1} \end{pmatrix} + \begin{pmatrix} 0 & \operatorname{div}_0 \\ \operatorname{grad} & 0 \end{pmatrix}} \right)^{-1} \begin{pmatrix} Q \\ 0 \end{pmatrix} \\ &\in \operatorname{dom}\left(\partial_{t,\nu}\right) \cap \operatorname{dom}\left(\begin{pmatrix} 0 & \operatorname{div}_0 \\ \operatorname{grad} & 0 \end{pmatrix} \right). \end{aligned} \tag{6.1}$$

Then, as in Remark 6.2.2, it follows that θ and q satisfy the heat flux balance and Fourier's law in the sense that $\theta \in \operatorname{dom}(\partial_{t,\nu}) \cap \operatorname{dom}(\operatorname{grad})$ and $q \in \operatorname{dom}(\operatorname{div}_0)$ and

$$\begin{aligned} \partial_t \theta + \operatorname{div}_0 q &= Q, \\ q &= -a \operatorname{grad} \theta. \end{aligned}$$

This regularity result is true even for $Q \in L_{2,\nu}(\mathbb{R}; L_2(\Omega))$; see [88] and Chap. 15, Theorem 15.2.3.

The Scalar Wave Equation

The classical scalar wave equation in a medium $\Omega \subseteq \mathbb{R}^d$ (think, for instance, of a vibrating string ($d = 1$) or membrane ($d = 2$)) consists of the equation of the balance of momentum where the acceleration of the (vertical) displacement, $u \colon \mathbb{R} \times$

$\Omega \to \mathbb{R}$, is balanced by external forces, $f : \mathbb{R} \times \Omega \to \mathbb{R}$, and the divergence of the stress, $\sigma : \mathbb{R} \times \Omega \to \mathbb{R}^d$, in such a way that

$$\partial_t^2 u - \operatorname{div} \sigma = f.$$

The stress is related to u via the following so-called stress-strain relation (here Hooke's law)

$$\sigma = T \operatorname{grad} u,$$

where the so-called elasticity tensor, $T : \Omega \to \mathbb{R}^{d \times d}$, is bounded, measurable, and satisfies

$$T(x) = T(x)^* \geqslant c$$

for some $c > 0$ uniformly in $x \in \Omega$. The quantity $\operatorname{grad} u$ is referred to as the strain. We think of u as being fixed at $\partial\Omega$ ("clamped boundary condition"). This is modelled by $u \in \operatorname{dom}(\operatorname{grad}_0)$.

Using $v := \partial_t u$ as an unknown, we can rewrite the balance of momentum and Hooke's law as 2×2-block-operator matrix equation

$$\left(\partial_t \begin{pmatrix} 1 & 0 \\ 0 & T^{-1} \end{pmatrix} - \begin{pmatrix} 0 & \operatorname{div} \\ \operatorname{grad}_0 & 0 \end{pmatrix}\right) \begin{pmatrix} v \\ \sigma \end{pmatrix} = \begin{pmatrix} f \\ 0 \end{pmatrix}.$$

The solution theory of evolutionary equations for the wave equation now reads as follows:

Theorem 6.2.6 *Let $\Omega \subseteq \mathbb{R}^d$ be open, and T as indicated above. Then, for all $\nu > 0$,*

$$\partial_{t,\nu} \begin{pmatrix} 1 & 0 \\ 0 & T^{-1} \end{pmatrix} - \begin{pmatrix} 0 & \operatorname{div} \\ \operatorname{grad}_0 & 0 \end{pmatrix}$$

is densely defined and closable in $L_{2,\nu}\left(\mathbb{R}; L_2(\Omega) \times L_2(\Omega)^d\right)$. The respective closure is continuously invertible with causal inverse being eventually independent of ν.

Proof We apply Theorem 6.2.1 to $A = -\begin{pmatrix} 0 & \operatorname{div} \\ \operatorname{grad}_0 & 0 \end{pmatrix}$, which is skew-selfadjoint by Proposition 6.2.3(a), and $M(z) = \begin{pmatrix} 1 & 0 \\ 0 & T^{-1} \end{pmatrix}$, which defines a material law with $s_b(M) = -\infty$. The positive definiteness constraint needed in Theorem 6.2.1 is satisfied by Proposition 6.2.3(b) on account of the selfadjointness of T, which implies the same for T^{-1}. Indeed, for $\nu_0 > 0$ and $z \in \mathbb{C}_{\mathrm{Re}\geqslant \nu_0}$ we estimate

$$\begin{aligned}\mathrm{Re}\,\langle (x,y), zM(z)(x,y)\rangle_{L_2(\Omega)\times L_2(\Omega)^d} &= \mathrm{Re}\,\langle x, zx\rangle_{L_2(\Omega)} + \mathrm{Re}\left\langle y, zT^{-1}y\right\rangle_{L_2(\Omega)^d} \\ &\geqslant \nu_0 \|x\|^2_{L_2(\Omega)} + \nu_0 \frac{c}{\|T\|^2}\|y\|^2_{L_2(\Omega)^d} \\ &\geqslant \nu_0 \min\{1, c/\|T\|^2\}\,\|(x,y)\|^2_{L_2(\Omega)\times L_2(\Omega)^d}\end{aligned}$$

for each $(x, y) \in L_2(\Omega) \times L_2(\Omega)^d$, where we used the selfadjointness of T^{-1} in the second line. □

Remark 6.2.7 Let $f \in L_{2,\nu}(\mathbb{R}; L_2(\Omega))$, $\nu > 0$, and define

$$\begin{pmatrix} u \\ \widetilde{\sigma}\end{pmatrix} = \left(\overline{\partial_{t,\nu}\begin{pmatrix} 1 & 0 \\ 0 & T^{-1}\end{pmatrix} - \begin{pmatrix} 0 & \mathrm{div} \\ \mathrm{grad}_0 & 0\end{pmatrix}}\right)^{-1}\begin{pmatrix}\partial_{t,\nu}^{-1} f \\ 0\end{pmatrix}.$$

By Theorem 6.2.1, we obtain $\begin{pmatrix} u \\ \widetilde{\sigma}\end{pmatrix} \in \mathrm{dom}(\partial_{t,\nu}) \cap \mathrm{dom}\left(\begin{pmatrix} 0 & \mathrm{div} \\ \mathrm{grad}_0 & 0\end{pmatrix}\right)$. Hence, we have

$$\begin{aligned}\partial_{t,\nu} u - \mathrm{div}\,\widetilde{\sigma} &= \partial_{t,\nu}^{-1} f \\ \partial_{t,\nu} T^{-1}\widetilde{\sigma} &= \mathrm{grad}_0\, u\end{aligned}$$

or

$$\begin{aligned}\partial_{t,\nu} u - \mathrm{div}\,\widetilde{\sigma} &= \partial_{t,\nu}^{-1} f \\ \widetilde{\sigma} &= T\partial_{t,\nu}^{-1}\,\mathrm{grad}_0\, u.\end{aligned}$$

Thus, formally, after another time-differentiation and the setting of $\sigma = \partial_{t,\nu}\widetilde{\sigma}$ we obtain a solution of the wave equation, (u, σ). Notice, however, that differentiating $\mathrm{div}\,\widetilde{\sigma}$ cannot be done without any additional knowledge of the regularity of $\widetilde{\sigma}$. In fact, in order to arrive at the balance of momentum equation, one would need to have $\mathrm{div}\,\widetilde{\sigma} \in \mathrm{dom}(\partial_{t,\nu})$. However, one only has $\widetilde{\sigma} \in \mathrm{dom}(\partial_{t,\nu}) \cap \mathrm{dom}(\mathrm{div})$. It is an elementary argument, see [110, Lemma 4.6], that we in fact have $\mathrm{div}\,\partial_{t,\nu}^{-1} = \overline{\partial_{t,\nu}^{-1}\,\mathrm{div}}$, which suggests that, in general, $\mathrm{div}\,\widetilde{\sigma} \notin \mathrm{dom}(\partial_{t,\nu})$, see Exercise 6.6.

Maxwell's Equations

The final example in this chapter forms the archetypical evolutionary equation—Maxwell's equations in a medium $\Omega \subseteq \mathbb{R}^3$. In order to identify the particular choices of $M(\partial_{t,\nu})$ and A in the present situation (and to finally conclude the 2×2-block matrix formulation historically due to the work of [59, 64, 102]), we start out with Faraday's law of induction, which relates the unknown electric field, $E\colon \mathbb{R}\times\Omega \to$

$\mathbb{R}^3$, to the magnetic induction, $B: \mathbb{R} \times \Omega \to \mathbb{R}^3$, via

$$\partial_t B + \operatorname{curl} E = 0.$$

We assume that the medium is contained in a perfect conductor, which is reflected in the so-called electric boundary condition which asks for the vanishing of the tangential component of E at the boundary. This is modelled by $E \in \operatorname{dom}(\operatorname{curl}_0)$. The next constituent of Maxwell's equations is Ampère's law

$$\partial_t D + j_c - \operatorname{curl} H = j_0,$$

which relates the unknown electric displacement, $D: \mathbb{R} \times \Omega \to \mathbb{R}^3$, (free) current (density), $j_c: \mathbb{R} \times \Omega \to \mathbb{R}^3$, and magnetic field, $H: \mathbb{R} \times \Omega \to \mathbb{R}^3$, to the (given) external currents, $j_0: \mathbb{R} \times \Omega \to \mathbb{R}^3$. Maxwell's equations are completed by constitutive relations specific to each material at hand. Indeed, the (bounded, measurable) dielectricity, $\varepsilon: \Omega \to \mathbb{R}^{3\times 3}$, and the (bounded, measurable) magnetic permeability, $\mu: \Omega \to \mathbb{R}^{3\times 3}$, are symmetric matrix-valued functions which couple the electric displacement to the electric field and the magnetic field to the magnetic induction via

$$D = \varepsilon E, \text{ and } B = \mu H.$$

Finally, Ohm's law relates the current to the electric field via the (bounded, measurable) electric conductivity, $\sigma: \Omega \to \mathbb{R}^{3\times 3}$, as

$$j_c = \sigma E.$$

All in all, in terms of (E, H), Maxwell's equations read

$$\left(\partial_t \begin{pmatrix} \varepsilon & 0 \\ 0 & \mu \end{pmatrix} + \begin{pmatrix} \sigma & 0 \\ 0 & 0 \end{pmatrix} + \begin{pmatrix} 0 & -\operatorname{curl} \\ \operatorname{curl}_0 & 0 \end{pmatrix}\right)\begin{pmatrix} E \\ H \end{pmatrix} = \begin{pmatrix} j_0 \\ 0 \end{pmatrix}.$$

For the time being, we shall assume that there exist $c > 0$ and $\nu_0 > 0$ such that for all $\nu \geqslant \nu_0$ we have

$$\nu\varepsilon(x) + \operatorname{Re}\sigma(x) \geqslant c, \quad \mu(x) \geqslant c \quad (x \in \Omega)$$

in the sense of positive definiteness. Note that the latter condition allows particularly for $\varepsilon = 0$ on certain regions, if $\operatorname{Re}\sigma$ compensates for this. To approximate small ε by 0 is referred to as the eddy current approximation in these regions. With the above preparations at hand, we may now formulate the well-posedness result concerning Maxwell's equations.

Theorem 6.2.8 *Let $\Omega \subseteq \mathbb{R}^3$ be open and $\nu \geqslant \nu_0$. Then*

$$\partial_{t,\nu} \begin{pmatrix} \varepsilon & 0 \\ 0 & \mu \end{pmatrix} + \begin{pmatrix} \sigma & 0 \\ 0 & 0 \end{pmatrix} + \begin{pmatrix} 0 & -\operatorname{curl} \\ \operatorname{curl}_0 & 0 \end{pmatrix}$$

is densely defined and closable in $L_{2,\nu}\left(\mathbb{R}; L_2(\Omega)^3 \times L_2(\Omega)^3\right)$. The respective closure is continuously invertible with causal inverse being eventually independent of ν.

Proof The assertion follows from Theorem 6.2.1 applied to the material law

$$M(z) = \begin{pmatrix} \varepsilon & 0 \\ 0 & \mu \end{pmatrix} + z^{-1} \begin{pmatrix} \sigma & 0 \\ 0 & 0 \end{pmatrix}$$

and the skew-selfadjoint operator

$$A = \begin{pmatrix} 0 & -\operatorname{curl} \\ \operatorname{curl}_0 & 0 \end{pmatrix}. \qquad \square$$

Remark 6.2.9 In the physics literature (see e.g. [40, Chapter 18]), Maxwell's equations are usually complemented by Gauss' law,

$$\operatorname{div}_0 B = 0,$$

as well as the introduction of the charge density, $\rho = \operatorname{div} \varepsilon E$, and the current, $j = j_0 - j_c$, by the continuity equation

$$\partial_t \rho = \operatorname{div} j.$$

We shall argue in the following that these equations are *automatically* satisfied if (E, H) is a solution to Maxwell's equation. Indeed, assuming $j_0 \in \operatorname{dom}(\partial_{t,\nu})$, then, as a consequence of Theorem 6.2.1, for

$$\begin{pmatrix} E \\ H \end{pmatrix} = \left(\overline{\partial_{t,\nu} \begin{pmatrix} \varepsilon & 0 \\ 0 & \mu \end{pmatrix} + \begin{pmatrix} \sigma & 0 \\ 0 & 0 \end{pmatrix} + \begin{pmatrix} 0 & -\operatorname{curl} \\ \operatorname{curl}_0 & 0 \end{pmatrix}}\right)^{-1} \begin{pmatrix} j_0 \\ 0 \end{pmatrix}$$

we observe $\begin{pmatrix} E \\ H \end{pmatrix} \in \operatorname{dom}\left(\partial_{t,\nu}\right) \cap \operatorname{dom}\left(\begin{pmatrix} 0 & -\operatorname{curl} \\ \operatorname{curl}_0 & 0 \end{pmatrix}\right)$. Reformulating the latter equation yields

$$\begin{aligned} B = \mu H &= -\partial_{t,\nu}^{-1} \operatorname{curl}_0 E, \\ \varepsilon E &= \partial_{t,\nu}^{-1} \left(-\sigma E + j_0 + \operatorname{curl} H\right) = \partial_{t,\nu}^{-1} j + \partial_{t,\nu}^{-1} \operatorname{curl} H. \end{aligned}$$

Since $\operatorname{curl}_0 E \in \operatorname{ran}(\operatorname{curl}_0)$, we have by Proposition 3.1.6(b) that $\partial_{t,\nu}^{-1} \operatorname{curl}_0 E \in \overline{\operatorname{ran}}(\operatorname{curl}_0)$. Thus, by Proposition 6.1.4, we obtain

$$\operatorname{div}_0 B = \operatorname{div}_0 \left(-\partial_{t,\nu}^{-1} \operatorname{curl}_0 E\right) = 0.$$

Similarly, we deduce that

$$\rho = \operatorname{div} \varepsilon E = \operatorname{div} \partial_{t,\nu}^{-1} j.$$

If, in addition, we have that $j \in \operatorname{dom}(\operatorname{div})$, we recover the continuity equation. In general, the continuity equation is satisfied in the integrated sense just derived.

We shall keep the list of examples to that for now. In the course of this book, we will see more (involved) examples. Furthermore, we will study the boundary conditions more deeply and shall relate the conditions introduced abstractly here to more classical formulations involving trace spaces.

6.3 Proof of Picard's Theorem

In this section we shall prove the well-posedness theorem. For this, we recall an elementary result from functional analysis. It is remindful of the Lax–Milgram lemma.

Proposition 6.3.1 *Let H be a Hilbert space and $B\colon \operatorname{dom}(B) \subseteq H \to H$ densely defined and closed. Assume there exists $c > 0$ such that*

$$\begin{aligned}\operatorname{Re}\langle \phi, B\phi\rangle_H &\geqslant c\,\|\phi\|_H^2 \quad (\phi \in \operatorname{dom}(B)),\\ \operatorname{Re}\left\langle \psi, B^*\psi\right\rangle_H &\geqslant c\,\|\psi\|_H^2 \quad (\psi \in \operatorname{dom}(B^*)).\end{aligned}$$

Then $B^{-1} \in L(H)$ and $\left\|B^{-1}\right\| \leqslant 1/c$.

Proof Since B is not necessarily bounded here, the present argument requires a refinement of the one in Proposition 6.2.3. In fact, the first assumed inequality implies closedness of the range of B as well as continuous invertibility with $B^{-1}\colon \operatorname{ran}(B) \to H$. The fact that $\operatorname{ran}(B)$ is dense in H follows from the second inequality. □

Remark 6.3.2 In the proof of Theorem 6.2.1, we will apply Proposition 6.3.1 in a situation, where $\operatorname{dom}(B^*) \subseteq \operatorname{dom}(B)$. In this case, the condition

$$\operatorname{Re}\langle \phi, B\phi\rangle_H \geqslant c\,\|\phi\|_H^2 \quad (\phi \in \operatorname{dom}(B))$$

readily implies

$$\operatorname{Re}\left\langle \psi, B^*\psi\right\rangle_H \geqslant c\,\|\psi\|_H^2 \quad (\psi \in \operatorname{dom}(B^*)).$$

Next, we turn to the proof of Picard's theorem. For this, we recall that we do not notationally distinguish between the operator A defined on H and its extension to H-valued functions. We leave it to the context, which realisation of A is considered, which will always be obvious; see also Exercises 6.1 and 6.2.

Proof of Theorem 6.2.1 Let $\nu \geqslant \nu_0$ and $z \in \mathbb{C}_{\operatorname{Re}\geqslant\nu}$. Define $B(z) := zM(z) + A$. Since $M(z) \in L(H)$ it follows from Theorem 2.3.2 that $B(z)^* = (zM(z))^* - A$ and $\operatorname{dom}(B(z)) = \operatorname{dom}(B(z)^*) = \operatorname{dom}(A)$. Moreover, for all $\phi \in \operatorname{dom}(A)$ we have

$$\operatorname{Re}\left\langle \phi, B(z)\phi\right\rangle_H = \operatorname{Re}\left\langle \phi, (zM(z) + A)\,\phi\right\rangle_H = \operatorname{Re}\left\langle \phi, zM(z)\phi\right\rangle_H \geqslant c\,\|\phi\|_H^2\,,$$

due to the skew-selfadjointness of A. Thus, by Proposition 6.3.1 (see also Remark 6.3.2) applied to $B(z)$ instead of B, we deduce that

$$S\colon \mathbb{C}_{\operatorname{Re}\geqslant\nu} \ni z \mapsto B(z)^{-1}$$

is bounded and assumes values in $L(H)$ with norm bounded by $1/c$. By Exercise 6.5, we have that S is holomorphic. Thus, S is a material law and $\left\|S(\partial_{t,\nu})\right\| \leqslant 1/c$ by Proposition 5.3.2. Moreover, Theorem 5.3.6 implies that $S(\partial_{t,\nu})$ is independent of ν and causal.

Next, if $f \in \operatorname{dom}(\partial_{t,\nu})$, it follows that $(\mathrm{i}m + \nu)\,\mathcal{L}_\nu f \in L_2(\mathbb{R}; H)$. Hence, for all $t \in \mathbb{R}$ we obtain

$$\begin{aligned} AS(\mathrm{i}t + \nu)\mathcal{L}_\nu f(t) &= A\big((\mathrm{i}t + \nu)\,M(\mathrm{i}t + \nu) + A\big)^{-1}\mathcal{L}_\nu f(t) \\ &= \mathcal{L}_\nu f(t) - (\mathrm{i}t + \nu)\,M(\mathrm{i}t + \nu)S(\mathrm{i}t + \nu)\mathcal{L}_\nu f(t). \end{aligned}$$

Thus, by the boundedness of M and S, we deduce $S(\mathrm{i}\cdot +\nu)\mathcal{L}_\nu f \in L_2(\mathbb{R}; \operatorname{dom}(A))$. This implies $S(\partial_{t,\nu})f \in L_{2,\nu}(\mathbb{R}; \operatorname{dom}(A))$ by Exercise 6.2. Similarly, but more easily, it follows that $(\mathrm{i}\cdot +\nu)\,S(\mathrm{i}\cdot +\nu)\mathcal{L}_\nu f \in L_2(\mathbb{R}; H)$ also, which shows $S(\partial_{t,\nu})f \in \operatorname{dom}(\partial_{t,\nu})$.

We now define the operator $B(\mathrm{i}m + \nu)$ by

$$\begin{aligned} \operatorname{dom}(B(\mathrm{i}m + \nu)) := \big\{f \in L_2(\mathbb{R}; H)\,;\, f(t) \in \operatorname{dom}(A) \text{ for a.e. } t \in \mathbb{R}, \\ (t \mapsto B(\mathrm{i}t + \nu)f(t)) \in L_2(\mathbb{R}; H)\big\} \end{aligned}$$

and

$$B(\mathrm{i}m + \nu)f := (t \mapsto B(\mathrm{i}t + \nu)f(t)) \quad (f \in \operatorname{dom}(B(\mathrm{i}m + \nu))).$$

Then one easily sees that $B(\mathrm{im}+\nu) = S(\mathrm{im}+\nu)^{-1}$ and since $S(\mathrm{im}+\nu)$ is closed, it follows that $B(\mathrm{im}+\nu)$ is closed as well. Moreover

$$(\mathrm{im}+\nu)M(\mathrm{im}+\nu) + A \subseteq B(\mathrm{im}+\nu)$$

and hence, the operator $(\mathrm{im}+\nu)M(\mathrm{im}+\nu) + A$ is closable, which also yields the closability of $\partial_{t,\nu}M(\partial_{t,\nu}) + A$ by unitary equivalence. To complete the proof, we have to show that

$$\overline{(\mathrm{im}+\nu)M(\mathrm{im}+\nu) + A} = B(\mathrm{im}+\nu),$$

as this equality implies $S(\partial_{t,\nu}) = \left(\overline{\partial_{t,\nu}M(\partial_{t,\nu}) + A}\right)^{-1}$ by unitary equivalence. For showing the asserted equality, let $f \in \mathrm{dom}(B(\mathrm{im}+\nu))$. For $n \in \mathbb{N}$ we define $f_n := \mathbb{1}_{[-n,n]}f$. Then $f_n \in \mathrm{dom}(\mathrm{im}+\nu) \cap \mathrm{dom}(A) \subseteq \mathrm{dom}\left((\mathrm{im}+\nu)M(\mathrm{im}+\nu)+A\right)$ for each $n \in \mathbb{N}$ and by dominated convergence, we have that $f_n \to f$ as $n \to \infty$ as well as

$$\begin{aligned}\left((\mathrm{im}+\nu)M(\mathrm{im}+\nu) + A\right)f_n &= B(\mathrm{im}+\nu)f_n \\ &= \mathbb{1}_{[-n,n]}B(\mathrm{im}+\nu)f \to B(\mathrm{im}+\nu)f\end{aligned}$$

$n \to \infty$. This shows that $f \in \mathrm{dom}\left(\overline{(\mathrm{im}+\nu)M(\mathrm{im}+\nu) + A}\right)$ and hence, the assertion follows. □

Remark 6.3.3 Note that Theorem 6.2.1 can partly be generalised in the following way (with the same proof). Let $M\colon \mathbb{C}_{\mathrm{Re}>\nu_0} \to L(H)$ be holomorphic and A a closed, densely defined operator in H such that $zM(z) + A$ is boundedly invertible for all $z \in \mathbb{C}_{\mathrm{Re}>\nu_0}$ and that $\sup_{z\in\mathbb{C}_{\mathrm{Re}>\nu_0}} \left\|(zM(z)+A)^{-1}\right\|_{L(H)} < \infty$. Then $S_\nu \in L(L_{2,\nu}(\mathbb{R}; H))$ is causal and eventually independent of ν.

Remark 6.3.4 As the proof of Theorem 6.2.1 shows, for $\nu \geqslant \nu_0$ we have that $S\colon \mathbb{C}_{\mathrm{Re}\geqslant\nu} \ni z \mapsto (zM(z)+A)^{-1} \in L(H)$ is a material law and $S_\nu = S(\partial_{t,\nu})$. Thus, the solution operator is a material law operator, and by Remark 5.3.3 applied to S and $z \mapsto \frac{1}{z}1_H$ we obtain

$$S_\nu \partial_{t,\nu} \subseteq \partial_{t,\nu} S_\nu.$$

6.4 Comments

The proof of Theorem 6.2.1 here is rather close to the strategy originally employed in [82], at least where existence and uniqueness are concerned. The causality part is a consequence of some observations detailed in [52, 131]. The original process of proving causality used the Theorem of Paley and Wiener, which we shall discuss later on.

The eddy current approximation has enjoyed great interest in the mathematical and physical community, in particular for the case when $\varepsilon = 0$ everywhere. The reason being that then Maxwell's equations are merely of parabolic type. We shall refer to [79] and the references therein for an extensive discussion.

Both Proposition 6.3.1 and the Lax–Milgram lemma have been put into a general perspective in [89].

Exercises

Exercise 6.1 Let (Ω, Σ, μ) be a σ-finite measure space and let H_0, H_1 be Hilbert spaces. Let $A\colon \operatorname{dom}(A) \subseteq H_0 \to H_1$ be densely defined and closed. Show that the operator

$$A_\mu\colon L_2(\mu; \operatorname{dom}(A)) \subseteq L_2(\mu; H_0) \to L_2(\mu; H_1)$$
$$f \mapsto \big(\omega \mapsto Af(\omega)\big)$$

is densely defined and closed. Moreover, show that $\big(A_\mu\big)^* = (A^*)_\mu$.

Exercise 6.2 In the situation of Exercise 6.1, if $(\Omega_1, \Sigma_1, \mu_1)$ is another σ-finite measure space and $\mathcal{F}\colon L_2(\mu) \to L_2(\mu_1)$ is unitary, show that for $j \in \{0, 1\}$ there exists a unique unitary operator $\mathcal{F}_{H_j}\colon L_2(\mu; H_j) \to L_2(\mu_1; H_j)$ such that

$$\mathcal{F}_{H_j}(\phi x) = (\mathcal{F}\phi)x \quad (\phi \in L_2(\mu), x \in H_j).$$

Furthermore, prove that

$$\mathcal{F}_{H_1} A_\mu \mathcal{F}_{H_0}^* = A_{\mu_1}.$$

Exercise 6.3 Show that for $\Omega \subseteq \mathbb{R}^d$ open, the set $C_c^\infty(\Omega) \subseteq L_2(\Omega)$ is dense.

Exercise 6.4 Prove Theorem 6.1.2.

Exercise 6.5 Let H be a Hilbert space, $A\colon \operatorname{dom}(A) \subseteq H \to H$ skew-selfadjoint, and $c > 0$. Moreover, let $M\colon \operatorname{dom}(M) \subseteq \mathbb{C} \to L(H)$ be holomorphic with

$$\operatorname{Re} M(z) \geqslant c \quad (z \in \operatorname{dom}(M)).$$

Show that $\operatorname{dom}(M) \ni z \mapsto (M(z) + A)^{-1}$ is holomorphic.

Exercise 6.6 Let $C\colon \operatorname{dom}(C) \subseteq H_0 \to H_1$ be a densely defined and closed linear operator acting in Hilbert spaces H_0 and H_1. For $\nu > 0$ show that

$$\overline{\partial_{t,\nu}^{-1} C} = C \partial_{t,\nu}^{-1}.$$

Hint: Apply Exercise 6.2 and show $\overline{(\mathrm{im}+\nu)^{-1}C} = C(\mathrm{im}+\nu)^{-1}$ with a suitable approximation argument.

Exercise 6.7 Let $\Omega \subseteq \mathbb{R}^d$ be open.

(a) Compute $H_0^1(\Omega)^\perp$ where the orthogonal complement is computed in $H^1(\Omega)$.
(b) Assume that

$$D := \left\{\phi \in H^1(\Omega)\,;\ \operatorname{grad}\phi \in \operatorname{dom}(\operatorname{div}), \phi = \operatorname{div}\operatorname{grad}\phi\right\} \subseteq C^\infty(\Omega).$$

and show that $C^\infty(\Omega) \cap H^1(\Omega) \subseteq H^1(\Omega)$ is dense.

Remark The regularity assumption in (b) always holds and is known as Weyl's Lemma, see e.g. [45, Corollary 8.11], where the more general situation of an elliptic operator with smooth coefficients is treated. See also [32, p.127], where the regularity is shown for harmonic distributions.

References

32. W.F. Donoghue Jr., *Distributions and Fourier transforms*, vol. 32. Pure and Applied Mathematics (Academic, New York, 1969)
40. R.P. Feynman, R.B. Leighton, M. Sands, *The Feynman Lectures on Physics. Vol. 2: Mainly Electromagnetism and Matter* (Addison-Wesley Publishing Co., Inc., Reading, MA, London, 1964)
45. D. Gilbarg, N.S. Trudinger, *Elliptic Partial Differential Equations of Second Order*, vol. 224, 2nd edn. Grundlehren der Mathematischen Wissenschaften [Fundamental Principles of Mathematical Sciences] (Springer, Berlin, 1983)
52. A. Kalauch et al., A Hilbert space perspective on ordinary differential equations with memory term. J. Dynam. Differ. Equ. **26**(2), 369–399 (2014)
59. R. Leis, Zur Theorie elektromagnetischer Schwingungen in anisotropen inhomogenen Medien. Math. Z. **106**, 213–224 (1968)
64. H. Minkowski, Die Grundgleichungen für die elektromagnetischen Vorgänge in bewegten Körpern. Math. Ann. **68**(4), 472–525 (1910)
79. D. Pauly, R. Picard, S. Trostorff, M. Waurick, On a class of degenerate abstract parabolic problems and applications to some Eddy current models. J. Funct. Anal. **280**(7), 108847 (2021)
82. R. Picard, A structural observation for linear material laws in classical mathematical physics. Math. Method Appl. Sci. **32**, 1768–1803 (2009)
88. R. Picard, S. Trostorff, M. Waurick, On maximal regularity for a class of evolutionary equations. J. Math. Anal. Appl. **449**(2), 1368–1381 (2017)
89. R. Picard, S. Trostorff, M. Waurick, Well-posedness via monotonicity. An overview, in *Operator Semigroups Meet Complex Analysis, Harmonic Analysis and Mathematical Physics*, vol. 250. Operator Theory: Advances and Applications (Birkhäuser, Cham, 2015), pp. 397–452
102. G. Schmidt, Spectral and scattering theory for Maxwell's equations in an exterior domain. Arch. Ration. Mech. Anal. **28**, 284–322 (1967/1968)
110. A. Süß, M. Waurick, A solution theory for a general class of SPDEs. Stoch. Part. Differ. Equ. Anal. Comput. **5**(2), 278–318 (2017)
131. M. Waurick, A note on causality in Banach spaces. Indagationes Mathematicae **26**(2), 404–412 (2015)

Chapter 7
Examples of Evolutionary Equations

This chapter is devoted to a small tour through a variety of evolutionary equations. More precisely, we shall look into the equations of poro-elastic media, (time-)fractional elasticity, thermodynamic media with delay as well as visco-elastic media. The discussion of these examples will be similar to that of the examples in the previous chapter in the sense that we shall present the equations first, reformulate them suitably and then apply the solution theory to them. The study of visco-elastic media within the framework of partial integro-differential equations will be carried out in the exercises section.

7.1 Poro-Elastic Deformations

In this section we will discuss the equations of poro-elasticity, which form a coupled system of equations. More precisely, the equations of (linearised) elasticity are coupled with the diffusion equation. Beforc properly writing these equations we introduce the following notation and differential operators.

Definition Let $\mathbb{K}^{d\times d}_{\mathrm{sym}} := \{A \in \mathbb{K}^{d\times d};\ A = A^\top\} \subseteq \mathbb{K}^{d\times d}$ be the (closed) subspace of symmetric $d \times d$ matrices. Let $\Omega \subseteq \mathbb{R}^d$ be open. Then define

$$L_2(\Omega)^{d\times d}_{\mathrm{sym}} := L_2(\Omega; \mathbb{K}^{d\times d}_{\mathrm{sym}})$$
$$= \left\{(\Phi_{jk})_{j,k\in\{1,\ldots,d\}} \in L_2(\Omega)^{d\times d};\ \forall j,k \in \{1,\ldots,d\}\colon \Phi_{jk} = \Phi_{kj}\right\}.$$

Analogously, we set $C_c^\infty(\Omega)^{d\times d}_{\mathrm{sym}} := C_c^\infty(\Omega; \mathbb{K}^{d\times d}_{\mathrm{sym}})$.

Note that the symmetry of a $d \times d$ matrix here means that the matrix elements are symmetric with respect to the main diagonal. For $\mathbb{K} = \mathbb{C}$, this does not

C. Seifert et al., *Evolutionary Equations*, Operator Theory: Advances and Applications 287, https://doi.org/10.1007/978-3-030-89397-2_7

correspond to the symmetry of the associated linear operator (which would rather be selfadjointness).

Definition Let $\Omega \subseteq \mathbb{R}^d$ be open. Then we define

$$\begin{aligned}\operatorname{Grad}_{\mathrm{c}} \colon C_{\mathrm{c}}^{\infty}(\Omega)^d \subseteq L_2(\Omega)^d &\to L_2(\Omega)^{d\times d}_{\mathrm{sym}} \\ \left(\phi_j\right)_{j\in\{1,\ldots,d\}} &\mapsto \frac{1}{2}\left(\partial_k\phi_j + \partial_j\phi_k\right)_{j,k\in\{1,\ldots,d\}},\end{aligned}$$

and

$$\begin{aligned}\operatorname{Div}_{\mathrm{c}} \colon C_{\mathrm{c}}^{\infty}(\Omega)^{d\times d}_{\mathrm{sym}} \subseteq L_2(\Omega)^{d\times d}_{\mathrm{sym}} &\to L_2(\Omega)^d \\ \left(\Phi_{jk}\right)_{j,k\in\{1,\ldots,d\}} &\mapsto \left(\sum_{k=1}^{d} \partial_k \Phi_{jk}\right)_{j\in\{1,\ldots,d\}}.\end{aligned}$$

Similarly to the definitions in the previous chapter, we put $\operatorname{Grad} := -\operatorname{Div}_{\mathrm{c}}^*$, $\operatorname{Div} := -\operatorname{Grad}_{\mathrm{c}}^*$ and $\operatorname{Grad}_0 := -\operatorname{Div}^*$, $\operatorname{Div}_0 := -\operatorname{Grad}^*$, where (analogously to the scalar-valued case) we observe that $\operatorname{Grad}_{\mathrm{c}} \subseteq -\operatorname{Div}_{\mathrm{c}}^*$ motivating the notation Grad and Grad_0.

Remark 7.1.1 Note that in the literature $\operatorname{Grad} u$ is also denoted by $\varepsilon(u)$ and is called the *strain tensor*. Due to the (obvious) similarity to the scalar case, we refrain from using ε in this context and prefer Grad instead. Again, the index 0 in the operators refers to generalised Dirichlet (for Grad_0) or Neumann (for Div_0) boundary conditions.

We are now properly equipped to formulate the equations of poro-elasticity; see also [69] and below for further details. In an elastic body $\Omega \subseteq \mathbb{R}^d$, the displacement field, $u\colon \mathbb{R}\times\Omega \to \mathbb{R}^d$, and the pressure field, $p\colon \mathbb{R}\times\Omega\to\mathbb{R}$, of a fluid diffusing through Ω satisfy the following two energy balance equations

$$\begin{aligned}\partial_t\rho\partial_t u - \operatorname{grad}\partial_t\lambda\operatorname{div} u - \operatorname{Div} C\operatorname{Grad} u + \operatorname{grad}\alpha^* p &= f, \\ \partial_t(c_0 p + \alpha\operatorname{div} u) - \operatorname{div} k\operatorname{grad} p &= g.\end{aligned}$$

The right-hand sides $f\colon \mathbb{R}\times\Omega\to\mathbb{R}^d$ and $g\colon\mathbb{R}\times\Omega\to\mathbb{R}$ describe some given external forcing. We assume homogeneous Neumann boundary conditions for the diffusing fluid as well as homogeneous Dirichlet (i.e. clamped) boundary conditions for the elastic body. The operator $\rho \in L(L_2(\Omega)^d)$ describes the density of the medium Ω (usually realised as a multiplication operator by a bounded, measurable, scalar function). The bounded linear operators $C \in L(L_2(\Omega)^{d\times d}_{\mathrm{sym}})$ and $k \in L(L_2(\Omega)^d)$ are the elasticity tensor and the hydraulic conductivity of the medium, whereas $c_0, \lambda \in L(L_2(\Omega))$ are the porosity of the medium and the compressibility of the fluid, respectively. The operator $\alpha \in L(L_2(\Omega))$ is the so-called Biot–Willis constant. Note that in many applications ρ, c_0, λ and α are just positive real numbers, and C and k are strictly positive definite tensors or matrices.

The reformulation of the equations for poro-elasticity involves several 'tricks'. One of these is to introduce the matrix trace as the operator

$$\begin{aligned} \operatorname{trace}\colon L_2(\Omega)^{d\times d}_{\mathrm{sym}} &\to L_2(\Omega) \\ (\Phi_{jk})_{j,k\in\{1,\ldots,d\}} &\mapsto \sum_{j=1}^{d} \Phi_{jj}. \end{aligned}$$

Note that the adjoint is given by $\operatorname{trace}^* f = \operatorname{diag}(f,\ldots,f) \in L_2(\Omega)^{d\times d}_{\mathrm{sym}}$. It is then elementary to obtain $\operatorname{trace}\operatorname{Grad} \subseteq \operatorname{div}$ as well as $\operatorname{grad} = \operatorname{Div}\operatorname{trace}^*$. Hence, we formally get

$$\begin{aligned} \partial_t \rho \partial_t u - \operatorname{Div}\big(\big(\partial_t \operatorname{trace}^* \lambda \operatorname{trace} + C\big)\operatorname{Grad} u - \operatorname{trace}^* \alpha^* p\big) &= f, \\ \partial_t\left(c_0 p + \alpha \operatorname{trace}\operatorname{Grad} u\right) - \operatorname{div} k \operatorname{grad} p &= g. \end{aligned}$$

Next, we introduce a new set of unknowns

$$\begin{aligned} v &:= \partial_t u, \\ T &:= C \operatorname{Grad} u, \\ \omega &:= \lambda \operatorname{trace}\operatorname{Grad} v - \alpha^* p, \\ q &:= -k \operatorname{grad} p. \end{aligned}$$

Here, v is the velocity, T is the stress tensor and q is the heat flux. The quantity ω is an additional variable, which helps to rewrite the system into the form of evolutionary equations.

In order to finalise the reformulation we shall assume some additional properties on the coefficients involved. Throughout the rest of this section, we assume that

$$\begin{aligned} \rho = \rho^* &\geqslant c, \\ c_0 = c_0^* &\geqslant c, \\ \operatorname{Re}\lambda &\geqslant c, \\ \operatorname{Re} k &\geqslant c, \text{ and} \\ C = C^* &\geqslant c \end{aligned}$$

for some $c > 0$, where all inequalities are thought of in the sense of positive definiteness (compare Chap. 6). As a consequence, we obtain

$$\operatorname{trace}\operatorname{Grad} v = \lambda^{-1}\omega + \lambda^{-1}\alpha^* p.$$

Rewriting the defining equations for T, ω, and q together with the two equations we started out with, we obtain the system

$$\begin{aligned}\partial_t \rho v - \operatorname{Div}\left(T + \operatorname{trace}^* \omega\right) &= f,\\ \partial_t c_0 p + \alpha\lambda^{-1}\omega + \alpha\lambda^{-1}\alpha^* p + \operatorname{div} q &= g,\\ \lambda^{-1}\omega + \lambda^{-1}\alpha^* p - \operatorname{trace}\operatorname{Grad} v &= 0,\\ \partial_t C^{-1} T - \operatorname{Grad} v &= 0,\\ k^{-1} q + \operatorname{grad} p &= 0.\end{aligned}$$

Note that at this stage of modelling we assumed that we can freely interchange the order of differentiation, so that $\operatorname{Grad}\partial_t u = \partial_t \operatorname{Grad} u$. Introducing

$$M_0 := \begin{pmatrix} \rho & 0 & 0 & 0 & 0\\ 0 & c_0 & 0 & 0 & 0\\ 0 & 0 & 0 & 0 & 0\\ 0 & 0 & 0 & C^{-1} & 0\\ 0 & 0 & 0 & 0 & 0\end{pmatrix}, \quad M_1 := \begin{pmatrix} 0 & 0 & 0 & 0 & 0\\ 0 & \alpha\lambda^{-1}\alpha^* & \alpha\lambda^{-1} & 0 & 0\\ 0 & \lambda^{-1}\alpha^* & \lambda^{-1} & 0 & 0\\ 0 & 0 & 0 & 0 & 0\\ 0 & 0 & 0 & 0 & k^{-1}\end{pmatrix}, \tag{7.1}$$

$$V := \begin{pmatrix} 1 & 0 & 0 & 0 & 0\\ 0 & 1 & 0 & 0 & 0\\ 0 & 0 & 1 & \operatorname{trace} & 0\\ 0 & 0 & 0 & 1 & 0\\ 0 & 0 & 0 & 0 & 1\end{pmatrix}, \quad A := \begin{pmatrix} 0 & 0 & 0 & -\operatorname{Div} & 0\\ 0 & 0 & 0 & 0 & \operatorname{div}_0\\ 0 & 0 & 0 & 0 & 0\\ -\operatorname{Grad}_0 & 0 & 0 & 0 & 0\\ 0 & \operatorname{grad} & 0 & 0 & 0\end{pmatrix}, \tag{7.2}$$

we obtain

$$\left(\partial_t M_0 + M_1 + VAV^*\right)\begin{pmatrix} v\\ p\\ \omega\\ T\\ q\end{pmatrix} = \begin{pmatrix} f\\ g\\ 0\\ 0\\ 0\end{pmatrix}.$$

This perspective enables us to prove well-posedness for the equations of poro-elasticity by applying Theorem 6.2.1.

Theorem 7.1.2 *Put* $H := L_2(\Omega)^d \times L_2(\Omega) \times L_2(\Omega) \times L_2(\Omega)^{d\times d}_{\mathrm{sym}} \times L_2(\Omega)^d$ *and let* $M_0, M_1, V \in L(H)$ *and* A *be given as in (7.1) and (7.2). Then there exists* $\nu_0 > 0$ *such that for all* $\nu \geqslant \nu_0$ *the operator* $\overline{\partial_{t,\nu} M_0 + M_1 + VAV^*}$ *is continuously invertible on* $L_{2,\nu}(\mathbb{R}; H)$. *The inverse* S_ν *of this operator is causal and eventually independent of* ν. *Moreover,* $\sup_{\nu\geqslant\nu_0}\|S_\nu\| < \infty$ *and* $F \in \operatorname{dom}(\partial_{t,\nu})$ *implies* $S_\nu F \in \operatorname{dom}(\partial_{t,\nu}) \cap \operatorname{dom}(VAV^*)$.

We will provide two prerequisites for the proof. We ask for the details of the proof of Theorem 7.1.2 in Exercise 7.1.

Proposition 7.1.3 *Let H_0, H_1 be Hilbert spaces, $B\colon \operatorname{dom}(B) \subseteq H_0 \to H_0$ skew-selfadjoint, $V \in L(H_0, H_1)$ bijective. Then $(VBV^*)^* = -VBV^*$.*

The proof of Proposition 7.1.3 is left as (part of) Exercise 7.1.

Proposition 7.1.4 *Let H be a Hilbert space, $N_0, N_1 \in L(H)$ with $N_0 = N_0^*$. Assume there exist $c_0, c_1 > 0$ such that $\langle x, N_0 x\rangle \geqslant c_0 \|x\|^2$ for all $x \in \operatorname{ran}(N_0)$ and $\operatorname{Re}\langle y, N_1 y\rangle \geqslant c_1 \|y\|^2$ for all $y \in \ker(N_0)$. Then for all $0 < c_1' < c_1$ there exists $\nu_0 > 0$ such that for all $\nu \geqslant \nu_0$ we have that*

$$\nu N_0 + \operatorname{Re} N_1 \geqslant c_1'.$$

Proof Note that by the selfadjointness of N_0 we can decompose $H = \overline{\operatorname{ran}}(N_0) \oplus \ker(N_0)$, see Corollary 2.2.6. Let $z \in H$, and $x \in \overline{\operatorname{ran}}(N_0)$, $y \in \ker(N_0)$ such that $z = x + y$. For $\varepsilon, \nu > 0$ we estimate

$$\begin{aligned}
&\nu \langle x+y, N_0(x+y)\rangle + \operatorname{Re}\langle x+y, N_1(x+y)\rangle \\
&= \nu\langle x, N_0 x\rangle + \operatorname{Re}\langle y, N_1 y\rangle + \operatorname{Re}\langle x, N_1 x\rangle + \operatorname{Re}\langle x, N_1 y\rangle + \operatorname{Re}\langle y, N_1 x\rangle \\
&\geqslant \nu c_0 \|x\|^2 + c_1\|y\|^2 - \|N_1\|\,\|x\|^2 - 2\|N_1\|\,\|x\|\,\|y\| \\
&\geqslant \nu c_0 \|x\|^2 + c_1\|y\|^2 - \|N_1\|\,\|x\|^2 - \frac{1}{\varepsilon}\|N_1\|^2\|x\|^2 - \varepsilon\|y\|^2 \\
&= \left(\nu c_0 - \frac{1}{\varepsilon}\|N_1\|^2 - \|N_1\|\right)\|x\|^2 + (c_1 - \varepsilon)\|y\|^2,
\end{aligned}$$

where we have used the Peter–Paul inequality (i.e., Young's inequality for products of non-negative numbers). For $0 < c_1' < c_1$ we find $\varepsilon > 0$ such that $c_1 - \varepsilon > c_1'$. Then we choose $\nu_0 > \frac{1}{c_0}\left(c_1' + \frac{1}{\varepsilon}\|N_1\|^2 + \|N_1\|\right)$. With this choice of ν_0 we deduce for all $\nu \geqslant \nu_0$ that

$$\nu\langle z, N_0 z\rangle + \operatorname{Re}\langle z, N_1 z\rangle \geqslant c_1'\left(\|x\|^2 + \|y\|^2\right) = c_1'\|z\|^2,$$

which yields the assertion. □

7.2 Fractional Elasticity

Let $\Omega \subseteq \mathbb{R}^d$ be open. In order to better fit to the experimental data of visco-elastic solids (i.e., to incorporate solids that 'memorise' previous force applied to them) the equations of linearised elasticity need to be extended in some way. The balance law for the momentum, however, is still satisfied; that is, for the displacement $u\colon \mathbb{R}\times\Omega\to\mathbb{R}^d$ we still have that

$$\partial_t \rho \partial_t u - \operatorname{Div} T = f,$$

where $\rho \in L(L_2(\Omega)^d)$ models the density and $f\colon \mathbb{R}\times\Omega\to\mathbb{R}^d$ is a given external forcing term. The stress tensor, $T\colon \mathbb{R}\times\Omega\to\mathbb{R}^{d\times d}_{\mathrm{sym}}$, does *not* follow the classical Hooke's law, which, if it did, would look like

$$T = C \operatorname{Grad} u$$

for $C \in L(L_2(\Omega)^{d\times d}_{\mathrm{sym}})$. Instead it is amended by another material dependent coefficient $D \in L(L_2(\Omega)^{d\times d}_{\mathrm{sym}})$ and a fractional time derivative; that is,

$$T = C \operatorname{Grad} u + D\partial_t^\alpha \operatorname{Grad} u,$$

for some $\alpha\in[0,1]$, where $\partial_t^\alpha := \partial_t\partial_t^{\alpha-1}$, see Example 5.3.1(e). We shall simplify the present consideration slightly and refer to Exercise 7.2 instead for a more involved example. Throughout this section, we shall assume that

$$C = 0,\ \ D = D^* \geqslant c,\ \text{and}\ \rho = \rho^* \geqslant c$$

for some $c > 0$. Thus, putting $v := \partial_t u$ and assuming the clamped boundary conditions again, we study well-posedness of

$$\partial_t \rho v - \operatorname{Div} T = f, \tag{7.3}$$

$$T = D\partial_t^\alpha \operatorname{Grad}_0 u. \tag{7.4}$$

In order to do that, we first rewrite the second equation. We will make use of the following proposition which will serve us to show bounded invertibility of ∂_t^α (in the space $L_{2,\nu}$), and which will also be employed to obtain well-posedness.

Proposition 7.2.1 *Let* $\nu > 0$, $z \in \mathbb{C}_{\mathrm{Re}\geqslant\nu}$, $\alpha\in[0,1]$. *Then*

$$\operatorname{Re} z^\alpha \geqslant (\operatorname{Re} z)^\alpha \geqslant \nu^\alpha.$$

Proof Let us prove the first inequality. Note that without loss of generality, we may assume that $\operatorname{Re} z = 1$. Let $\varphi := \arg z \in \left(-\frac{\pi}{2},\frac{\pi}{2}\right)$. Since $\ln\circ\cos$ is concave on $\left(-\frac{\pi}{2},\frac{\pi}{2}\right)$ (as $(\ln\circ\cos)' = -\tan$ is decreasing) and $(\ln\circ\cos)(0) = 0$, we obtain

$$\ln\cos(\alpha\varphi) = \ln\cos(\alpha\varphi + (1-\alpha)0) \geqslant \alpha\ln\cos(\varphi) + (1-\alpha)\ln\cos(0) = \ln\big(\cos(\varphi)^\alpha\big),$$

and therefore $\cos(\alpha\varphi) \geqslant \cos(\varphi)^\alpha$. Since $\operatorname{Re} z = 1$ implies $|z| = \frac{1}{\cos(\varphi)}$, we obtain

$$\operatorname{Re} z^\alpha = \frac{\cos(\alpha\varphi)}{(\cos\varphi)^\alpha} \geqslant 1 = (\operatorname{Re} z)^\alpha.$$

The second inequality follows from monotonicity of $x \mapsto x^\alpha$. □

Applying Proposition 7.2.1 and noting that D is boundedly invertible we can reformulate (7.4) as

$$\partial_{t,\nu}^{-\alpha} D^{-1} T - \operatorname{Grad}_0 u = 0,$$

so that (7.4) and (7.3) read

$$\left(\partial_{t,\nu} \begin{pmatrix} \rho & 0 \\ 0 & \partial_{t,\nu}^{-\alpha} D^{-1} \end{pmatrix} - \begin{pmatrix} 0 & \operatorname{Div} \\ \operatorname{Grad}_0 & 0 \end{pmatrix}\right) \begin{pmatrix} v \\ T \end{pmatrix} = \begin{pmatrix} f \\ 0 \end{pmatrix}.$$

A solution theory for the latter equation, thus, reads as follows, where again $v := \partial_{t,\nu} u$.

Theorem 7.2.2 *Put $H := L_2(\Omega)^d \times L_2(\Omega)^{d\times d}_{\mathrm{sym}}$. Then for all $\nu > 0$ the operator*

$$\partial_{t,\nu} \begin{pmatrix} \rho & 0 \\ 0 & \partial_{t,\nu}^{-\alpha} D^{-1} \end{pmatrix} - \begin{pmatrix} 0 & \operatorname{Div} \\ \operatorname{Grad}_0 & 0 \end{pmatrix}$$

is densely defined and closable in $L_{2,\nu}(\mathbb{R}; H)$. The inverse of the closure is continuous, causal and eventually independent of ν.

Proof The proof rests on Theorem 6.2.1. Since $\begin{pmatrix} 0 & \operatorname{Div} \\ \operatorname{Grad}_0 & 0 \end{pmatrix}$ is skew-selfadjoint by Proposition 6.2.3(a), it suffices to confirm the positive definiteness condition for the material law. For this let $z \in \mathbb{C}_{\operatorname{Re}\geqslant\nu}$ and compute for $x \in L_2(\Omega)^{d\times d}_{\mathrm{sym}}$, using Proposition 7.2.1 and Proposition 6.2.3(b),

$$\operatorname{Re}\left\langle x, zz^{-\alpha} D^{-1} x\right\rangle = \operatorname{Re}\left\langle x, z^{1-\alpha} D^{-1} x\right\rangle \geqslant \nu^{1-\alpha}\left\langle x, D^{-1} x\right\rangle \geqslant \nu^{1-\alpha} \frac{c}{\|D\|^2} \|x\|^2.$$

This yields the assertion. □

7.3 The Heat Equation with Delay

Let $\Omega \subseteq \mathbb{R}^d$ be open. In this section we concentrate on a generalisation of the heat equation discussed in the previous chapter. Although we keep the heat flux balance in the sense that

$$\partial_t \theta + \operatorname{div} q = Q,$$

with $q\colon \mathbb{R} \times \Omega \to \mathbb{R}^d$ being the heat flux and $\theta\colon \mathbb{R} \times \Omega \to \mathbb{R}$ being the heat, we shall now modify Fourier's law to the extent that

$$q = -a \operatorname{grad} \theta - b\tau_{-h} \operatorname{grad} \theta$$

for some $a, b \in L(L_2(\Omega)^d)$ with $\operatorname{Re} a \geqslant c$ for some $c > 0$, and $h > 0$. We shall again assume homogeneous Neumann boundary conditions for q. Written in the now standard block operator matrix form, this modified heat equation reads

$$\left(\partial_{t,\nu} \begin{pmatrix} 1 & 0 \\ 0 & 0 \end{pmatrix} + \begin{pmatrix} 0 & 0 \\ 0 & (a + b\tau_{-h})^{-1} \end{pmatrix} + \begin{pmatrix} 0 & \operatorname{div}_0 \\ \operatorname{grad} & 0 \end{pmatrix}\right) \begin{pmatrix} \theta \\ q \end{pmatrix} = \begin{pmatrix} Q \\ 0 \end{pmatrix}.$$

In order to actually justify the existence of the operator $(a + b\tau_{-h})^{-1}$ as a bounded linear operator, we provide the following lemma.

Lemma 7.3.1 *Let $h > 0$.*

(a) *There exists $\nu_0 > 0$ such that for all $\nu \geqslant \nu_0$ the operator $a + b\tau_{-h}$ is continuously invertible on $L_{2,\nu}(\mathbb{R}; L_2(\Omega)^d)$.*
(b) *For all $0 < c' < c/\|a\|^2$ there is $\nu_1 \geqslant \nu_0$ such that for all $z \in \mathbb{C}_{\operatorname{Re} \geqslant \nu_1}$ we have $\operatorname{Re}\left(a + b\mathrm{e}^{-zh}\right)^{-1} \geqslant c'$.*

Proof Note that a is invertible with $\left\|a^{-1}\right\| \leqslant \frac{1}{c}$ and $\operatorname{Re} a^{-1} \geqslant \frac{c}{\|a\|^2}$ by Proposition 6.2.3(b).

(a) By Example 5.3.4(c), for all $\nu > 0$ we obtain

$$\|b\tau_{-h}\|_{L(L_{2,\nu})} \leqslant \|b\|_{L(L_2(\Omega)^d)} \sup_{t \in \mathbb{R}} \left|\mathrm{e}^{-(it+\nu)h}\right| = \|b\|_{L(L_2(\Omega)^d)} \mathrm{e}^{-h\nu}.$$

Thus, we find $\nu_0 > 0$ such that for all $\nu \geqslant \nu_0$ we obtain $\left\|b\tau_{-h} a^{-1}\right\|_{L(L_{2,\nu})} \leqslant \frac{1}{c} \|b\tau_{-h}\|_{L(L_{2,\nu})} < 1$. Thus,

$$a + b\tau_{-h} = \left(1 + b\tau_{-h} a^{-1}\right) a$$

is continuously invertible by a Neumann series argument.

(b) Let $0 < c' < c/\|a\|^2$, and set $d(z) := -be^{-zh}a^{-1}$. Moreover, we choose $\nu_1 \geqslant \nu_0$ such that $\|d(z)\|_{L(L_2(\Omega)^d)} \leqslant \min\{\frac{1}{2}, \varepsilon\}$ for all $z \in \mathbb{C}_{\mathrm{Re} \geqslant \nu_1}$, where $0 < \varepsilon \leqslant \frac{1}{2}c\left(\frac{c}{\|a\|^2} - c'\right)$. For $z \in \mathbb{C}_{\mathrm{Re} \geqslant \nu_1}$ we compute

$$
\begin{aligned}
\operatorname{Re}\left(a + be^{-zh}\right)^{-1} &= \operatorname{Re} a^{-1}(1 - d(z))^{-1} = \operatorname{Re}\left(a^{-1}\sum_{k=0}^{\infty} d(z)^k\right) \\
&= \operatorname{Re}\left(a^{-1} + \sum_{k=1}^{\infty} a^{-1}d(z)^k\right) \\
&\geqslant \frac{c}{\|a\|^2} - \left\|\sum_{k=1}^{\infty} a^{-1}d(z)^k\right\| \geqslant \frac{c}{\|a\|^2} - \frac{1}{c}\sum_{k=1}^{\infty}\|d(z)\|^k \\
&= \frac{c}{\|a\|^2} - \frac{1}{c}\frac{\|d(z)\|}{1 - \|d(z)\|} \geqslant \frac{c}{\|a\|^2} - \frac{1}{c}2\varepsilon \geqslant c'.
\end{aligned}
$$

□

With this lemma we are in the position to provide the well-posedness for the modified heat equation.

Theorem 7.3.2 *Let $H = L_2(\Omega) \times L_2(\Omega)^d$. There exists $\nu_0 > 0$ such that for all $\nu \geqslant \nu_0$ the operator*

$$
\partial_{t,\nu}\begin{pmatrix} 1 & 0 \\ 0 & 0 \end{pmatrix} + \begin{pmatrix} 0 & 0 \\ 0 & (a + b\tau_{-h})^{-1} \end{pmatrix} + \begin{pmatrix} 0 & \operatorname{div}_0 \\ \operatorname{grad} & 0 \end{pmatrix}
$$

is densely defined and closable with continuously invertible closure on $L_{2,\nu}(\mathbb{R}; H)$. The inverse of the closure is causal and eventually independent of ν.

Proof The proof rests on Theorem 6.2.1 and Lemma 7.3.1. □

7.4 Dual Phase Lag Heat Conduction

The last example is concerned with a different modification of Fourier's law. The heat flux balance

$$
\partial_t \theta + \operatorname{div} q = Q \tag{7.5}
$$

is accompanied by the modified Fourier's law

$$
\left(1 + s_q\partial_t + \frac{1}{2}s_q^2\partial_t^2\right)q = -(1 + s_\theta\partial_t)\operatorname{grad}\theta, \tag{7.6}
$$

where $s_q \in \mathbb{R}$, $s_\theta > 0$ are given numbers, which are called 'phases'.

Remark 7.4.1 The modified Fourier's law in (7.6) is an attempt to resolve the problem of infinite propagation speed which stems from a truncated Taylor series expansion of a model given by

$$\tau_{s_q} q = -\tau_{s_\theta} \operatorname{grad} \theta.$$

Note that it can be shown that such a model would even be ill-posed, see [34].

Let us turn back to the system (7.5) and (7.6). Notice, since $s_\theta > 0$, and due to a strictly positive real part of the derivative in our functional analytic setting, we deduce that $(1 + s_\theta \partial_{t,\nu})$ is continuously invertible for $\nu \geqslant 0$. Thus, we obtain

$$\partial_{t,\nu}\big(\partial_{t,\nu}^{-1} + s_q + \frac{1}{2}s_q^2\partial_{t,\nu}\big)(1 + s_\theta\partial_{t,\nu})^{-1} q = -\operatorname{grad}\theta.$$

The block operator matrix formulation of the dual phase lag heat conduction model is thus

$$\left(\partial_{t,\nu}\begin{pmatrix}1 & 0\\ 0 & (\partial_{t,\nu}^{-1} + s_q + \frac{1}{2}s_q^2\partial_{t,\nu})(1 + s_\theta\partial_{t,\nu})^{-1}\end{pmatrix} + \begin{pmatrix}0 & \operatorname{div}_0\\ \operatorname{grad} & 0\end{pmatrix}\right)\begin{pmatrix}\theta\\ q\end{pmatrix} = \begin{pmatrix}Q\\ 0\end{pmatrix}.$$

Theorem 7.4.2 *Let $H = L_2(\Omega) \times L_2(\Omega)^d$. Assume $s_q \in \mathbb{R}\setminus\{0\}$, $s_\theta > 0$. Then there exists $\nu_0 > 0$ such that for all $\nu \geqslant \nu_0$ the operator*

$$\partial_{t,\nu}\begin{pmatrix}1 & 0\\ 0 & (\partial_{t,\nu}^{-1} + s_q + \frac{1}{2}s_q^2\partial_{t,\nu})(1 + s_\theta\partial_{t,\nu})^{-1}\end{pmatrix} + \begin{pmatrix}0 & \operatorname{div}_0\\ \operatorname{grad} & 0\end{pmatrix}$$

is densely defined and closable with continuously invertible closure on $L_{2,\nu}(\mathbb{R}; H)$. The inverse of the closure is causal and eventually independent of ν.

The proof of Theorem 7.4.2 is again based on Theorem 6.2.1. Thus, we shall only record the decisive observation in the next result. For this, we define

$$M(z) := \frac{z^{-1} + s_q + \frac{1}{2}s_q^2 z}{1 + s_\theta z} \in \mathbb{C} \quad (z \in \mathbb{C}\setminus\{0, -\tfrac{1}{s_\theta}\}).$$

Lemma 7.4.3 *Let $s_q \in \mathbb{R}\setminus\{0\}$, $s_\theta > 0$. Then there exist $\nu_0 \in \mathbb{R}$ and $c > 0$ such that for all $z \in \mathbb{C}_{\mathrm{Re}\geqslant\nu_0}$ we have*

$$\operatorname{Re} zM(z) \geqslant c.$$

Proof We put $\sigma := \frac{s_q}{s_\theta}$. Let $z \in \mathbb{C}\setminus\{0, -\frac{1}{s_\theta}\}$. We compute

$$zM(z) = \frac{1 + s_q z + \frac{1}{2}s_q^2 z^2}{1 + s_\theta z} = \frac{1}{2}s_q z\sigma + \sigma\left(1 - \frac{1}{2}\sigma\right) + \frac{1 - \sigma\left(1 - \frac{1}{2}\sigma\right)}{1 + s_\theta z}$$

and therefore

$$\operatorname{Re} zM(z) = \frac{1}{2}s_q\sigma \operatorname{Re} z + \sigma\left(1 - \frac{1}{2}\sigma\right) + \frac{\left(1-\sigma\left(1-\frac{1}{2}\sigma\right)\right)(1+s_\theta \operatorname{Re} z)}{|1+s_\theta z|^2}.$$

By assumption

$$0 < \frac{s_q^2}{s_\theta} = s_q\sigma,$$

and since

$$\frac{\left(1-\sigma\left(1-\frac{1}{2}\sigma\right)\right)(1+s_\theta \operatorname{Re} z)}{|1+s_\theta z|^2} \to 0$$

as $\operatorname{Re} z \to \infty$, we obtain

$$\operatorname{Re} zM(z) \geqslant \frac{1}{2}s_q\sigma \operatorname{Re} z - \delta$$

for some $\delta > 0$ and all $z \in \mathbb{C}$ with $\operatorname{Re} z$ large enough. □

7.5 Comments

The equations of poro-elasticity have been proposed in [69] and were mathematically studied in [63, 103].

Equations of fractional elasticity are discussed in [20, 73, 87, 134]. The well-posedness conditions stated here and in Exercise 7.2 can be generalised as it is outlined in [87] to the case where both C and D are non-negative, selfadjoint operators so that C and D satisfy the conditions imposed on N_1 and N_0 in Proposition 7.1.4. We refrained from presenting this argument here, as it seemed too technical for the time being. Note however that the proof is neither fundamentally different nor considerably less elementary.

The heat equation with delay has also been studied in [55] with an entirely different strategy; the dual phase lag models have been dealt with in [68, 127].

Other ideas to rectify infinite propagation speed of the heat equation can be found in [3], where nonlinear models for heat conduction are being discussed.

The visco-elastic equations discussed in Exercise 7.6 are studied with convolution operators more general than below in [119]; see also [19, 27, 95, 116].

Exercises

Exercise 7.1 (Solutions to the Equations of Poro-Elasticity)

(a) Prove Proposition 7.1.3.
(b) Prove Theorem 7.1.2.
(c) Let $\Omega \subseteq \mathbb{R}^d$ be open, $\nu > 0$, $f \in H_\nu^1(\mathbb{R}; L_2(\Omega)^d)$ and $g \in H_\nu^1(\mathbb{R}; L_2(\Omega))$. With the help of Theorem 7.1.2 show that for large enough $\nu > 0$ there exist a unique $u \in \operatorname{dom}\left(\partial_{t,\nu}^2\right) \cap \operatorname{dom}\left(\operatorname{grad} \lambda \operatorname{div} \partial_{t,\nu}\right) \cap \operatorname{dom}\left(\operatorname{Div} C \operatorname{Grad}_0\right)$ and $p \in \operatorname{dom}(\partial_{t,\nu}) \cap \operatorname{dom}(\operatorname{grad} \alpha^*) \cap \operatorname{dom}(\operatorname{div}_0 k \operatorname{grad})$ such that

$$\begin{aligned} \partial_{t,\nu} \rho \partial_{t,\nu} u - \operatorname{grad} \lambda \operatorname{div} \partial_{t,\nu} u - \operatorname{Div} C \operatorname{Grad}_0 u + \operatorname{grad} \alpha^* p &= f \\ \partial_{t,\nu} c_0 p + \alpha \operatorname{div} \partial_{t,\nu} u - \operatorname{div}_0 k \operatorname{grad} p &= g. \end{aligned}$$

Exercise 7.2 Let $\Omega \subseteq \mathbb{R}^d$ be open, $C, D \in L(L_2(\Omega)_{\mathrm{sym}}^{d \times d})$, $D = D^* \geqslant c$ for some $c > 0$ and $\alpha \in [\frac{1}{2}, 1]$. Show that there exists $\nu_0 > 0$ such that for all $\nu \geqslant \nu_0$ the system

$$\begin{aligned} \partial_{t,\nu} \rho v - \operatorname{Div} T &= f, \\ T &= \left(C + D \partial_{t,\nu}^\alpha\right) \operatorname{Grad}_0 u, \end{aligned}$$

where $v = \partial_{t,\nu} u$, admits a unique solution $(v, T) \in L_{2,\nu}(\mathbb{R}; L_2(\Omega)^d \times L_2(\Omega)_{\mathrm{sym}}^{d \times d})$ for all $f \in H_\nu^1(\mathbb{R}; L_2(\Omega)^d)$.

The following exercises are devoted to showing the well-posedness of certain equations in visco-elasticity, where the 'viscous part' is modelled by convolution with certain integral kernels. The proof of the positive definiteness property requires some preliminary results. We assume the reader to be equipped with the basics from the theory of functions of one complex variable.

For $U \subseteq \mathbb{C}$ open write $\widetilde{U} := \left\{(x, y) \in \mathbb{R}^2;\ x + \mathrm{i}y \in U\right\}$, and for $u \colon U \to \mathbb{C}$ holomorphic, define $f_{\mathrm{Re}\, u} \colon \widetilde{U} \to \mathbb{R}$ by $f_{\mathrm{Re}\, u}(x, y) := \operatorname{Re} u(x + \mathrm{i}y)$ for $(x, y) \in \widetilde{U}$. We put

$$H_{\mathrm{Re}}(U) := \left\{ f_{\mathrm{Re}\, u}\, ;\ u \colon U \to \mathbb{C} \text{ holomorphic} \right\}.$$

Exercise 7.3 Let $U \subseteq \mathbb{C}$ be open.

(a) Let $f \in H_{\mathrm{Re}}(U)$. Show that f satisfies the *mean value property*; that is, for all $(x, y) \in \widetilde{U}$ and $r > 0$ with $\overline{B\left((x, y), r\right)} \subseteq \widetilde{U}$ we have

$$f(x, y) = \frac{1}{2\pi} \int_0^{2\pi} f(x + r\cos\theta, y + r\sin\theta)\, \mathrm{d}\theta.$$

(b) Let $U := \mathbb{C}_{\mathrm{Im}>0}$ and $f \in H_{\mathrm{Re}}(U) \cap C(\mathbb{R} \times \mathbb{R}_{\geqslant 0})$. Moreover, assume that $f(x, 0) = 0$ for each $x \in \mathbb{R}$ and $f(x, y) \to 0$ as $|(x, y)| \to \infty$. Show that $f = 0$ on $\mathbb{R} \times \mathbb{R}_{\geqslant 0}$.

Exercise 7.4 In this exercise we show a version of *Poisson's formula*. Let $U := \mathbb{C}_{\mathrm{Im}>0}$ and $f \in H_{\mathrm{Re}}(U) \cap C(\mathbb{R} \times \mathbb{R}_{\geqslant 0})$.

(a) Assume that $f(\cdot, 0) \in L_p(\mathbb{R})$ for some $1 \leqslant p < \infty$. Show that $\mathbb{C}_{\mathrm{Im}>0} \ni z \mapsto \frac{1}{\pi} \int_{\mathbb{R}} \frac{\mathrm{Im}\, z' + \mathrm{i}(\mathrm{Re}\, z - x')}{(\mathrm{Re}\, z - x')^2 + (\mathrm{Im}\, z)^2} f(x', 0)\, \mathrm{d}x'$ is holomorphic.
(b) Assume that $f(\cdot, 0) \in L_\infty(\mathbb{R})$. Show that $\frac{1}{\pi} \int_{\mathbb{R}} \frac{y}{(x - x')^2 + y^2)} f(x', 0)\, \mathrm{d}x' \to f(x_0, 0)$ as $x \to x_0$ and $y \to 0+$.
(c) (Poisson's formula) Assume that $f(\cdot, 0) \in L_p(\mathbb{R})$ for some $1 \leqslant p < \infty$ and $f(x, y) \to 0$ as $|(x, y)| \to \infty$ in $\mathbb{R} \times \mathbb{R}_{\geqslant 0}$. Show that

$$f(x, y) = \frac{1}{\pi} \int_{\mathbb{R}} \frac{y}{(x - x')^2 + y^2} f(x', 0)\, \mathrm{d}x' \quad ((x, y) \in \mathbb{R} \times \mathbb{R}_{>0}).$$

Hint: Apply Exercise 7.3(b).

Exercise 7.5 Let $\nu_0 \in \mathbb{R}$ and $k \in L_{1,\nu_0}(\mathbb{R}; \mathbb{R})$ with $\operatorname{spt} k \subseteq \mathbb{R}_{\geqslant 0}$.

(a) Show that for all $(x, \nu) \in \mathbb{R} \times \mathbb{R}_{>\nu_0}$ we have

$$\mathrm{Im}(\mathcal{L}k)(\mathrm{i}x + \nu) = \frac{1}{\pi} \int_{\mathbb{R}} \frac{\nu - \nu_0}{(x - x')^2 + (\nu - \nu_0)^2} \mathrm{Im}(\mathcal{L}k)(\mathrm{i}x' + \nu_0)\, \mathrm{d}x'.$$

Hint: Approximate k by functions in $C_c^\infty(\mathbb{R}_{\geqslant 0}; \mathbb{R})$ and use Poisson's formula (see Exercise 7.4).
(b) Assume there exists $d \geqslant 0$ such that for all $x \in \mathbb{R}$

$$x\, \mathrm{Im}(\mathcal{L}k)(\mathrm{i}x + \nu_0) \leqslant d.$$

Show that for all $\nu \geqslant \nu_0$ and $x \in \mathbb{R}$ we have

$$x\, \mathrm{Im}(\mathcal{L}k)(\mathrm{i}x + \nu) \leqslant 4d.$$

Hint: Use the formula in (a) and split the integral into positive and negative part of $\mathbb{R}$; use symmetry of $(\mathcal{L}k)$ under conjugation due to the realness of k.

Exercise 7.6 Let $\Omega \subseteq \mathbb{R}^d$ be open, $\nu_0 \in \mathbb{R}$ and $k \in L_{1,\nu_0}(\mathbb{R}; \mathbb{R})$ with $\operatorname{spt} k \subseteq \mathbb{R}_{\geqslant 0}$. Assume there exists $d \geqslant 0$ such that

$$x\, \mathrm{Im}(\mathcal{L}k)(\mathrm{i}x + \nu_0) \leqslant d \quad (x \in \mathbb{R}).$$

Show that there exists $\nu_1 \geqslant \nu_0$ such that for all $\nu \geqslant \nu_1$ the operator

$$\partial_{t,\nu}\begin{pmatrix}1 & 0\\ 0 & (1-k*)^{-1}\end{pmatrix}+\begin{pmatrix}0 & \operatorname{Div}\\ \operatorname{Grad}_0 & 0\end{pmatrix}$$

is well-defined, densely defined and closable in $L_{2,\nu}(\mathbb{R}; H)$ with $H = L_2(\Omega)^d \times L_2(\Omega)^{d\times d}_{\mathrm{sym}}$. Further, show that its closure is continuously invertible, and that the corresponding inverse is causal and eventually independent of ν.

Exercise 7.7 Let $\nu_0 \in \mathbb{R}$ and $k \in L_{1,\nu_0}(\mathbb{R}; \mathbb{R})$ with $\operatorname{spt} k \subseteq \mathbb{R}_{\geqslant 0}$.

(a) Assume that k is absolutely continuous with $k' \in L_{1,\nu_0}(\mathbb{R}; \mathbb{R})$. Show that there exist $\nu_1 \geqslant \nu_0$ and $d \geqslant 0$ with

$$x \operatorname{Im}(\mathcal{L}k)(\mathrm{i}x + \nu_1) \leqslant d \quad (x \in \mathbb{R}).$$

(b) Assume that $k(t) \geqslant 0$ for all $t \in \mathbb{R}$ and that $k(t) \leqslant k(s)$, whenever $s \leqslant t$. Show that there exists $\nu_1 \geqslant \nu_0$ with

$$x \operatorname{Im}(\mathcal{L}k)(\mathrm{i}x + \nu_1) \leqslant 0 \quad (x \in \mathbb{R}).$$

Hint: For part (b) use the explicit formula for $\operatorname{Im}(\mathcal{L}k)$ as an integral and the periodicity of sin.
Remark: The condition in (a) is a standard assumption for convolution kernels in the framework of visco-elastic equations; the condition in (b) is from [95].

References

3. F. Andreu et al., Finite propagation speed for limited flux diffusion equations. Arch. Ration. Mech. Anal. **182**(2), 269 (2006)
19. P. Cannarsa, D. Sforza, Global solutions of abstract semilinear parabolic equations with memory terms. Nonlinear Differ. Equ. Appl. **10**(4), 399–430 (2003)
20. K. Cherednichenko, M. Waurick, Resolvent estimates in homogenisation of periodic problems of fractional elasticity. J. Differ. Equ. **264**(6), 3811–3835 (2018)
27. C.M. Dafermos, An abstract Volterra equation with applications to linear viscoelasticity. J. Differ. Equ. **7**, 554–569 (1970)
34. M. Dreher, R. Quintanilla, R. Racke, Ill-posed problems in thermomechanics. Appl. Math. Lett. **22**(9), 1374–1379 (2009)
55. D. Khusainov, M. Pokojovy, R. Racke, Strong and mild extrapolated L^2- solutions to the heat equation with constant delay. SIAM J. Math. Anal. **47**(1), 427–454 (2015)
63. D.F. McGhee, R. Picard, A note on anisotropic, inhomogeneous, poro-elastic media. Math. Methods Appl. Sci. **33**(3), 313–322 (2010)
68. S. Mukhopadyay et al., On some models in linear thermo-elasticity with rational material laws. Math. Mech. Solids **21**(9), 1149–1163 (2016)
69. M.A. Murad, J.H. Cushman, Multiscale flow and deformation in hydrophilic swelling porous media. Int. J. Eng. Sci. **34**(3), 313–338 (1996)

73. B. Nolte, S. Kempfle, I. Schäfer, Does a real material behave fractionally? Applications of fractional differential operators to the damped structure borne sound in viscoelastic solids. J. Comput. Acoust. **11**(03), 451–489 (2003). eprint: https://doi.org/10.1142/S0218396X03002024
87. R. Picard, S. Trostorff, M. Waurick, On evolutionary equations with material laws containing fractional integrals. Math. Method Appl. Sci. **38**(15), 3141–3154 (2015)
95. J. Prüss, Decay properties for the solutions of a partial differential equation with memory. Archiv der Mathematik **92**(2), 158–173 (2009)
103. R.E. Showalter, Diffusion in poro-elastic media. J. Math. Anal. Appl. **251**(1), 310–340 (2000)
116. S. Trostorff, Exponential Stability and Initial Value Problems for Evolutionary Equations. Habilitation Thesis. TU Dresden, 2018
119. S. Trostorff, On integro-differential inclusions with operator-valued kernels. Math. Method Appl. Sci. **38**(5), 834–850 (2015)
127. D. Tzou, A unified field approach for heat conduction from macro-to microscales. J. Heat Transfer **117**(1), 8–16 (1995)
134. M. Waurick, Homogenization in fractional elasticity. SIAM J. Math. Anal. **46**(2), 1551–1576 (2014)

Chapter 8
Causality and a Theorem of Paley and Wiener

In this chapter we turn our focus back to causal operators. In Chap. 5 we found out that material laws provide a class of causal and autonomous bounded operators. In this chapter we will present another proof of this fact, which rests on a result which characterises functions in $L_2(\mathbb{R}; H)$ with support contained in the non-negative reals; the celebrated Theorem of Paley and Wiener. With the help of this theorem, which is interesting in its own right, the proof of causality for material laws becomes very easy. At a first glance it seems that holomorphy of a material law is a rather strong assumption. In the second part of this chapter, however, we shall see that in designing autonomous and causal solution operators, there is no way of circumventing holomorphy.

In the following, let H be a Hilbert space, and we consider $L_{2,\nu}(\mathbb{R}_{\geqslant 0}; H)$ as the subspace of functions in $L_{2,\nu}(\mathbb{R}; H)$ vanishing on $(-\infty, 0)$.

8.1 A Theorem of Paley and Wiener

We start with the following lemma, for which we need the notion of locally integrable functions. We define

$$\begin{aligned} L_{1,\mathrm{loc}}(\mathbb{R}, H) &:= \{f\,;\, \forall K \subseteq \mathbb{R} \text{ compact} : \mathbb{1}_K f \in L_1(\mathbb{R}; H)\} \\ &= \left\{f\,;\, \forall \varphi \in C_c^\infty(\mathbb{R}) : \varphi f \in L_1(\mathbb{R}; H)\right\}. \end{aligned}$$

Lemma 8.1.1 *Let $f \in L_{1,\mathrm{loc}}(\mathbb{R}; H)$. Then we have $f \in L_2(\mathbb{R}_{\geqslant 0}; H)$ if and only if $f \in \bigcap_{\nu>0} L_{2,\nu}(\mathbb{R}; H)$ with $\sup_{\nu>0} \|f\|_{L_{2,\nu}(\mathbb{R};H)} < \infty$. In the latter case we have that*

$$\|f\|_{L_2(\mathbb{R}_{\geqslant 0};H)} = \lim_{\nu\to 0+} \|f\|_{L_{2,\nu}(\mathbb{R};H)} = \sup_{\nu>0} \|f\|_{L_{2,\nu}(\mathbb{R};H)}\,.$$

C. Seifert et al., *Evolutionary Equations*, Operator Theory: Advances and Applications 287, https://doi.org/10.1007/978-3-030-89397-2_8

Proof Let $f \in L_2(\mathbb{R}_{\geqslant 0}; H)$ and $\nu > 0$. Then we estimate

$$\int_{\mathbb{R}} \|f(t)\|_H^2 \,\mathrm{e}^{-2\nu t}\,\mathrm{d}t = \int_{\mathbb{R}_{\geqslant 0}} \|f(t)\|_H^2 \,\mathrm{e}^{-2\nu t}\,\mathrm{d}t \leqslant \int_{\mathbb{R}_{\geqslant 0}} \|f(t)\|_H^2 \,\mathrm{d}t = \|f\|^2_{L_2(\mathbb{R}_{\geqslant 0}; H)},$$

which proves that $f \in L_{2,\nu}(\mathbb{R}; H)$ with $\|f\|_{L_{2,\nu}(\mathbb{R};H)} \leqslant \|f\|_{L_2(\mathbb{R}_{\geqslant 0};H)}$ for each $\nu > 0$. Moreover, $\|f\|_{L_{2,\nu}(\mathbb{R};H)} \to \|f\|_{L_2(\mathbb{R}_{\geqslant 0};H)}$ as $\nu \to 0$ by monotone convergence and since clearly $\|f\|_{L_{2,\nu}(\mathbb{R};H)} \leqslant \|f\|_{L_{2,\mu}(\mathbb{R};H)}$ for $0 < \mu \leqslant \nu$ we obtain

$$\|f\|_{L_2(\mathbb{R}_{\geqslant 0};H)} = \lim_{\nu \to 0+} \|f\|_{L_{2,\nu}(\mathbb{R};H)} = \sup_{\nu > 0} \|f\|_{L_{2,\nu}(\mathbb{R};H)}.$$

Assume now that $f \in \bigcap_{\nu > 0} L_{2,\nu}(\mathbb{R}; H)$ with $C := \sup_{\nu > 0} \|f\|_{L_{2,\nu}(\mathbb{R};H)} < \infty$. This inequality yields

$$\sup_{\nu \in (0,\infty)} \int_{(-\infty,0)} \|f(t)\|^2 \,\mathrm{e}^{-2\nu t}\,\mathrm{d}t \leqslant C^2.$$

Hence, the monotone convergence theorem yields that $g(t) := \lim_{\nu \to \infty} \|f(t)\|^2 \,\mathrm{e}^{-2\nu t}$ for $t \in (-\infty, 0)$ defines a function $g \in L_1(-\infty, 0)$. Thus, $[g = \infty]$ is a set of measure zero and thus $[f = 0] \cap (-\infty, 0) = (-\infty, 0) \setminus [g = \infty]$ has full measure in $(-\infty, 0)$ implying that $\operatorname{spt} f \subseteq \mathbb{R}_{\geqslant 0}$.

Finally, from

$$\sup_{\nu \in (0,\infty)} \int_{(0,\infty)} \|f(t)\|^2 \,\mathrm{e}^{-2\nu t}\,\mathrm{d}t \leqslant C^2.$$

we infer again by the monotone convergence theorem that $t \mapsto \lim_{\nu \to 0} \|f(t)\|^2 \,\mathrm{e}^{-2\nu t} = \|f(t)\|^2$ defines a function in $L_1(0, \infty)$, showing the remaining assertion. □

For the proof of the Paley–Wiener theorem we need a suitable space of holomorphic functions on the right half-plane, the so-called *Hardy space* $\mathcal{H}_2(\mathbb{C}_{\mathrm{Re}>\nu}; H)$, which we introduce in the following.

Definition For $\nu \in \mathbb{R}$ we define the *Hardy space*

$$\mathcal{H}_2(\mathbb{C}_{\mathrm{Re}>\nu}; H) := \left\{ g \colon \mathbb{C}_{\mathrm{Re}>\nu} \to H\,;\ g \text{ holomorphic, } \sup_{\rho > \nu} \int_{\mathbb{R}} \|g(\mathrm{i}t + \rho)\|_H^2 \,\mathrm{d}t < \infty \right\}$$

and equip it with the norm $\|\cdot\|_{\mathcal{H}_2(\mathbb{C}_{\mathrm{Re}>\nu};H)}$ defined by

$$\|g\|_{\mathcal{H}_2(\mathbb{C}_{\mathrm{Re}>\nu};H)} := \sup_{\rho>\nu} \left(\int_{\mathbb{R}} \|g(\mathrm{i}t+\rho)\|_H^2 \,\mathrm{d}t \right)^{\frac{1}{2}}.$$

We motivate the Theorem of Paley–Wiener first. For this, let $f \in L_{2,\nu}(\mathbb{R}_{\geqslant 0}; H)$ and define its *Laplace transform* as

$$\mathbb{C}_{\mathrm{Re}>\nu} \ni z \mapsto \mathcal{L}f(z) := \frac{1}{\sqrt{2\pi}} \int_0^\infty f(t)\mathrm{e}^{-zt}\,\mathrm{d}t. \tag{8.1}$$

Note that $\mathcal{L}f(z) = \mathcal{L}_{\mathrm{Re}\,z} f(\mathrm{Im}\,z)$ for all $z \in \mathbb{C}_{\mathrm{Re}>\nu}$ due to the support constraint on f. Moreover, it is not difficult to see that the integral on the right-hand side of (8.1) exists as $\left(t \mapsto \mathrm{e}^{-\rho t} f(t)\right) \in L_1(\mathbb{R}_{\geqslant 0}; H) \cap L_2(\mathbb{R}_{\geqslant 0}; H)$ for all $\rho > \nu$. Hence, $\mathcal{L}f\colon \mathbb{C}_{\mathrm{Re}>\nu} \to H$ is holomorphic (cf. Exercise 5.6). Moreover, by Lemma 8.1.1

$$\begin{aligned}\sup_{\rho>\nu} \|\mathcal{L}f(\mathrm{i}\cdot+\rho)\|_{L_2(\mathbb{R};H)} &= \sup_{\rho>\nu} \left\|\mathcal{L}_\rho f\right\|_{L_2(\mathbb{R};H)} = \sup_{\rho>\nu} \|f\|_{L_{2,\rho}(\mathbb{R};H)} \\ &= \sup_{\rho>0} \left\|\mathrm{e}^{-\nu\cdot} f\right\|_{L_{2,\rho}(\mathbb{R};H)} \\ &= \left\|\mathrm{e}^{-\nu\cdot} f\right\|_{L_2(\mathbb{R};H)} = \|f\|_{L_{2,\nu}(\mathbb{R};H)},\end{aligned}$$

which proves that

$$\begin{aligned}\mathcal{L}\colon L_{2,\nu}(\mathbb{R}_{\geqslant 0}; H) &\to \mathcal{H}_2(\mathbb{C}_{\mathrm{Re}>\nu}; H) \\ f &\mapsto \left(z \mapsto (\mathcal{L}_{\mathrm{Re}\,z} f)(\mathrm{Im}\,z)\right)\end{aligned}$$

is well-defined and isometric. It turns out that $\mathcal{L}$ is actually surjective, see Corollary 8.1.3 below. The surjectivity statement is contained in the following Theorem of Paley–Wiener, [78]. We mainly follow the proof given in [101, 19.2 Theorem].

Theorem 8.1.2 (Paley–Wiener) *Let $g \in \mathcal{H}_2(\mathbb{C}_{\mathrm{Re}>0}, H)$. Then there exists an $f \in L_2(\mathbb{R}_{\geqslant 0}; H)$ such that*

$$\mathcal{L}_\nu f = g(\mathrm{i}\cdot+\nu) \quad (\nu > 0).$$

Proof For $\nu > 0$ we set $g_\nu := g(\mathrm{i}\cdot+\nu) \in L_2(\mathbb{R}; H)$ and $f_\nu := \mathcal{F}^* g_\nu \in L_2(\mathbb{R}; H)$. Moreover, we set $f := \mathrm{e}^{(\cdot)} f_1$. We first prove that $f \in \bigcap_{\nu>0} L_{2,\nu}(\mathbb{R}; H)$ with $\sup_{\nu>0} \|f\|_{L_{2,\nu}(\mathbb{R};H)} < \infty$. For doing so, let $a > 0$, $\rho > 0$ and $x \in \mathbb{R}$. Applying

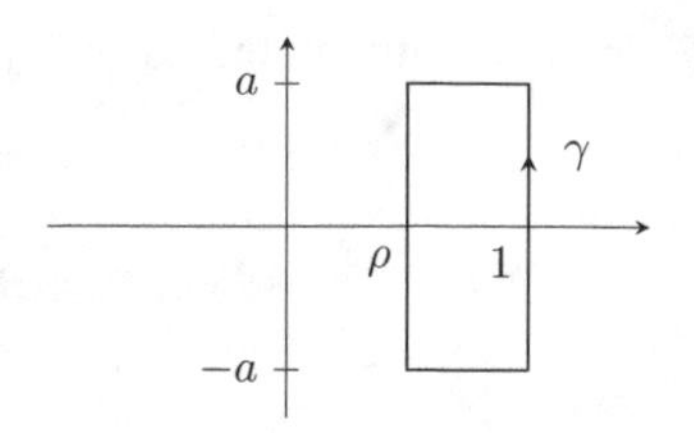

Fig. 8.1 Curve γ

Cauchy's integral theorem to the function $z \mapsto \mathrm{e}^{zx} g(z)$ and the curve γ, as indicated in Fig. 8.1, we obtain

$$\begin{aligned} 0 = \mathrm{i} \int_{-a}^{a} \mathrm{e}^{(\mathrm{i}t+1)x} g(\mathrm{i}t+1)\,\mathrm{d}t - \int_{\rho}^{1} \mathrm{e}^{(\mathrm{i}a+\kappa)x} g(\mathrm{i}a+\kappa)\,\mathrm{d}\kappa \\ - \mathrm{i} \int_{-a}^{a} \mathrm{e}^{(\mathrm{i}t+\rho)x} g(\mathrm{i}t+\rho)\,\mathrm{d}t + \int_{\rho}^{1} \mathrm{e}^{(-\mathrm{i}a+\kappa)x} g(-\mathrm{i}a+\kappa)\,\mathrm{d}\kappa. \end{aligned} \tag{8.2}$$

Moreover, since

$$\begin{aligned} \int_{\mathbb{R}} \left\| \int_{\rho}^{1} \mathrm{e}^{(\pm\mathrm{i}a+\kappa)x} g(\pm\mathrm{i}a+\kappa)\,\mathrm{d}\kappa \right\|_H^2 \mathrm{d}a &\leqslant \int_{\mathbb{R}} \left| \int_{\rho}^{1} \left|\mathrm{e}^{(\pm\mathrm{i}a+\kappa)x}\right|^2 \mathrm{d}\kappa \int_{\rho}^{1} \|g(\pm\mathrm{i}a+\kappa)\|_H^2 \,\mathrm{d}\kappa \right| \mathrm{d}a \\ &\leqslant \left| \int_{\rho}^{1} \mathrm{e}^{2\kappa x}\,\mathrm{d}\kappa \right| \left| \int_{\rho}^{1} \int_{\mathbb{R}} \|g(\pm\mathrm{i}a+\kappa)\|_H^2 \,\mathrm{d}a\,\mathrm{d}\kappa \right| \\ &\leqslant \left| \int_{\rho}^{1} \mathrm{e}^{2\kappa x}\,\mathrm{d}\kappa \right| |1-\rho|\, \|g\|^2_{\mathcal{H}_2(\mathbb{C}_{\mathrm{Re}>0};H)} < \infty, \end{aligned}$$

we infer that $\left(a \mapsto \int_{\rho}^{1} \mathrm{e}^{(\pm\mathrm{i}a+\kappa)x} g(\pm\mathrm{i}a+\kappa)\,\mathrm{d}\kappa\right) \in L_2(\mathbb{R}; H)$ and thus, we find a sequence $(a_n)_{n\in\mathbb{N}}$ in $\mathbb{R}_{>0}$ such that $a_n \to \infty$ and

$$\int_{\rho}^{1} \mathrm{e}^{(\pm\mathrm{i}a_n+\kappa)x} g(\pm\mathrm{i}a_n+\kappa)\,\mathrm{d}\kappa \to 0$$

as $n \to \infty$. Hence, using (8.2) with a replaced by a_n and letting n tend to infinity, we derive that

$$\int_{-a_n}^{a_n} \mathrm{e}^{(\mathrm{i}t+1)x} g(\mathrm{i}t+1)\,\mathrm{d}t - \int_{-a_n}^{a_n} \mathrm{e}^{(\mathrm{i}t+\rho)x} g(\mathrm{i}t+\rho)\,\mathrm{d}t \to 0 \quad (n \to \infty).$$

Noting that for each $\mu > 0$ we have

$$\int_{-a_n}^{a_n} \mathrm{e}^{(\mathrm{i}t+\mu)x} g(\mathrm{i}t+\mu)\,\mathrm{d}t = \sqrt{2\pi}\mathrm{e}^{\mu x}\mathcal{F}^*(\mathbb{1}_{[-a_n,a_n]}g_\mu)(x) \quad (x \in \mathbb{R})$$

and that $\mathbb{1}_{[-a_n,a_n]}g_\mu \to g_\mu$ in $L_2(\mathbb{R}; H)$ as $n \to \infty$, we may choose a subsequence (again denoted by $(a_n)_n$) such that

$$\begin{aligned} 0 &= \lim_{n\to\infty} \left(\int_{-a_n}^{a_n} \mathrm{e}^{(\mathrm{i}t+1)x} g(\mathrm{i}t+1)\,\mathrm{d}t - \int_{-a_n}^{a_n} \mathrm{e}^{(\mathrm{i}t+\rho)x} g(\mathrm{i}t+\rho)\,\mathrm{d}t \right) \\ &= \lim_{n\to\infty} \left(\sqrt{2\pi}\mathrm{e}^{x}\mathcal{F}^*(\mathbb{1}_{[-a_n,a_n]}g_1)(x) - \sqrt{2\pi}\mathrm{e}^{\rho x}\mathcal{F}^*(1_{[-a_n,a_n]}g_\rho)(x) \right) \\ &= \sqrt{2\pi}\left(\mathrm{e}^x f_1(x) - \mathrm{e}^{\rho x} f_\rho(x)\right) \end{aligned}$$

for almost every $x \in \mathbb{R}$. Hence, $f = \mathrm{e}^{(\cdot)} f_1 = \exp(\rho \mathrm{m}) f_\rho$ for each $\rho > 0$ and thus,

$$\int_{\mathbb{R}} \|f(t)\|_H^2 \,\mathrm{e}^{-2\rho t}\,\mathrm{d}t = \int_{\mathbb{R}} \left\|f_\rho(t)\right\|_H^2 \,\mathrm{d}t < \infty$$

which shows $f \in \bigcap_{\rho>0} L_{2,\rho}(\mathbb{R}; H)$ with

$$\sup_{\rho>0} \|f\|_{L_{2,\rho}(\mathbb{R};H)} = \sup_{\rho>0} \left\|f_\rho\right\|_{L_2(\mathbb{R};H)} = \sup_{\rho>0} \left\|g_\rho\right\|_{L_2(\mathbb{R};H)} = \|g\|_{\mathcal{H}_2(\mathbb{C}_{\mathrm{Re}>0};H)}\,.$$

Thus, $f \in L_2(\mathbb{R}_{\geqslant 0}; H)$ with $\|f\|_{L_2(\mathbb{R}_{\geqslant 0};H)} = \|g\|_{\mathcal{H}_2(\mathbb{C}_{\mathrm{Re}>0};H)}$ by Lemma 8.1.1. Moreover,

$$\mathcal{L}_\nu f = \mathcal{F}\exp(-\nu \mathrm{m}) f = \mathcal{F}\exp(-\nu m)\exp(\nu \mathrm{m}) f_\nu = \mathcal{F} f_\nu = g_\nu = g(\mathrm{i}\cdot+\nu)$$

for each $\nu > 0$, which shows the representation formula for g. □

Summarising the results of Theorem 8.1.2 and the arguments carried out just before Theorem 8.1.2, we obtain the following statement.

Corollary 8.1.3 *Let $\nu \in \mathbb{R}$. Then the mapping*

$$\begin{aligned} \mathcal{L}\colon L_{2,\nu}(\mathbb{R}_{\geqslant 0}; H) &\to \mathcal{H}_2(\mathbb{C}_{\mathrm{Re}>\nu}; H) \\ f &\mapsto \left(z \mapsto (\mathcal{L}_{\mathrm{Re}\,z} f)(\mathrm{Im}\,z)\right) \end{aligned}$$

is an isometric isomorphism. In particular, $\mathcal{H}_2(\mathbb{C}_{\mathrm{Re}>\nu}; H)$ is a Hilbert space.

Proof We have argued already that $\mathcal{L}$ is well-defined and isometric. Thus, we show that $\mathcal{L}$ is onto, next. For this, let $g \in \mathcal{H}_2(\mathbb{C}_{\mathrm{Re}>\nu}; H)$ and define $\widetilde{g}(z) := g(z+\nu)$ for $z \in \mathbb{C}_{\mathrm{Re}>0}$. Then $\widetilde{g} \in \mathcal{H}_2(\mathbb{C}_{\mathrm{Re}>0}; H)$ and thus, Theorem 8.1.2 yields the existence of $\widetilde{f} \in L_2(\mathbb{R}_{\geqslant 0}; H)$ with

$$g(\mathrm{i}\cdot+\rho) = \widetilde{g}(\mathrm{i}\cdot+\rho-\nu) = \mathcal{L}_{\rho-\nu}\widetilde{f} = \mathcal{L}_\rho\big(\mathrm{e}^{\nu\cdot}\widetilde{f}\big) \quad (\rho > \nu).$$

Hence, setting $f := \mathrm{e}^{\nu\cdot}\widetilde{f} \in L_{2,\nu}(\mathbb{R}_{\geqslant 0}; H)$, we obtain $\mathcal{L}f = g$. □

We can now provide an alternative proof of Theorem 5.3.6 by proving causality with the help of the Theorem of Paley–Wiener.

Proposition 8.1.4 *Let $M\colon \operatorname{dom}(M) \subseteq \mathbb{C} \to L(H)$ be a material law. Then for $\nu > \mathrm{s_b}(M)$ we have $M(\partial_{t,\nu}) \in L(L_{2,\nu}(\mathbb{R}; H))$ and $M(\partial_{t,\nu})$ is causal and autonomous (see Exercise 5.7).*

Proof Let $\nu > \mathrm{s_b}(M)$. Then $M\colon \mathbb{C}_{\mathrm{Re}\geqslant\nu} \to L(H)$ is bounded and holomorphic on $\mathbb{C}_{\mathrm{Re}>\nu}$. Hence, by unitary equivalence, $M(\partial_{t,\nu}) \in L(L_{2,\nu}(\mathbb{R}; H))$. Moreover, $M(\partial_{t,\nu})$ is autonomous by Exercise 5.7. Thus, for causality it suffices to check that $\operatorname{spt} M(\partial_{t,\nu})f \subseteq \mathbb{R}_{\geqslant 0}$ whenever $f \in L_{2,\nu}(\mathbb{R}_{\geqslant 0}; H)$. So let $f \in L_{2,\nu}(\mathbb{R}_{\geqslant 0}; H)$. Then $\mathcal{L}f \in \mathcal{H}_2(\mathbb{C}_{\mathrm{Re}>\nu}; H)$ by Corollary 8.1.3 and since M is bounded and holomorphic on $\mathbb{C}_{\mathrm{Re}>\nu}$, we infer also that

$$\big(z \mapsto M(z)\,(\mathcal{L}f)\,(z)\big) \in \mathcal{H}_2(\mathbb{C}_{\mathrm{Re}>\nu}; H).$$

Again by Corollary 8.1.3 there exists $g \in L_{2,\nu}(\mathbb{R}_{\geqslant 0}; H)$ such that

$$\mathcal{L}g(z) = M(z)\,(\mathcal{L}f)\,(z) \quad (z \in \mathbb{C}_{\mathrm{Re}>\nu}).$$

Thus, in particular

$$\mathcal{L}_\rho g = M(\mathrm{im}+\rho)\mathcal{L}_\rho f \quad (\rho > \nu).$$

Since $f, g \in L_{2,\nu}(\mathbb{R}_{\geqslant 0}; H)$ we infer that $\mathcal{L}_\rho g \to \mathcal{L}_\nu g$ and $\mathcal{L}_\rho f \to \mathcal{L}_\nu f$ in $L_2(\mathbb{R}; H)$ as $\rho \to \nu$ by dominated convergence. Moreover, $M(\mathrm{im}+\rho) \to M(\mathrm{im}+\nu)$ strongly on $L_2(\mathbb{R}; H)$ as $\rho \to \nu$ (cf. Exercise 8.2). Hence, we derive

$$\mathcal{L}_\nu g = M(\mathrm{im}+\nu)\mathcal{L}_\nu f,$$

and thus, $g = M(\partial_{t,\nu})f$ which shows causality. □

8.2 A Representation Result

In this section we argue that our solution theory needs holomorphy as a central property for the material law. There are two key properties for rendering $T \in L(L_{2,\nu_0}(\mathbb{R}; H))$ a material law operator. The first one is causality (i.e., $\mathbb{1}_{(-\infty,a]}(\mathrm{m})T\mathbb{1}_{(-\infty,a]}(\mathrm{m}) = \mathbb{1}_{(-\infty,a]}(\mathrm{m})T$ for all $a \in \mathbb{R}$) and, secondly, T needs to be autonomous (i.e., $\tau_h T = T\tau_h$ for all $h \in \mathbb{R}$ where $\tau_h f = f(\cdot + h)$). The main theorem of this section reads as follows:

Theorem 8.2.1 *Let $\nu_0 \in \mathbb{R}$ and let $T \in L(L_{2,\nu_0}(\mathbb{R}; H))$ be causal and autonomous. Then $T|_{L_{2,\nu_0} \cap L_{2,\nu}}$ has a unique extension $T_\nu \in L(L_{2,\nu}(\mathbb{R}; H))$ for each $\nu > \nu_0$ and there exists a unique $M : \mathbb{C}_{\mathrm{Re}>\nu_0} \to L(H)$ holomorphic and bounded such that $T_\nu = M(\partial_{t,\nu})$ for each $\nu > \nu_0$.*

We consider the following (shifted) variant of Theorem 8.2.1 first.

Theorem 8.2.2 *Let $T \in L(L_2(\mathbb{R}; H))$ be causal and autonomous. Then there exists $M : \mathbb{C}_{\mathrm{Re}>0} \to L(H)$, a material law (i.e., holomorphic and bounded), such that*

$$(\mathcal{L}Tf)(z) = M(z)(\mathcal{L}f)(z) \quad (f \in L_2(\mathbb{R}_{\geqslant 0}; H), z \in \mathbb{C}_{\mathrm{Re}>0}).$$

Proof For $s > 0$ and $x \in H$ we define $f_{x,s} := \mathbb{1}_{(0,s)}x$ and compute

$$\mathcal{L}f_{x,s}(z) = \frac{1}{\sqrt{2\pi}} \int_0^s \mathrm{e}^{-zt} x \, \mathrm{d}t = \frac{1}{\sqrt{2\pi}} \frac{1 - \mathrm{e}^{-zs}}{z} x \quad (z \in \mathbb{C}_{\mathrm{Re}>0}). \tag{8.3}$$

We define $M : \mathbb{C}_{\mathrm{Re}>0} \to L(H)$ via

$$M(z)x := \frac{\sqrt{2\pi}z}{1 - \mathrm{e}^{-z}} \mathcal{L}Tf_{x,1}(z),$$

which is well-defined since $\operatorname{spt} Tf_{x,1} \subseteq [0, \infty)$ (use causality of T); $M(z) \in L(H)$, since T is bounded. Also, $M(\cdot)x$ is evidently holomorphic for every $x \in H$ as a product of two holomorphic mappings and thus by Exercise 5.3, M is holomorphic itself. Next, we show that for all $z \in \mathbb{C}_{\mathrm{Re}>0}$ and $f \in L_2(\mathbb{R}_{\geqslant 0}; H)$, we have

$$(\mathcal{L}Tf)(z) = M(z)(\mathcal{L}f)(z). \tag{8.4}$$

By definition of M, the equality is true for f replaced by $f_{x,1}$, $x \in H$. Next, observe that $\operatorname{lin}\left\{\mathbb{1}_{(a,a+1/n)}x \,;\, a \geqslant 0, n \in \mathbb{N}, x \in H\right\}$ is dense in $L_2(\mathbb{R}_{\geqslant 0}; H)$. Hence, for (8.4), it suffices to show

$$\left(\mathcal{L}T\mathbb{1}_{(a,a+1/n)}x\right)(z) = M(z)\left(\mathcal{L}\mathbb{1}_{(a,a+1/n)}x\right)(z) \tag{8.5}$$

for all $a \geqslant 0$, $n \in \mathbb{N}$, $x \in H$, and $z \in \mathbb{C}_{\mathrm{Re}>0}$. Next, using that T is autonomous in the situation of (8.5), we see $\left(T\mathbb{1}_{(a,a+1/n)}x\right) = \left(T\tau_{-a}\mathbb{1}_{(0,1/n)}x\right) = \tau_{-a}\left(T\mathbb{1}_{(0,1/n)}x\right)$ and, by a straightforward computation, $(\mathcal{L}\tau_{-a}f)(z) = \mathrm{e}^{-za}\mathcal{L}f(z)$ for all $f \in L_2(\mathbb{R}_{\geqslant 0}; H)$. Thus,

$$\left(\mathcal{L}T\mathbb{1}_{(a,a+1/n)}x\right)(z) = \mathrm{e}^{-za}\left(\mathcal{L}T\mathbb{1}_{(0,1/n)}x\right)(z),$$

which yields that it suffices to show (8.5) for $a = 0$ only, that is, for $f = f_{x,1/n}$. Furthermore, we compute for $n \in \mathbb{N}$ and $z \in \mathbb{C}_{\mathrm{Re}>0}$

$$\begin{aligned}\mathcal{L}Tf_{x,1}(z) &= \sum_{k=0}^{n-1}(\mathcal{L}T\mathbb{1}_{(k/n,(k+1)/n)}x)(z) = \sum_{k=0}^{n-1}\mathrm{e}^{-zk/n}(\mathcal{L}T\mathbb{1}_{(0,1/n)}x)(z)\\ &= \frac{1-\mathrm{e}^{-z}}{1-\mathrm{e}^{-z/n}}(\mathcal{L}Tf_{x,1/n})(z).\end{aligned}$$

Thus, using (8.3) for $s = 1/n$, we deduce from the definition of M,

$$\begin{aligned}\mathcal{L}Tf_{x,1/n}(z) &= \frac{1-\mathrm{e}^{-z/n}}{\sqrt{2\pi}z}\frac{\sqrt{2\pi}z}{1-\mathrm{e}^{-z}}\mathcal{L}Tf_{x,1}(z) = \frac{1-\mathrm{e}^{-z/n}}{\sqrt{2\pi}z}M(z)x\\ &= M(z)\mathcal{L}f_{x,1/n}(z).\end{aligned}$$

Hence, (8.4) holds for all $f \in L_2(\mathbb{R}_{\geqslant 0}; H)$. It remains to show boundedness of M. For this, let $z \in \mathbb{C}_{\mathrm{Re}>0}$ and $x \in H$. Set $f := \mathbb{1}_{[0,\infty)}\mathrm{e}^{-z^*}x$ as well as $c := 2\,\mathrm{Re}\,z\sqrt{2\pi}$. Then

$$\mathcal{L}f(z) = \frac{1}{\sqrt{2\pi}}\int_0^\infty \mathrm{e}^{-zt-z^*t}x\,\mathrm{d}t = \frac{x}{c}.$$

By virtue of (8.4), we get $\mathcal{L}Tf(z) = M(z)\mathcal{L}f(z)$ and thus $M(z)x = c\mathcal{L}Tf(z)$. This leads to

$$\begin{aligned}\|M(z)x\| &\leqslant \frac{c}{\sqrt{2\pi}}\int_0^\infty \left\|\mathrm{e}^{-zt}Tf(t)\right\|\,\mathrm{d}t \leqslant \frac{c}{\sqrt{2\pi}}\left\|\mathbb{1}_{[0,\infty)}\mathrm{e}^{-z(\cdot)}\right\|_{L_2(\mathbb{R})}\|Tf\|_{L_2(\mathbb{R})}\\ &\leqslant \frac{c}{\sqrt{2\pi}}\left\|\mathbb{1}_{[0,\infty)}\mathrm{e}^{-z(\cdot)}\right\|_{L_2(\mathbb{R})}^2\|T\|_{L(L_2(\mathbb{R};H))}\|x\|_H = \|T\|_{L(L_2(\mathbb{R};H))}\|x\|_H,\end{aligned}$$

where we used that $\|f\|_{L_2(\mathbb{R};H)} = \left\|\mathbb{1}_{[0,\infty)}\mathrm{e}^{-z(\cdot)}\right\|_{L_2(\mathbb{R})}\|x\|_H$. Thus, $\|M(z)\| \leqslant \|T\|$, which yields boundedness of M and the assertion of the theorem. □

We can now prove our main result of this section.

Proof of Theorem 8.2.1 We just prove the existence of a function M. The proof of its uniqueness is left as Exercise 8.3.

We first prove the assertion for $\nu_0 = 0$. So, let $T \in L(L_2(\mathbb{R}; H))$ be causal and autonomous. According to Theorem 8.2.2 we find $M\colon \mathbb{C}_{\mathrm{Re}>0} \to L(H)$ holomorphic and bounded such that

$$(\mathcal{L}Tf)(z) = M(z)(\mathcal{L}f)(z) \quad (f \in L_2(\mathbb{R}_{\geqslant 0}; H), z \in \mathbb{C}_{\mathrm{Re}>0}).$$

Let now $\varphi \in C_{\mathrm{c}}^{\infty}(\mathbb{R}; H)$ and set $a := \inf \operatorname{spt} \varphi$. Then $\tau_a \varphi \in L_2(\mathbb{R}_{\geqslant 0}; H)$, and for $\nu > 0$ we compute

$$\begin{aligned}\mathcal{L}_\nu T\varphi &= \mathcal{L}_\nu \tau_{-a} T \tau_a \varphi = \mathrm{e}^{-(\mathrm{im}+\nu)a} \mathcal{L}_\nu T \tau_a \varphi = \mathrm{e}^{-(\mathrm{im}+\nu)a} M(\mathrm{im}+\nu) \mathcal{L}_\nu \tau_a \varphi \\ &= M(\mathrm{im}+\nu)\mathcal{L}_\nu \varphi. \end{aligned} \tag{8.6}$$

The latter implies

$$\begin{aligned}\|T\varphi\|_{L_{2,\nu}(\mathbb{R};H)} = \|\mathcal{L}_\nu T\varphi\|_{L_2(\mathbb{R};H)} &= \|M(\mathrm{im}+\nu)\mathcal{L}_\nu\varphi\|_{L_2(\mathbb{R};H)} \\ &\leqslant \|M\|_{\infty,\mathbb{C}_{\mathrm{Re}>0}} \|\varphi\|_{L_{2,\nu}(\mathbb{R};H)}\end{aligned}$$

and hence, $T|_{C_{\mathrm{c}}^\infty(\mathbb{R};H)}$ has a unique continuous extension $T_\nu \in L(L_{2,\nu}(\mathbb{R}; H))$. Using (8.6) we obtain

$$T_\nu = \mathcal{L}_\nu^* M(\mathrm{im}+\nu)\mathcal{L}_\nu = M(\partial_{t,\nu})$$

by approximation.

Let now $\nu_0 \in \mathbb{R}$. Then the operator

$$\widetilde{T} := \mathrm{e}^{-\nu_0 \mathrm{m}} T \mathrm{e}^{\nu_0 \mathrm{m}} \in L(L_2(\mathbb{R}; H))$$

is causal and autonomous as well. Thus, $\widetilde{T}|_{C_{\mathrm{c}}^\infty(\mathbb{R};H)}$ has continuous extensions $\widetilde{T}_\rho \in L(L_{2,\rho}(\mathbb{R}; H))$ for each $\rho > 0$ and there is $\widetilde{M}\colon \mathbb{C}_{\mathrm{Re}>0} \to L(H)$ holomorphic and bounded such that $\widetilde{T}_\rho = \widetilde{M}(\partial_{t,\rho})$ for each $\rho > 0$. Using $T|_{C_{\mathrm{c}}^\infty(\mathbb{R};H)} = \mathrm{e}^{\nu_0 \mathrm{m}} \widetilde{T}|_{C_{\mathrm{c}}^\infty(\mathbb{R};H)} \mathrm{e}^{-\nu_0 \mathrm{m}}$, we derive that $T|_{C_{\mathrm{c}}^\infty(\mathbb{R};H)}$ has the unique continuous extension $T_\nu = \mathrm{e}^{\nu_0 \mathrm{m}} \widetilde{T}_{\nu-\nu_0} \mathrm{e}^{-\nu_0 \mathrm{m}} \in L(L_{2,\nu}(\mathbb{R}; H))$ for each $\nu > \nu_0$ and

$$\begin{aligned}\mathcal{L}_\nu T_\nu = \mathcal{L}_\nu \mathrm{e}^{\nu_0 \mathrm{m}} \widetilde{T}_{\nu-\nu_0} \mathrm{e}^{-\nu_0 \mathrm{m}} &= \mathcal{L}_{\nu-\nu_0} \widetilde{T}_{\nu-\nu_0} \mathrm{e}^{-\nu_0 \mathrm{m}} = \widetilde{M}(\mathrm{im}+\nu-\nu_0)\mathcal{L}_{\nu-\nu_0} \mathrm{e}^{-\nu_0 \mathrm{m}} \\ &= \widetilde{M}(\mathrm{im}+\nu-\nu_0)\mathcal{L}_\nu.\end{aligned}$$

Hence,

$$T_\nu = M(\partial_{t,\nu})$$

for the holomorphic and bounded function M given by $M(z) := \widetilde{M}(z - \nu_0)$ for $z \in \mathbb{C}_{\mathrm{Re}>\nu_0}$. □

8.3 Comments

The stated Theorem of Paley and Wiener is of course not the only theorem characterising properties of the support of L_2-functions in terms of their Fourier or Laplace transform. For instance, a similar result holds for functions having compact support, see e.g. [101, 19.3 Theorem] and Exercise 8.7. These theorems provide a nice connection between L_2-functions and spaces of holomorphic functions in form of Hardy spaces. In this chapter we just introduced the Hardy space $\mathcal{H}_2$ and it is not surprising that there are also the Hardy spaces $\mathcal{H}_p$ for $1 \leqslant p \leqslant \infty$. We refer to [35] for this topic.

The representation result presented in the second part of this chapter was originally proved by Fourès and Segal in 1955, [41]. In this article the authors prove an analogous representation result for causal operators on $L_2(\mathbb{R}^d; H)$, where causality is defined with respect to a closed and convex cone on $\mathbb{R}^d$. The quite elementary proof of Theorem 8.2.2 for $d = 1$ presented here was kindly communicated to us by Hendrik Vogt.

Exercises

Exercise 8.1 Let $\Lambda \subseteq \mathbb{R}_{>0}$ be a set with an accumulation point in $\mathbb{R}_{>0}$. Prove that $\{(x \mapsto \mathrm{e}^{-\lambda x}) \,;\, \lambda \in \Lambda\}$ is a total set in $L_1(\mathbb{R}_{\geqslant 0})$.
Hint: Use that the set is total if and only if

$$\forall f \in L_\infty(\mathbb{R}_{\geqslant 0}) : \left(\forall \lambda \in \Lambda : \int_{\mathbb{R}_{\geqslant 0}} \mathrm{e}^{-\lambda x} f(x)\,\mathrm{d}x = 0 \Rightarrow f = 0 \right).$$

Exercise 8.2 Let $M \colon \mathrm{dom}(M) \subseteq \mathbb{C} \to L(H)$ be a material law. Moreover, let $\nu > \mathrm{s_b}(M)$. Show that $\lim_{\rho \to \nu+} M(\mathrm{im} + \rho) = M(\mathrm{im} + \nu)$ where the limit is meant in the strong operator topology on $L_2(\mathbb{R}; H)$.

Exercise 8.3 Prove the uniqueness statement in Theorem 8.2.1.

Exercise 8.4 Give an example of a continuous and bounded function $M \colon \mathbb{C}_{\mathrm{Re}>0} \to L(H)$ such that the corresponding operator $M(\partial_{t,\nu})$ is not causal for any $\nu > 0$.

Exercise 8.5 Prove the following distributional variant of the Paley–Wiener theorem: Let $\nu_0 > 0$, $k \in \mathbb{N}$, $f \colon \mathbb{C}_{\mathrm{Re}>\nu_0} \to \mathbb{C}$, and set $h(z) := \frac{1}{z^k} f(z)$ for $z \in \mathbb{C}_{\mathrm{Re}>\nu_0}$.

We assume that $h \in \mathcal{H}_2(\mathbb{C}_{\mathrm{Re}>\nu_0}; \mathbb{C})$. For $\nu > \nu_0$ we define the distribution $u\colon C_c^\infty(\mathbb{R}) \to \mathbb{C}$ by

$$u(\psi) := \left\langle \mathcal{L}_\nu^* h(\mathrm{i}\cdot +\nu), (\partial_{t,\nu}^*)^k \psi \right\rangle_{L_{2,\nu}(\mathbb{R};\mathbb{C})} \quad (\psi \in C_c^\infty(\mathbb{R}; \mathbb{C})).$$

Prove that $\operatorname{spt} u \subseteq \mathbb{R}_{\geqslant 0}$, where

$$\operatorname{spt} u := \mathbb{R} \setminus \bigcup \left\{U \subseteq \mathbb{R} \text{ open}\,;\ \forall\, \psi \in C_c^\infty(U; \mathbb{C}) : u(\psi) = 0\right\}.$$

What is u if $f = \mathbb{1}_{\mathbb{C}_{\mathrm{Re}>\nu_0}}$?

Exercise 8.6 Let $g \in L_2(\mathbb{R})$, $a > 0$ such that $\operatorname{spt} g \subseteq [-a, a]$. Show that $f := \mathcal{F}g$ extends to a holomorphic function $\widetilde{f}\colon \mathbb{C} \to \mathbb{C}$ with $\widetilde{f}(\mathrm{i}t) = f(t)$ for each $t \in \mathbb{R}$ such that

$$\exists C \geqslant 0\, \forall z \in \mathbb{C} : |f(z)| \leqslant C\mathrm{e}^{a|\operatorname{Re} z|}.$$

Exercise 8.7 Let $f : \mathbb{C} \to \mathbb{C}$ be holomorphic such that

(a) $\exists C \geqslant 0,\ a > 0\, \forall z \in \mathbb{C} : |f(z)| \leqslant C\mathrm{e}^{a|\operatorname{Re} z|}$,
(b) $f(\mathrm{i}\cdot) \in L_2(\mathbb{R})$.

Prove that $g := \mathcal{F}^* f(\mathrm{i}\cdot)$ satisfies $\operatorname{spt} g \subseteq [-a, a]$.
Hint: Apply Theorem 8.1.2 to the function $h : \mathbb{C}_{\mathrm{Re}>0} \to \mathbb{C}$ given by

$$h(z) := \mathrm{e}^{-za} \frac{f(z)}{z+1} \quad (z \in \mathbb{C}_{\mathrm{Re}>0})$$

to derive that $\operatorname{spt} g \subseteq \mathbb{R}_{\geqslant -a}$.
Remark: The assertion even holds true if one replaces condition (a) by

$$\exists C \geqslant 0,\ a > 0\, \forall z \in \mathbb{C} : |f(z)| \leqslant C\mathrm{e}^{a|z|}.$$

References

35. P.L. Duren, *Theory of H^p Spaces*, vol. XII (Academic, New York, London, 1970), 258 p. 1970.
41. Y. Fourès, I. Segal, Causality and analyticity. Trans. Am. Math. Soc. **78**, 385–405 (1955)
78. R.E. Paley, N. Wiener, *Fourier Transforms in the Complex Domain*, vols. 19, VIII. American Mathematical Society Colloquium publications (American Mathematical Society, New York, 1934)
101. W. Rudin, *Real and Complex Analysis*. Mathematics Series (McGraw-Hill, New York, 1987)

Chapter 9
Initial Value Problems and Extrapolation Spaces

Up until now we have dealt with evolutionary equations of the form

$$\left(\overline{\partial_{t,\nu}M(\partial_{t,\nu}) + A}\right)U = F$$

for some given $F \in L_{2,\nu}(\mathbb{R}; H)$ for some Hilbert space H, a skew-selfadjoint operator A in H and a material law M defined on a suitable half-plane satisfying an appropriate positive definiteness condition with $\nu \in \mathbb{R}$ chosen suitably large. Under these conditions, we established that the solution operator, $S_\nu := \left(\overline{\partial_{t,\nu}M(\partial_{t,\nu}) + A}\right)^{-1} \in L(L_{2,\nu}(\mathbb{R}; H))$, is eventually independent of ν and causal; that is, if $F = 0$ on $(-\infty, a]$ for some $a \subset \mathbb{R}$, then so too is U.

To solve for $U \in L_{2,\nu}(\mathbb{R}; H)$ for some non-negative ν penalises U having support on $\mathbb{R}_{\leq 0}$. This might be interpreted as an implicit *initial condition at* $-\infty$. In this chapter, we shall study how to obtain a solution for initial value problems with an initial condition at 0, based on the solution theory developed in the previous chapters.

9.1 What are Initial Values?

This section is devoted to the motivation of the framework to follow in the subsequent section. Let us consider the following, arguably easiest but not entirely trivial, initial value problem: find a 'causal' $u\colon \mathbb{R} \to \mathbb{R}$ such that for $u_0 \in \mathbb{R}$ we have

$$\begin{cases} u'(t) = 0 & (t > 0), \\ u(0) = u_0. \end{cases} \tag{9.1}$$

C. Seifert et al., *Evolutionary Equations*, Operator Theory: Advances and Applications 287, https://doi.org/10.1007/978-3-030-89397-2_9

First of all note that there is no condition for u on $(-\infty, 0)$. Since, there is no source term or right-hand side supported on $(-\infty, 0)$, causality would imply that $u = 0$ on $(-\infty, 0)$. Moreover, $u = c$ for some constant $c \in \mathbb{R}$ on $(0, \infty)$. Thus, in order to match with the initial condition,

$$u(t) = u_0 \mathbb{1}_{[0,\infty)}(t) \quad (t \in \mathbb{R}).$$

Notice also that u is not continuous. Hence, by the Sobolev embedding theorem (Theorem 4.1.2), $u \notin \bigcup_{\nu>0} \operatorname{dom}(\partial_{t,\nu})$.

Proposition 9.1.1 *Let H be a Hilbert space, $u_0 \in H$. Define*

$$\begin{aligned} \delta_0 u_0 \colon C_c^\infty(\mathbb{R}; H) &\to \mathbb{K} \\ f &\mapsto \langle u_0, f(0)\rangle_H . \end{aligned}$$

Then, for all $\nu \in \mathbb{R}_{>0}$, $\delta_0 u_0$ extends to a continuous linear functional on $\operatorname{dom}(\partial_{t,\nu})$. *Re-using the notation for this extension, for all $f \in \operatorname{dom}(\partial_{t,\nu})$ we have*

$$(\delta_0 u_0)(f) = -\left\langle \mathbb{1}_{[0,\infty)} u_0, \left(\partial_{t,\nu} - 2\nu\right) f\right\rangle_{L_{2,\nu}(\mathbb{R};H)} . \tag{9.2}$$

Proof The equality (9.2) is obvious for $f \in C_c^\infty(\mathbb{R}; H)$ as it is a direct consequence of the fundamental theorem of calculus (look at the right-hand side first). The continuity of $\delta_0 u_0$ follows from the Cauchy–Schwarz inequality applied to the right-hand side of (9.2). Note that $\mathbb{1}_{[0,\infty)} u_0 \in L_{2,\nu}(\mathbb{R}; H)$. □

Recall from Corollary 3.2.6 that

$$\partial_{t,\nu}^* = -\partial_{t,\nu} + 2\nu.$$

Hence, if we *formally* apply this formula to (9.2), we obtain

$$\left\langle \partial_{t,\nu} \mathbb{1}_{[0,\infty)} u_0, f\right\rangle = \left\langle \mathbb{1}_{[0,\infty)} u_0, \partial_{t,\nu}^* f\right\rangle_{L_{2,\nu}(\mathbb{R};H)} = (\delta_0 u_0)(f).$$

Therefore, in order to use the introduced time derivative operator for the above initial value problem, we need to extend the time derivative to a broader class of functions than just $\operatorname{dom}(\partial_{t,\nu})$. To utilise the adjoint operator in this way will be central to the construction to follow. It will turn out that indeed

$$\partial_{t,\nu} \mathbb{1}_{[0,\infty)} u_0 = \delta_0 u_0.$$

Moreover, we shall show below that

$$\partial_{t,\nu} u = \delta_0 u_0$$

considered on the full time-line $\mathbb{R}$ is one possible replacement of the initial value problem (9.1).

9.2 Extrapolating Operators

Since we are dealing with functionals, let us recall the definition of the dual space. Throughout this section let H, H_0, H_1 be Hilbert spaces.

Definition The space

$$H' := \{\varphi\colon H \to \mathbb{K};\ \varphi \text{ linear and bounded}\}$$

is called the *dual space of* H. We equip H' with the linear structure

$$(\lambda \odot \varphi + \psi)(x) := \lambda^* \varphi(x) + \psi(x) \quad (\lambda \in \mathbb{K}, \varphi, \psi \in H', x \in H).$$

Remark 9.2.1 Note that H' is a Hilbert space itself, since by the Riesz representation theorem for each $\varphi \in H'$ we find a unique element $R_H\varphi \in H$ such that

$$\forall x \in H:\ \varphi(x) = \langle R_H\varphi, x\rangle\,.$$

Due to the linear structure on H', the so induced mapping $R_H\colon H' \to H$ (which is one-to-one and onto) becomes linear and

$$H' \times H' \ni (\varphi, \psi) \mapsto \langle R_H\varphi, R_H\psi\rangle$$

defines an inner product on H', which induces the usual norm on functionals.

From now on we will identify elements $x \in H$ with their representatives in H'; that is, we identify x with $R_H^{-1}x$.

Let $C\colon \operatorname{dom}(C) \subseteq H_0 \to H_1$ be linear, densely defined and closed. We recall that in this case $\operatorname{dom}(C)$ endowed with the graph inner product

$$(u, v) \mapsto \langle u, v\rangle_{H_0} + \langle Cu, Cv\rangle_{H_1}$$

becomes a Hilbert space. Clearly, $\operatorname{dom}(C) \hookrightarrow H_0$ is continuous with dense range. Moreover, we see that $\operatorname{dom}(C) \ni x \mapsto Cx \in H_1$ is continuous. We define

$$\begin{aligned} C^\diamond\colon H_1 &\to \operatorname{dom}(C)' =: H^{-1}(C), \\ (C^\diamond\phi)(x) &:= \langle \phi, Cx\rangle_{H_1} \quad (\phi \in H_1, x \in \operatorname{dom}(C)). \end{aligned}$$

Note that $C^\diamond$ is related to the dual operator C' of C considered as a bounded operator from $\operatorname{dom}(C)$ to H_1 by

$$C^\diamond = C' R_{H_1}^{-1}.$$

Proposition 9.2.2 *With the notions and definitions from this section, the following statements hold:*

(a) $C^\diamond$ *is continuous and linear.*
(b) $C^* \subseteq C^\diamond$.
(c) $\ker(C^*) = \ker(C^\diamond)$.
(d) $C \subseteq (C^*)^\diamond : H_0 \to \operatorname{dom}(C^*)' = H^{-1}(C^*)$.
(e) $H_0 \cong H_0' \hookrightarrow H^{-1}(C)$ *densely and continuously.*

Proof

(a) Let $\phi, \psi \in H_1$, $\lambda \in \mathbb{K}$. Then

$$C^\diamond(\lambda\phi + \psi)(x) = \lambda^*(C^\diamond\phi)(x) + (C^\diamond\psi)(x) = (\lambda \odot C^\diamond\phi + C^\diamond\psi)(x) \quad (x \in \operatorname{dom}(C)).$$

To show continuity, let $\phi \in H_1$ and $x \in \operatorname{dom}(C)$. Then

$$|C^\diamond(\phi)(x)| = \left|\langle \phi, Cx\rangle_{H_1}\right| \leqslant \|\phi\|_{H_1} \|Cx\|_{H_1} \leqslant \|\phi\|_{H_1} \|x\|_{\operatorname{dom}(C)}\,.$$

Hence, $\left\|C^\diamond\right\| = \sup_{\phi \in H_1, \|\phi\|_{H_1} \leqslant 1} \left\|C^\diamond\phi\right\|_{\operatorname{dom}(C)'} \leqslant 1$.

(b) Let $\phi \in \operatorname{dom}(C^*)$. Then we have for all $x \in \operatorname{dom}(C)$

$$\left(C^\diamond\phi\right)(x) = \langle \phi, Cx\rangle_{H_1} = \left\langle C^*\phi, x\right\rangle_{H_0} = \left(C^*\phi\right)(x).$$

We obtain $C^\diamond\phi = C^*\phi$ (note that a functional on H_0 is uniquely determined by its values on $\operatorname{dom}(C)$).

(c) Using (b), we are left with showing $\ker(C^\diamond) \subseteq \ker(C^*)$. So, let $\phi \in \ker(C^\diamond)$. Then for all $x \in \operatorname{dom}(C)$ we have

$$0 = \left(C^\diamond\phi\right)(x) = \langle \phi, Cx\rangle_{H_1}\,,$$

which leads to $\phi \in \operatorname{dom}(C^*)$ and $\phi \in \ker(C^*)$.

(d) is a direct consequence of (b) applied to C^*.

(e) Since $\operatorname{dom}(C) \hookrightarrow H_0$ is dense and continuous, so is that $H_0' \hookrightarrow \operatorname{dom}(C)'$; cf. Exercise 9.2.

□

We will also write $C_{-1} := (C^*)^\diamond$ for the so-called *extrapolated operator* of C. Then $(C^*)_{-1} = C^\diamond$. We will record the index -1 at the beginning, but in order to avoid too much clutter in the notation we will drop this index again, bearing in mind that $C_{-1} \supseteq C$ and $(C^*)_{-1} \supseteq C^*$.

Example 9.2.3 We have shown that for all $\nu \in \mathbb{R}$ the operator $\partial_{t,\nu}$ is densely defined and closed. Then for $f \in L_{2,\nu}(\mathbb{R})$ we have for all $\phi \in C_c^\infty(\mathbb{R})$

$$\left((\partial_{t,\nu})_{-1} f\right)(\phi) = \left\langle f, \partial_{t,\nu}^*\phi\right\rangle_{L_{2,\nu}} = \left\langle f, \left(-\partial_{t,\nu} + 2\nu\right)\phi\right\rangle_{L_{2,\nu}} = -\int_{\mathbb{R}} \left\langle f, \left(\mathrm{e}^{-2\nu\cdot}\phi\right)'\right\rangle_{\mathbb{C}}\,.$$

Hence, $(\partial_{t,\nu})_{-1} f$ acts as the 'usual' distributional derivative taking into account the exponential weight in the scalar product.

With this observation we deduce that for $\nu > 0$ we have

$$\left(\partial_{t,\nu}\right)_{-1} \mathbb{1}_{[0,\infty)} = \partial_{t,\nu} \mathbb{1}_{[0,\infty)} = \delta_0.$$

Hence, the initial value problem from the beginning reads: find u such that

$$(\partial_{t,\nu})_{-1} u = \delta_0 u_0.$$

Example 9.2.4 Let $\Omega \subseteq \mathbb{R}^d$ be open. Consider $\operatorname{grad}_0 \colon H_0^1(\Omega) \subseteq L_2(\Omega) \to L_2(\Omega)^d$. We compute $\operatorname{div}_{-1} \colon L_2(\Omega)^d \to H^{-1}(\Omega)$ with $H^{-1}(\Omega) := H_0^1(\Omega)'$. For $q \in L_2(\Omega)^d$ we obtain for all $\phi \in H_0^1(\Omega)$

$$(\operatorname{div}_{-1} q)(\phi) = \langle q, \operatorname{div}^* \phi\rangle_{L_2(\Omega)^d} = -\langle q, \operatorname{grad}_0 \phi\rangle_{L_2(\Omega)^d}.$$

Also, with similar arguments, we see that

$$\left(\operatorname{grad}_{-1} f\right)(q) = -\langle f, \operatorname{div}_0 q\rangle_{L_2(\Omega)}$$

for all $f \in L_2(\Omega)$ and $q \in H_0(\operatorname{div}, \Omega)$.

We consider a case of particular interest within the framework of evolutionary equations.

Proposition 9.2.5 *Let* $A \colon \operatorname{dom}(C) \times \operatorname{dom}(C^*) \subseteq H_0 \times H_1 \to H_0 \times H_1$ *be given by*

$$A\begin{pmatrix}\phi\\ \psi\end{pmatrix} = \begin{pmatrix}0 & C^*\\ -C & 0\end{pmatrix}\begin{pmatrix}\phi\\ \psi\end{pmatrix} = \begin{pmatrix}C^*\psi\\ -C\phi\end{pmatrix}.$$

Then $A_{-1} \colon H_0 \times H_1 \to H^{-1}(C) \times H^{-1}(C^*)$ *acts as*

$$A_{-1}\begin{pmatrix}\phi\\ \psi\end{pmatrix} = \begin{pmatrix}0 & (C^*)_{-1}\\ -C_{-1} & 0\end{pmatrix}\begin{pmatrix}\phi\\ \psi\end{pmatrix} = \begin{pmatrix}(C^*)_{-1}\psi\\ -C_{-1}\phi\end{pmatrix}.$$

Next, we will look at the solution theory when carried over to distributional right-hand sides.

An immediate consequence of the introduction of extrapolated operators, however, is that we are now in the position to omit the closure bar for the operator sum in an evolutionary equation, which we will see in an abstract version in Theorem 9.2.6 and for evolutionary equations in Theorem 9.3.2. The main advantage is that we can calculate an operator sum much easier than the closure of it. The price we have to pay is that we have to work in a larger space H^{-1} rather than in the original Hilbert space $L_{2,\nu}(\mathbb{R}; H)$. Put differently, this provides another notion of "solutions" for evolutionary equations. For this, we need to introduce the set

$$\mathrm{Fun}(H) := \{\phi\colon \mathrm{dom}(\phi) \subseteq H \to \mathbb{K};\ \phi \text{ linear}\}$$

of not necessarily everywhere defined linear functionals on H. Any $u \in H$ is thus identified with an element in $\mathrm{Fun}(H)$ via $\psi \mapsto \langle u, \psi\rangle_H$. Note that we can add and scalarly multiply elements in $\mathrm{Fun}(H)$ with respect to the same addition and multiplication defined on H' and with their natural domains. As usual, we will use the $\subseteq$-sign for extension/restriction of mappings.

Theorem 9.2.6 *Let $A\colon \mathrm{dom}(A) \subseteq H \to H$, $B\colon \mathrm{dom}(B) \subseteq H \to H$ be densely defined and closed such that $A + B$ is closable, and assume that there exists $(T_n)_{n\in\mathbb{N}}$ in $L(H)$ such that $T_n \to 1_H$ in the strong operator topology with* $\mathrm{ran}(T_n) \subseteq \mathrm{dom}(B)$ *and*

$$T_n A \subseteq AT_n, \quad T_n B \subseteq BT_n \text{ for all } n \in \mathbb{N}.$$

Then $T_n^* A^* \subseteq A^* T_n^*$ and $T_n^* B^* \subseteq B^* T_n^*$ for each $n \in \mathbb{N}$ and $\mathrm{ran}(T_n^*) \subseteq \mathrm{dom}(B^*)$. *Moreover, for $x, f \in H$ the following conditions are equivalent:*

(i) $x \in \mathrm{dom}(\overline{A + B})$ *and* $(\overline{A + B})x = f$.
(ii) $A_{-1}x + B_{-1}x \subseteq f \in \mathrm{Fun}(H)$.

Proof Let $n \in \mathbb{N}$. Taking adjoints in the inclusion $T_n A \subseteq AT_n$, we derive $(AT_n)^* \subseteq (T_n A)^*$. By Theorem 2.3.4 and Remark 2.3.7 we obtain

$$T_n^* A^* \subseteq \overline{T_n^* A^*} = (AT_n)^* \subseteq (T_n A)^* = A^* T_n^*.$$

The same argument shows the claim for B^*. Moreover, since BT_n is a closed linear operator defined on the whole space H, it follows that $BT_n \in L(H)$ by the closed graph theorem. Hence, $(BT_n)^*$ is bounded by Lemma 2.2.9 and since $(BT_n)^* \subseteq (T_n B)^* = B^* T_n^*$, we derive that $\mathrm{dom}(B^* T_n^*) = H$, showing that $\mathrm{ran}(T_n^*) \subseteq \mathrm{dom}(B^*)$.
We now prove the asserted equivalence.
(i)$\Rightarrow$(ii): By definition, there exists $(x_n)_n$ in $\mathrm{dom}(A) \cap \mathrm{dom}(B)$ such that $x_n \to x$ in H and $Ax_n + Bx_n \to f$. By continuity, we obtain $A_{-1}x_n \to A_{-1}x$ and $B_{-1}x_n \to B_{-1}x$ in $H^{-1}(A^*)$ and $H^{-1}(B^*)$, respectively. Thus, we have

$$\begin{aligned}(A_{-1}x + B_{-1}x)(y) &= \lim_{n\to\infty} (A_{-1}x_n + B_{-1}x_n)(y) = \lim_{n\to\infty} \langle Ax_n + Bx_n, y\rangle \\ &= \langle f, y\rangle ,\end{aligned}$$

for each $y \in \mathrm{dom}(A^*) \cap \mathrm{dom}(B^*)$, which shows the asserted inclusion.

(ii)$\Rightarrow$(i): For $n \in \mathbb{N}$ we put $x_n := T_n x$. Then $x_n \in \operatorname{dom}(B)$ and for all $y \in \operatorname{dom}(A^*) \cap \operatorname{dom}(B^*)$, we obtain

$$\begin{aligned}\langle T_n f - Bx_n, y\rangle &= \langle T_n f, y\rangle - \langle T_n x, B^* y\rangle = \langle f, T_n^* y\rangle - \langle x, T_n^* B^* y\rangle \\ &= \langle f, T_n^* y\rangle - \langle x, B^* T_n^* y\rangle = f(T_n^* y) - (B_{-1}x)(T_n^* y) \\ &= (A_{-1}x)(T_n^* y) = \langle x, A^* T_n^* y\rangle = \langle x, T_n^* A^* y\rangle = \langle x_n, A^* y\rangle,\end{aligned}$$

where we have used that $T_n^* y \in \operatorname{dom}(A^*) \cap \operatorname{dom}(B^*)$. Let now $y \in \operatorname{dom}(A^*)$. Then $T_k^* y \in \operatorname{dom}(A^*) \cap \operatorname{dom}(B^*)$ for each $k \in \mathbb{N}$ and thus, by what we have shown above

$$\begin{aligned}\langle T_k(T_n f - Bx_n), y\rangle &= \langle T_n f - Bx_n, T_k^* y\rangle = \langle x_n, A^* T_k^* y\rangle \\ &= \langle x_n, T_k^* A^* y\rangle = \langle T_k x_n, A^* y\rangle\end{aligned}$$

for each $k \in \mathbb{N}$. Letting k tend to infinity, we derive

$$\langle T_n f - Bx_n, y\rangle = \langle x_n, A^* y\rangle.$$

Since this holds for each $y \in \operatorname{dom}(A^*)$, this implies that we have $x_n \in \operatorname{dom}(A)$ and $Ax_n + Bx_n = T_n f$. Letting $n \to \infty$, we deduce $x_n \to x$ and $Ax_n + Bx_n \to f$; that is, (i). □

Lemma 9.2.7 *Let $T\colon \operatorname{dom}(T) \subseteq H \to H$ be densely defined and closed with $0 \in \rho(T)$. Then $T_{-1}\colon H \to H^{-1}(T^*)$ is an isomorphsim. In particular, the norms $\left\|(T_{-1})^{-1}\cdot\right\|_H$ and $\|\cdot\|_{H^{-1}(T^*)}$ are equivalent.*

Proof Note that since $0 \in \rho(T)$ we obtain $\{0\} = \ker(T) = \ker((T^*)^\diamond) = \ker(T_{-1})$, see Proposition 9.2.2(c). Thus, T_{-1} is one-to-one. Next, let $f \in H^{-1}(T^*)$. Since $0 \in \rho(T)$, we obtain $0 \in \rho(T^*)$ by Exercise 2.4, which implies that $\langle T^*\cdot, T^*\cdot\rangle$ defines an equivalent scalar product on $\operatorname{dom}(T^*)$. Thus, by the Riesz representation theorem, we find $\phi \in \operatorname{dom}(T^*)$ such that for all $\psi \in \operatorname{dom}(T^*)$ we have

$$f(\psi) = \langle T^*\phi, T^*\psi\rangle = \left((T^*)^\diamond \left(T^*\phi\right)\right)(\psi).$$

Hence, $f \in \operatorname{ran}((T^*)^\diamond) = \operatorname{ran}(T_{-1})$, thus proving that T_{-1} is onto. □

The following alternative description of $H^{-1}(T^*)$ is content of Exercise 9.5.

Proposition 9.2.8 *Let $T\colon \operatorname{dom}(T) \subseteq H \to H$ be densely defined and closed with $0 \in \rho(T)$. Then*

$$H^{-1}(T^*) \cong \widetilde{\left(H, \left\|T^{-1}\cdot\right\|_H\right)},$$

where $\cong$ means isomorphic as Banach spaces and $\widetilde{(\cdot)}$ denotes the completion.

Proposition 9.2.9 *Let $B \in L(H)$. Assume that $T\colon \operatorname{dom}(T) \subseteq H \to H$ is densely defined and closed with $0 \in \rho(T)$ and $T^{-1}B = BT^{-1}$. Then B admits a unique continuous extension $\overline{B} \in L(H^{-1}(T^*))$.*

Proof By Proposition 9.2.2(e), $\operatorname{dom}(B) = H$ is dense in $H^{-1}(T^*)$. Thus, it suffices to show that $B\colon H \subseteq H^{-1}(T^*) \to H^{-1}(T^*)$ is continuous. For this, let $\phi \in H$ and compute for all $q \in \operatorname{dom}(T^*)$

$$\begin{aligned}|(B\phi)(q)| = |\langle B\phi, q\rangle| &= \left|\left\langle B\phi, \left(T^*\right)^{-1} T^*q\right\rangle\right| = \left|\left\langle T^{-1}B\phi, T^*q\right\rangle\right| \\ &= \left|\left\langle BT^{-1}\phi, T^*q\right\rangle\right| \leqslant \|B\| \left\|T^{-1}\phi\right\| \|q\|_{\operatorname{dom}(T^*)} .\end{aligned}$$

The statement now follows upon invoking Lemma 9.2.7. □

The abstract notions and concepts just developed will be applied to evolutionary equations next.

9.3 Evolutionary Equations in Distribution Spaces

In this section, we will specialise the results from the previous section and provide an extension of the solution theory in $L_{2,\nu}(\mathbb{R}; H)$. For this, and throughout this whole section, we let H be a Hilbert space, $\mu \in \mathbb{R}$ and $M\colon \mathbb{C}_{\mathrm{Re}>\mu} \to L(H)$ be a material law. Furthermore, let $\nu > \max\{\mathrm{s_b}(M), 0\}$ and $A\colon \operatorname{dom}(A) \subseteq H \to H$ be skew-selfadjoint. In order to keep track of the Hilbert spaces involved, we shall put

$$\begin{aligned}H_\nu^1(\mathbb{R}; H) &:= \operatorname{dom}(\partial_{t,\nu}), \\ H_\nu^{-1}(\mathbb{R}; H) &:= \operatorname{dom}(\partial_{t,\nu})' \cong \operatorname{dom}(\partial_{t,\nu}^*)'.\end{aligned}$$

Proposition 9.3.1 *Let $D\colon \operatorname{dom}(D) \subseteq H \to H$ be densely defined and closed and $B \in L(H)$. Assume that DB is densely defined. Then for all $\phi \in H$, $(DB)_{-1}(\phi) = (D_{-1}B)(\phi)$ on* $\operatorname{dom}(D^*)$.

Proof First of all, note that $(DB)^* = \overline{B^*D^*}$, by Theorem 2.3.4. Next, let $\phi \in H$ and $x \in \operatorname{dom}(D^*)$. Then

$$\begin{aligned}((DB)_{-1}\phi)(x) &= \left\langle \phi, (DB)^*x\right\rangle = \left\langle \phi, \overline{B^*D^*}x\right\rangle \\ &= \left\langle \phi, B^*D^*x\right\rangle = \left\langle B\phi, D^*x\right\rangle = (D_{-1}B\phi)(x).\end{aligned}$$ □

The first application of the theory developed in the previous section reads as follows.

Theorem 9.3.2 *Let $U, F \in L_{2,\nu}(\mathbb{R}; H)$. Then the following statements are equivalent:*

(i) $U \in \operatorname{dom}(\overline{\partial_{t,\nu}M(\partial_{t,\nu}) + A})$ *and* $\left(\overline{\partial_{t,\nu}M(\partial_{t,\nu}) + A}\right)U = F$.

(ii) $\partial_{t,\nu}M(\partial_{t,\nu})U + AU \subseteq F$ *where the left-hand side is considered as an element of* $H_\nu^{-1}(\mathbb{R}; H) \cap L_{2,\nu}(\mathbb{R}; H^{-1}(A)) \subseteq \mathrm{Fun}(L_{2,\nu}(\mathbb{R}; H))$.

Before we come to the proof, we state the following lemma, the proof of which is left as Exercise 9.7.

Lemma 9.3.3 *Let H be a Hilbert space.*

(a) *Let $B\colon \mathrm{dom}(B) \subseteq H \to H$ and $C\colon \mathrm{dom}(C) \subseteq H \to H$ be densely defined closed linear operators. Moreover, let $\lambda, \mu \in \rho(C)$ be in the same connected component of $\rho(C)$ and*

$$(\mu - C)^{-1}B \subseteq B(\mu - C)^{-1}.$$

Then $(\lambda - C)^{-1}B \subseteq B(\lambda - C)^{-1}$.

(b) *For $\nu > 0$ we have $(1+\varepsilon\partial_{t,\nu})^{-1} \to 1_{L_{2,\nu}(\mathbb{R};H)}$ and $(1+\varepsilon\partial_{t,\nu}^*)^{-1} \to 1_{L_{2,\nu}(\mathbb{R};H)}$ strongly as $\varepsilon \to 0+$.*

Proof of Theorem 9.3.2 At first, we want to apply Theorem 9.2.6 from above to the case $L_{2,\nu}(\mathbb{R}; H)$ being the Hilbert space, A the operator in $L_{2,\nu}(\mathbb{R}; H)$, $B = \partial_{t,\nu}M(\partial_{t,\nu})$, and $T_n := \left(1 + \frac{1}{n}\partial_{t,\nu}\right)^{-1}$, $n \in \mathbb{N}$. The operators A and B are densely defined. Indeed, A is skew-selfadjoint and $\mathrm{dom}(B) \supseteq \mathrm{dom}(\partial_{t,\nu})$. Next, by Theorems 2.3.2 and 2.3.4,

$$(B + A)^* \supseteq B^* + A^* = (\partial_{t,\nu}M(\partial_{t,\nu}))^* - A \supseteq M(\partial_{t,\nu})^*\partial_{t,\nu}^* - A.$$

In consequence, $\mathrm{dom}((A + B)^*) \supseteq \mathrm{dom}(\partial_{t,\nu}) \cap \mathrm{dom}(A)$ is dense. Thus, $B + A$ is closable by Lemma 2.2.7.

By Lemma 9.3.3 we obtain $T_n, T_n^* \to 1_{L_{2,\nu}(\mathbb{R};H)}$ strongly in $L_{2,\nu}(\mathbb{R}; H)$ as $n \to \infty$. Moreover, by Hille's theorem (see Proposition 3.1.6) we have $\partial_{t,\nu}^{-1}A \subseteq A\partial_{t,\nu}^{-1}$ and thus, $T_nA \subseteq AT_n$ for each $n \in \mathbb{N}$ by Lemma 9.3.3, which also yields $T_n^*A \subseteq AT_n^*$ for each $n \in \mathbb{N}$ by Theorem 9.2.6. The latter, together with the strong convergence of $(T_n)_n$ and $(T_n^*)_n$, yields that $T_n, T_n^* \to 1_{L_{2,\nu}(\mathbb{R};\mathrm{dom}(A))}$ strongly in $L_{2,\nu}(\mathbb{R}; \mathrm{dom}(A))$ as $n \to \infty$.

Next, we infer $\mathrm{ran}(T_n) \subseteq \mathrm{dom}(\partial_{t,\nu}) \subseteq \mathrm{dom}(B)$ and

$$T_nB \subseteq BT_n$$

for all $n \in \mathbb{N}$ by using the Fourier–Laplace transformation, see also Theorem 5.2.3.

Hence, by Theorem 9.2.6, condition (i) is equivalent to

$$(\partial_{t,\nu}M(\partial_{t,\nu}))_{-1}U + A_{-1}U \subseteq F. \tag{9.3}$$

It remains to show that (9.3) is equivalent to (ii): We apply Proposition 9.3.1 to the case $D = \partial_{t,\nu}$, $B = M(\partial_{t,\nu})$. For this assume that (9.3) holds. By Proposition 9.3.1, we deduce that for all $\varphi \in \operatorname{dom}(\partial_{t,\nu}^*) \cap \operatorname{dom}(A)$ we have that (use $\operatorname{dom}(A) = \operatorname{dom}(A^*)$)

$$((\partial_{t,\nu} M(\partial_{t,\nu}))_{-1} U + A_{-1} U)(\varphi) = ((\partial_{t,\nu})_{-1} M(\partial_{t,\nu}) U + A_{-1} U)(\varphi)$$

Thus, (9.3) implies (ii).

Now, assume that (ii) holds. Let $\phi \in \operatorname{dom}((\partial_{t,\nu} M(\partial_{t,\nu}))^*) \cap L_{2,\nu}(\mathbb{R}; \operatorname{dom}(A))$. Then, for $n \in \mathbb{N}$, $\phi_n := T_n^* \phi \to \phi$ as $n \to \infty$ in $L_{2,\nu}(\mathbb{R}; \operatorname{dom}(A))$ and

$$(\partial_{t,\nu} M(\partial_{t,\nu}))^* \phi_n = T_n^* (\partial_{t,\nu} M(\partial_{t,\nu}))^* \phi \to (\partial_{t,\nu} M(\partial_{t,\nu}))^* \phi \quad (n \to \infty)$$

in $L_{2,\nu}(\mathbb{R}; H)$. By (ii) we obtain

$$((\partial_{t,\nu})_{-1} M(\partial_{t,\nu}) U + A_{-1} U)(\phi_n) = F(\phi_n).$$

Using Proposition 9.3.1, we infer

$$((\partial_{t,\nu} M(\partial_{t,\nu}))_{-1} U + A_{-1} U)(\phi_n) = F(\phi_n).$$

Letting $n \to \infty$, we deduce (9.3). □

Assume now that there exists $c > 0$ such that

$$\operatorname{Re} z M(z) \geqslant c \quad (z \in \mathbb{C}_{\operatorname{Re} \geqslant \nu}).$$

We recall from Theorem 6.2.1 that the operator $\overline{\partial_{t,\nu} M(\partial_{t,\nu}) + A}$ is continuously invertible in $L_{2,\nu}(\mathbb{R}; H)$.

Theorem 9.3.4 *The operator $S_\nu := \left(\overline{\partial_{t,\nu} M(\partial_{t,\nu}) + A}\right)^{-1} \in L(L_{2,\nu}(\mathbb{R}; H))$ admits a continuous extension to $L(H_\nu^{-1}(\mathbb{R}; H))$.*

Proof We apply Proposition 9.2.9 to $L_{2,\nu}(\mathbb{R}; H)$ being the Hilbert space, $T = \partial_{t,\nu}$ and $B = S_\nu$. For this, it remains to prove that $T^{-1} S_\nu = S_\nu T^{-1}$. This however follows from the fact that $z \mapsto S(z) := (zM(z) + A)^{-1}$ is a material law and $S(\partial_{t,\nu}) = S_\nu$. □

9.4 Initial Value Problems for Evolutionary Equations

Let H be a Hilbert space, $\mu \in \mathbb{R}$, $M : \mathbb{C}_{\operatorname{Re} > \mu} \to L(H)$ a material law, $\nu > \max\{\mathrm{s_b}(M), 0\}$ and $A : \operatorname{dom}(A) \subseteq H \to H$ skew-selfadjoint. In this section we shall focus on the implementation of initial value problems for evolutionary equations. A priori there is no explicit initial condition implemented in the theory established in $L_{2,\nu}(\mathbb{R}; H)$. Indeed, choosing $\nu > 0$ we have only an implicit exponential decay condition at $-\infty$. For initial values at 0, we would rather want to solve the following type of equation. In the situation of the previous section, for a given initial value $U_0 \in H$ we seek to solve the initial value problem

$$\begin{cases} \big(\partial_{t,\nu} M(\partial_{t,\nu}) + A\big) U = 0 & \text{on } (0,\infty), \\ U(0+) = U_0. \end{cases} \tag{9.4}$$

In this generality the initial value problem cannot be solved. Indeed, for $U \in L_{2,\nu}(\mathbb{R}; H)$ evaluation at 0 is not well-defined. A way to overcome this difficulty is to weaken the attainment of the initial value. For this, we specialise to the case when

$$M(\partial_{t,\nu}) = M_0 + \partial_{t,\nu}^{-1} M_1$$

with $M_0, M_1 \in L(H)$.

We start with two lemmas, the second of which will also be useful in the next chapter.

Lemma 9.4.1 *Let H_0, H_1 be Hilbert spaces and assume that $H_1 \hookrightarrow H_0$ continuously and densely. Then $C_c^\infty(\mathbb{R}; H_1) \subseteq L_{2,\nu}(\mathbb{R}; H_1) \cap H_\nu^1(\mathbb{R}; H_0)$ is dense.*

Proof By Proposition 3.2.4, $C_c^\infty(\mathbb{R}; H_1) \subseteq H_\nu^1(\mathbb{R}; H_1)$ is dense. Since the embedding $H_\nu^1(\mathbb{R}; H_1) \hookrightarrow L_{2,\nu}(\mathbb{R}; H_1) \cap H_\nu^1(\mathbb{R}; H_0)$ is continuous, it thus suffices to show that this embedding is also dense. For this, let $f \in L_{2,\nu}(\mathbb{R}; H_1) \cap H_\nu^1(\mathbb{R}; H_0)$. For $\varepsilon > 0$ small enough, we define

$$f_\varepsilon := (1 + \varepsilon \partial_{t,\nu})^{-1} f \in H_\nu^1(\mathbb{R}; H_1).$$

By Lemma 9.3.3(b), $f_\varepsilon \to f$ in $L_{2,\nu}(\mathbb{R}; H_1)$ as $\varepsilon \to 0$. It remains to show that $\partial_{t,\nu} f_\varepsilon \to \partial_{t,\nu} f$ in $L_{2,\nu}(\mathbb{R}; H_0)$ as $\varepsilon \to 0$. For this, by definition of $H_\nu^1(\mathbb{R}; H_0)$, we find $g \in L_{2,\nu}(\mathbb{R}; H_0)$ such that $f = \partial_{t,\nu}^{-1} g$. Using again Lemma 9.3.3(b), we infer

$$\partial_{t,\nu} f_\varepsilon = \partial_{t,\nu}(1 + \varepsilon \partial_{t,\nu})^{-1} f = (1 + \varepsilon \partial_{t,\nu})^{-1} g \to g = \partial_{t,\nu} f$$

in $L_{2,\nu}(\mathbb{R}; H_0)$ as $\varepsilon \to 0$. This concludes the proof. □

Lemma 9.4.2 *Let $U_0 \in \operatorname{dom}(A)$, $U \in L_{2,\nu}(\mathbb{R}; H)$ such that $M_0 U - \mathbb{1}_{[0,\infty)} M_0 U_0 : \mathbb{R} \to H^{-1}(A)$ is continuous, $\operatorname{spt} U \subseteq [0,\infty)$ and*

$$\begin{cases} \partial_{t,\nu} M_0 U + M_1 U + AU = 0 & \textit{on } (0,\infty), \\ (M_0 U)(0+) = M_0 U_0 & \textit{in } H^{-1}(A), \end{cases}$$

where the first equality is meant in the sense that for all $\varphi \in H_\nu^1(\mathbb{R}; H) \cap L_{2,\nu}(\mathbb{R}; \operatorname{dom}(A))$ with $\operatorname{spt} \varphi \subseteq [0,\infty)$

$$\big(\partial_{t,\nu} M_0 U + M_1 U + AU\big)(\varphi) = 0.$$

Then $U - \mathbb{1}_{[0,\infty)}U_0 \in \operatorname{dom}(\overline{\partial_{t,\nu}M_0 + M_1 + A})$ *and*

$$\left(\overline{\partial_{t,\nu}M_0 + M_1 + A}\right)(U - \mathbb{1}_{[0,\infty)}U_0) = -(M_1 + A)U_0\mathbb{1}_{[0,\infty)}.$$

Proof We apply Theorem 9.3.2 for showing the claim; that is, we show that

$$\big((\partial_{t,\nu}M_0+M_1)(U-\mathbb{1}_{[0,\infty)}U_0)+A(U-\mathbb{1}_{[0,\infty)}U_0)\big)(\psi) = (-(M_1+A)U_0\mathbb{1}_{[0,\infty)})(\psi)$$

for each $\psi \in H_\nu^1(\mathbb{R}; H) \cap L_{2,\nu}(\mathbb{R}; \operatorname{dom}(A))$. Note that by continuity (use Lemma 9.4.1 with $H_0 = H$ and $H_1 = \operatorname{dom}(A)$), it suffices to show the equality for $\psi \in C_c^\infty(\mathbb{R}; \operatorname{dom}(A))$. So, let $\psi \in C_c^\infty(\mathbb{R}; \operatorname{dom}(A))$ and for $n \in \mathbb{N}$ we define the function $\varphi_n \in H_\nu^1(\mathbb{R})$ by

$$\varphi_n(t) := \begin{cases} 0 & \text{if } t \leqslant 0, \\ nt & \text{if } t \in (0, 1/n), \\ 1 & \text{if } t \geqslant 1/n. \end{cases}$$

Note that $\varphi_n\psi \in H_\nu^1(\mathbb{R}; H) \cap L_{2,\nu}(\mathbb{R}; \operatorname{dom}(A))$ and $\operatorname{spt}(\varphi_n\psi) \subseteq [0, \infty)$ for each $n \in \mathbb{N}$. Thus, we obtain

$$\begin{aligned}
&\big((\partial_{t,\nu}M_0 + M_1 + A)(U - \mathbb{1}_{[0,\infty)}U_0)\big)(\psi) \\
&= \big((\partial_{t,\nu}M_0 + M_1 + A)U\big)(\psi) - \big((\partial_{t,\nu}M_0 + M_1 + A)(\mathbb{1}_{[0,\infty)}U_0)\big)(\psi) \\
&= \big((\partial_{t,\nu}M_0 + M_1 + A)U\big)(\varphi_n\psi) + \big((\partial_{t,\nu}M_0 + M_1 + A)U\big)((1 - \varphi_n)\psi) \\
&\quad - \big((\partial_{t,\nu}M_0 + M_1 + A)(\mathbb{1}_{[0,\infty)}U_0)\big)(\psi) \\
&= \big((\partial_{t,\nu}M_0 + M_1 + A)U\big)((1 - \varphi_n)\psi) - (\delta_0M_0U_0)(\psi) \\
&\quad - \big((M_1 + A)(\mathbb{1}_{[0,\infty)}U_0)\big)(\psi)
\end{aligned}$$

for each $n \in \mathbb{N}$. Thus, the claim follows if we can show that

$$\big((\partial_{t,\nu}M_0 + M_1 + A)U\big)((1 - \varphi_n)\psi) - (\delta_0M_0U_0)(\psi) \to 0 \quad (n \to \infty).$$

For doing so, we first observe that for all $n \in \mathbb{N}$ we have

$$(\delta_0M_0U_0)(\psi) = (\delta_0M_0U_0)((1 - \varphi_n)\psi) = (\partial_{t,\nu}M_0\mathbb{1}_{[0,\infty)}U_0)((1 - \varphi_n)\psi),$$

since $\varphi_n(0) = 0$. Moreover,

$$\big((M_1 + A)U\big)((1 - \varphi_n)\psi) = \langle U, (1 - \varphi_n)(M_1^* + A^*)\psi\rangle_{L_{2,\nu}} \to 0 \quad (n \to \infty),$$

since $1-\varphi_n(\mathrm{m}) \to \mathbb{1}_{(-\infty,0]}(\mathrm{m})$ strongly in $L_{2,\nu}(\mathbb{R};H)$ and $\operatorname{spt} U \subseteq [0,\infty)$. Thus, it remains to show that

$$\left(\partial_{t,\nu} M_0(U-\mathbb{1}_{[0,\infty)}U_0)\right)((1-\varphi_n)\psi) \to 0 \quad (n\to\infty).$$

We compute

$$\begin{aligned}
&\left(\partial_{t,\nu} M_0(U-\mathbb{1}_{[0,\infty)}U_0)\right)((1-\varphi_n)\psi)\\
&= \left\langle M_0(U-\mathbb{1}_{[0,\infty)}U_0), \partial_{t,\nu}^*((1-\varphi_n)\psi)\right\rangle_{L_{2,\nu}}\\
&= \left\langle M_0(U-\mathbb{1}_{[0,\infty)}U_0), n\mathbb{1}_{[0,1/n]}\psi\right\rangle_{L_{2,\nu}} - \left\langle M_0(U-\mathbb{1}_{[0,\infty)}U_0), (1-\varphi_n)\partial_{t,\nu}\psi\right\rangle_{L_{2,\nu}}\\
&+ 2\nu\left\langle M_0(U-\mathbb{1}_{[0,\infty)}U_0), (1-\varphi_n)\psi\right\rangle_{L_{2,\nu}}.
\end{aligned}$$

Note that the last two terms on the right-hand side tend to 0 as $n\to\infty$ since, as above, $1-\varphi_n(\mathrm{m}) \to \mathbb{1}_{(-\infty,0]}(\mathrm{m})$ strongly in $L_{2,\nu}(\mathbb{R};H)$ and $\operatorname{spt} U \subseteq [0,\infty)$. For the first term, we observe that

$$\begin{aligned}
&\left|\left\langle M_0(U-\mathbb{1}_{[0,\infty)}U_0), n\mathbb{1}_{[0,1/n]}\psi\right\rangle_{L_{2,\nu}}\right|\\
&\leqslant n\int_0^{1/n} \left|\langle M_0(U(t)-U_0), \psi(t)\rangle_H\right| \mathrm{e}^{-2\nu t}\,\mathrm{d}t\\
&\leqslant n\int_0^{1/n} \|M_0(U(t)-U_0)\|_{H^{-1}(A)}\,\|\psi(t)\|_{\mathrm{dom}(A^*)}\,\mathrm{e}^{-2\nu t}\,\mathrm{d}t \to 0 \quad (n\to\infty),
\end{aligned}$$

by the fundamental theorem of calculus, since $(M_0U)(t) \to M_0U_0$ in $H^{-1}(A)$ as $t\to 0+$. □

Assume now additionally that there exists $c>0$ such that

$$zM_0 + M_1 \geqslant c \quad (z\in\mathbb{C}_{\mathrm{Re}\geqslant\nu}).$$

Then we can actually prove a stronger result than in the previous lemma.

Theorem 9.4.3 *Let $U_0\in\operatorname{dom}(A)$, $U\in L_{2,\nu}(\mathbb{R};H)$. Then the following statements are equivalent:*

(i) *$M_0U-\mathbb{1}_{[0,\infty)}M_0U_0\colon \mathbb{R}\to H^{-1}(A)$ is continuous,* $\operatorname{spt} U\subseteq[0,\infty)$ *and*

$$\begin{cases}\partial_{t,\nu}M_0U + M_1U + AU = 0 & \text{on } (0,\infty),\\ M_0U(0+) = M_0U_0 & \text{in } H^{-1}(A),\end{cases}$$

where the first equality is meant as in Lemma 9.4.2.

(ii) $U - \mathbb{1}_{[0,\infty)}U_0 \in \operatorname{dom}(\overline{\partial_{t,\nu}M_0 + M_1 + A})$, *and we have that* $\left(\overline{\partial_{t,\nu}M_0 + M_1 + A}\right)(U - \mathbb{1}_{[0,\infty)}U_0) = -(M_1 + A)U_0\mathbb{1}_{[0,\infty)}$.

(iii) $U = S_\nu\delta_0 M_0 U_0$, *with* $S_\nu \in L(H_\nu^{-1}(\mathbb{R}; H))$ *as in Theorem 9.3.4.*

Moreover, in either case we have $M_0U - \mathbb{1}_{[0,\infty)}M_0U_0 \in H_\nu^1(\mathbb{R}; H^{-1}(A))$.

Proof (i)⇒(ii): This was shown in Lemma 9.4.2.
(ii)⇒(iii): We have that

$$U - \mathbb{1}_{[0,\infty)}U_0 = -S_\nu((M_1 + A)\mathbb{1}_{[0,\infty)}U_0).$$

Applying $\partial_{t,\nu}^{-1}$ to both sides of this equality we infer that

$$\begin{aligned}\partial_{t,\nu}^{-1}(U - \mathbb{1}_{[0,\infty)}U_0) &= -S_\nu((M_1 + A)\partial_{t,\nu}^{-1}\mathbb{1}_{[0,\infty)}U_0)\\ &= -\partial_{t,\nu}^{-1}\mathbb{1}_{[0,\infty)}U_0 + S_\nu(\partial_{t,\nu}M_0\partial_{t,\nu}^{-1}\mathbb{1}_{[0,\infty)}U_0),\end{aligned}$$

which gives

$$\partial_{t,\nu}^{-1}U = S_\nu(\partial_{t,\nu}M_0\partial_{t,\nu}^{-1}\mathbb{1}_{[0,\infty)}U_0) = S_\nu(M_0\mathbb{1}_{[0,\infty)}U_0).$$

Applying $\partial_{t,\nu}$ to both sides and taking into account Theorem 9.3.4, we derive the claim.
(iii)⇒(ii): We do the argument in the proof of (ii)⇒(iii) backwards. First, we apply $\partial_{t,\nu}^{-1}$ to $U = S_\nu(\delta_0 M_0 U_0)$, which yields

$$\partial_{t,\nu}^{-1}U = \partial_{t,\nu}^{-1}S_\nu(\delta_0 M_0 U_0) = S_\nu(M_0\mathbb{1}_{[0,\infty)}U_0) = S_\nu(\partial_{t,\nu}M_0\partial_{t,\nu}^{-1}\mathbb{1}_{[0,\infty)}U_0).$$

Thus,

$$\begin{aligned}\partial_{t,\nu}^{-1}(U - \mathbb{1}_{[0,\infty)}U_0) &= S_\nu(\partial_{t,\nu}M_0\partial_{t,\nu}^{-1}\mathbb{1}_{[0,\infty)}U_0) - \partial_{t,\nu}^{-1}\mathbb{1}_{[0,\infty)}U_0\\ &= -S_\nu((M_1 + A)\partial_{t,\nu}^{-1}\mathbb{1}_{[0,\infty)}U_0).\end{aligned}$$

An application of $\partial_{t,\nu}$ yields the claim.
(ii),(iii)⇒(i): Since $U = S_\nu(\delta_0 M_0 U_0)$, we derive that

$$(\partial_{t,\nu}M_0 + M_1 + A)U \subseteq \delta_0 M_0 U_0,$$

which in particular yields $(\partial_{t,\nu}M_0 + M_1 + A)U = 0$ on $(0,\infty)$. By (ii) we infer

$$U - \mathbb{1}_{[0,\infty)}U_0 = -S_\nu((M_1 + A)\mathbb{1}_{[0,\infty)}U_0),$$

which shows that $\operatorname{spt}(U - \mathbb{1}_{[0,\infty)}U_0) \subseteq [0,\infty)$ due to causality and hence, $\operatorname{spt} U \subseteq [0,\infty)$. It remains to show that $M_0(U - \mathbb{1}_{[0,\infty)}U_0) \in H_\nu^1(\mathbb{R}; H^{-1}(A))$, since this would imply the continuity of $M_0(U - \mathbb{1}_{[0,\infty)}U_0)$ with values in $H^{-1}(A)$ by Theorem 4.1.2 and thus,

$$M_0(U - \mathbb{1}_{[0,\infty)}U_0)(0+) = M_0(U - \mathbb{1}_{[0,\infty)}U_0)(0-) = 0 \text{ in } H^{-1}(A)$$

since the function is supported on $[0, \infty)$ only. We compute

$$\begin{aligned}
&M_0(U - \mathbb{1}_{[0,\infty)}U_0)\\
&= -M_0 S_\nu((M_1 + A)\mathbb{1}_{[0,\infty)}U_0)\\
&= -\partial_{t,\nu} M_0 S_\nu(\partial_{t,\nu}^{-1}(M_1 + A)\mathbb{1}_{[0,\infty)}U_0)\\
&= -\partial_{t,\nu}^{-1}(M_1 + A)\mathbb{1}_{[0,\infty)}U_0 + (M_1 + A)S_\nu(\partial_{t,\nu}^{-1}(M_1 + A)\mathbb{1}_{[0,\infty)}U_0),
\end{aligned}$$

and since the right-hand side belongs to $H_\nu^1(\mathbb{R}; H^{-1}(A))$, the assertion follows. □

Remark 9.4.4 By Theorem 9.3.4, we always have $U = S_\nu \delta_0 M_0 U_0 \in H_\nu^{-1}(\mathbb{R}; H)$. This then serves as our generalisation for the initial value problem even if $U_0 \notin \operatorname{dom}(A)$.

The upshot of Theorem 9.4.3(ii) is that, provided $U_0 \in \operatorname{dom}(A)$, we can reformulate initial value problems with the help of our theory as evolutionary equations with $L_{2,\nu}$-right-hand sides. Thus, we do not need the detour to extrapolation spaces for being able to solve the initial value problem (9.4) (with an adapted initial condition as in (i)) in this situation.

Also note that it may seem that U does depend on the 'full information' of U_0 as it is indicated in (ii). In fact, U only depends on the values of U_0 orthogonal to the kernel of M_0 as it is seen in (iii). We conclude this chapter with two examples; the first one is the heat equation, the second example considers Maxwell's equations.

Example 9.4.5 (Initial Value Problems for the Heat Equation) We recall the setting for the heat equation outlined in Theorem 6.2.4. This time, we will use homogeneous Dirichlet boundary conditions for the heat distribution θ. Let $\Omega \subseteq \mathbb{R}^d$ be open and bounded, $a \in L_\infty(\Omega)^{d\times d}$ with $\operatorname{Re} a(x) \geqslant c > 0$ for a.e. $x \in \Omega$ for some $c > 0$. In this case, we have

$$M_0 = \begin{pmatrix} 1 & 0 \\ 0 & 0 \end{pmatrix}, \quad M_1 = \begin{pmatrix} 0 & 0 \\ 0 & a^{-1} \end{pmatrix}, \quad A = \begin{pmatrix} 0 & \operatorname{div} \\ \operatorname{grad}_0 & 0 \end{pmatrix}.$$

For the unknown heat distribution, θ, we ask it to have the initial value $\theta_0 \in \operatorname{dom}(\operatorname{grad}_0)$. Let $\nu > 0$ and $V \in L_{2,\nu}\big(\mathbb{R}; L_2(\Omega) \times L_2(\Omega)^d\big)$ be the unique solution of

$$\big(\overline{\partial_{t,\nu} M_0 + M_1 + A}\big)V = -\left(M_1 + A\right)\mathbb{1}_{[0,\infty)}\begin{pmatrix} \theta_0 \\ 0 \end{pmatrix} = -\mathbb{1}_{[0,\infty)}\begin{pmatrix} 0 \\ \operatorname{grad}_0 \theta_0 \end{pmatrix}.$$

Then $(\theta, q) := U := V + \mathbb{1}_{[0,\infty)} \begin{pmatrix} \theta_0 \\ 0 \end{pmatrix} \in L_{2,\nu}\big(\mathbb{R}; L_2(\Omega) \times L_2(\Omega)^d\big)$ satisfies (ii) from Theorem 9.4.3. Hence, on $(0, \infty)$ we have

$$\begin{pmatrix} \partial_{t,\nu}\theta \\ a^{-1}q \end{pmatrix} + \begin{pmatrix} \operatorname{div} q \\ \operatorname{grad}_0 \theta \end{pmatrix} = 0$$

and the initial value is attained in the sense that

$$(M_0\,(\theta, q))\,(0+) = \begin{pmatrix} \theta(0+) \\ 0 \end{pmatrix} = \begin{pmatrix} \theta_0 \\ 0 \end{pmatrix} \quad \text{in} \quad H^{-1}(A) = H^{-1}(\operatorname{grad}_0) \times H^{-1}(\operatorname{div}),$$

which follows from Proposition 9.2.5 where we computed $H^{-1}(A)$. Let us have a closer look at the attainment of the initial value. As a particular consequence of strong convergence in $H^{-1}(\operatorname{grad}_0)$, we obtain for all $\phi \in \operatorname{dom}(\operatorname{div})$

$$\langle \theta(t), \operatorname{div} \phi \rangle \to \langle \theta_0, \operatorname{div} \phi \rangle$$

as $t \to 0+$. Since grad_0 is one-to-one and has closed range (see Corollary 11.3.2), we see that div has dense and closed range. Hence div is onto. This implies that for all $\psi \in L_2(\Omega)$

$$\langle \theta(t), \psi \rangle \to \langle \theta_0, \psi \rangle \quad (t \to 0+).$$

We deduce that the initial value is attained weakly. This might seem a bit unsatisfactory, however, we shall see stronger assertions for more particular cases in the next chapter.

Next, we have a look at Maxwell's equations.

Example 9.4.6 (Initial Value Problems for Maxwell's Equations) We briefly recall the situation of Maxwell's equations from Theorem 6.2.8. Let $\varepsilon, \mu, \sigma \colon \Omega \to \mathbb{R}^{3\times 3}$ satisfy the assumptions in Theorem 6.2.8 and let $(E_0, H_0) \in \operatorname{dom}(\operatorname{curl}_0) \times \operatorname{dom}(\operatorname{curl})$. Let $(\widehat{E}, \widehat{H}) \in L_{2,\nu}(\mathbb{R}; L_2(\Omega)^6)$ satisfy

$$\overline{\left(\partial_{t,\nu} \begin{pmatrix} \varepsilon & 0 \\ 0 & \mu \end{pmatrix} + \begin{pmatrix} \sigma & 0 \\ 0 & 0 \end{pmatrix} + \begin{pmatrix} 0 & -\operatorname{curl} \\ \operatorname{curl}_0 & 0 \end{pmatrix}\right)} \begin{pmatrix} \widehat{E} \\ \widehat{H} \end{pmatrix}$$
$$= -\left(\begin{pmatrix} \sigma & 0 \\ 0 & 0 \end{pmatrix} + \begin{pmatrix} 0 & -\operatorname{curl} \\ \operatorname{curl}_0 & 0 \end{pmatrix}\right) \mathbb{1}_{[0,\infty)} \begin{pmatrix} E_0 \\ H_0 \end{pmatrix} = \mathbb{1}_{[0,\infty)} \begin{pmatrix} -\sigma E_0 + \operatorname{curl} H_0 \\ -\operatorname{curl}_0 E_0 \end{pmatrix}.$$

Then, as we have argued for the heat equation,

$$\begin{pmatrix} E \\ H \end{pmatrix} := \begin{pmatrix} \widehat{E} \\ \widehat{H} \end{pmatrix} + \mathbb{1}_{[0,\infty)} \begin{pmatrix} E_0 \\ H_0 \end{pmatrix}$$

satisfies a corresponding initial value problem. We note here that although often the second component in the right-hand side is set to 0, as there are 'no magnetic monopoles', in the theory of evolutionary equations the second component of the right-hand side does appear as an initial value in disguise.

9.5 Comments

There are many ways to define spaces generalising the action of an operator to a bigger class of elements; both in a concrete setting and in abstract situations; see e.g. [22, 38]. People have also taken into account simultaneous extrapolation spaces for operators that commute, see e.g. [77, 93].

These spaces are particularly useful for formulating initial value problems as was exemplified above; see also the concluding chapter of [84] for more insight. Yet there is more to it as one can in fact generalise the equation under consideration or even force the attainment of the initial value in a stronger sense. These issues, however, imply that either the initial value is attained in a much weaker sense, or that there are other structural assumptions needed to be imposed on the material law M (as well as on the operator A).

In fact, quite recently, it was established that a particular proper subclass of evolutionary equations can be put into the framework of C_0-semigroups. The conditions required to allow for statements in this direction are, on the other hand, rather hard to check in practice; see [116, 120].

Exercises

Exercise 9.1 Let H_0 be a Hilbert space, $T \in L(H_0)$. Compute $H^{-1}(T)$ and $H^{-1}(T^*)$.

Exercise 9.2 Let H_0, H_1 be Hilbert spaces such that $H_0 \hookrightarrow H_1$ is dense and continuous. Prove that $H_1' \hookrightarrow H_0'$ is dense and continuous as well.

Exercise 9.3 Prove the following statement which generalises Proposition 9.2.9 from above: Let H_0 be a Hilbert space, $A \in L(H_0)$. Assume that $T\colon \operatorname{dom}(T) \subseteq H_0 \to H_0$ is densely defined and closed with $0 \in \rho(T)$ and $T^{-1}A = AT^{-1} + T^{-1}BT^{-1}$ for some bounded $B \in L(H_0)$. Then A admits a unique continuous extension, $\overline{A} \in L(H^{-1}(T^*))$.

Exercise 9.4 Let H_0 be a Hilbert space, $N\colon \operatorname{dom}(N) \subseteq H_0 \to H_0$ be a *normal* operator; that is, N is densely defined and closed and $NN^* = N^*N$. Show that $H^{-1}(N) \cong H^{-1}(N^*)$ and deduce $H^{-1}(\partial_{t,\nu}) \cong H^{-1}(\partial_{t,\nu}^*)$.

Exercise 9.5 Prove Proposition 9.2.8.

Exercise 9.6 Let H_0 be a Hilbert space, $n \in \mathbb{N}$ and $T : \operatorname{dom}(T) \subseteq H_0 \to H_0$ be a densely defined, closed linear operator with $0 \in \rho(T)$. We define $H^n(T) := \operatorname{dom}(T^n)$ and $H^{-n}(T) := H^{-1}(T^n)$. Show that for all $k \in \mathbb{N}$ and $\ell \in \mathbb{Z}$ we have that $H^{k+\ell}(T) \hookrightarrow H^\ell(T)$ continuously and densely. Also show that $\mathcal{D} := \bigcap_{n\in\mathbb{N}} \operatorname{dom}(T^n)$ is dense in $H^\ell(T)$ and dense in $H^{-\ell}(T^*)$ for all $\ell \in \mathbb{N}$ and that $T|_{\mathcal{D}}$ can be continuously extended to a topological isomorphism $H^\ell(T) \to H^{\ell-1}(T)$ and to an isomorphism $H^{-\ell+1}(T^*) \to H^{-\ell}(T^*)$ for each $\ell \in \mathbb{N}$.

Exercise 9.7 Prove Lemma 9.3.3.
Hint: Prove a similar equality with $\partial_{t,\nu}^{-1}$ formally replaced by $z \in \partial B(r, r) \subseteq \mathbb{C}$ and deduce the assertion with the help of Theorem 5.2.3.

References

22. P. Cojuhari, A. Gheondea, Closed embeddings of Hilbert spaces. J. Math. Anal. Appl. **369**(1), 60–75 (2010)
38. K.-J. Engel, R. Nagel, *One-Parameter Semigroups for Linear Evolution Equations*, vol. 194. Graduate Texts in Mathematics. With contributions by S. Brendle, M. Campiti, T. Hahn, G. Metafune, G. Nickel, D. Pallara, C. Perazzoli, A. Rhandi, S. Romanelli, R. Schnaubelt (Springer, New York, 2000)
77. R.S. Palais, *Seminar on the Atiyah-Singer Index Theorem*. With contributions by M.F. Atiyah, A. Borel, E.E. Floyd, R.T. Seeley, W. Shih, R. Solovay. Annals of Mathematics Studies, vol. 57 (Princeton University Press, Princeton, NJ, 1965)
84. R. Picard, D. McGhee, *Partial Differential Equations: A Unified Hilbert Space Approach*, vol. 55. Expositions in Mathematics (DeGruyter, Berlin, 2011)
93. R. Picard, Evolution equations as operator equations in lattices of Hilbert spaces. Glas. Mat. Ser. III **35**(55), 1 (2000). Dedicated to the memory of Branko Najman, pp. 111–136
116. S. Trostorff, Exponential Stability and Initial Value Problems for Evolutionary Equations. Habilitation Thesis. TU Dresden, 2018
120. S. Trostorff, Semigroups and evolutionary equations. Semigr. Forum **103**(2), 661–699 (2021)

Chapter 10
Differential Algebraic Equations

Let H be a Hilbert space and $\nu \in \mathbb{R}$. We saw in the previous chapter how initial value problems can be formulated within the framework of evolutionary equations. More precisely, we have studied problems of the form

$$\begin{cases} \left(\partial_{t,\nu} M_0 + M_1 + A\right) U = 0 \quad \text{on } (0,\infty), \\ M_0 U(0+) = M_0 U_0 \end{cases} \tag{10.1}$$

for $U_0 \in H$, $M_0, M_1 \in L(H)$ and $A\colon \operatorname{dom}(A) \subseteq H \to H$ skew-selfadjoint; that is, we have considered material laws of the form

$$M(z) := M_0 + z^{-1} M_1 \quad (z \in \mathbb{C} \setminus \{0\}).$$

Here, the initial value is attained in a weak sense as an equality in the extrapolation space $H^{-1}(A)$. The first line is also meant in a weak sense since the left-hand side turned out to be a functional in $H_\nu^{-1}(\mathbb{R}; H) \cap L_{2,\nu}(\mathbb{R}; H^{-1}(A))$. In Theorem 9.4.3 it was shown that the latter problem can be rewritten as

$$\left(\partial_{t,\nu} M_0 + M_1 + A\right) U = \delta_0 M_0 U_0.$$

In this chapter we aim to inspect initial value problems a little closer but in the particularly simple case when $A = 0$. However, we want to impose the initial condition for U and not just $M_0 U$. Thus, we want to deal with the problem

$$\begin{cases} \left(\partial_{t,\nu} M_0 + M_1\right) U = 0 \quad \text{on } (0,\infty), \\ U(0+) = U_0 \end{cases} \tag{10.2}$$

C. Seifert et al., *Evolutionary Equations*, Operator Theory: Advances and Applications 287, https://doi.org/10.1007/978-3-030-89397-2_10

for two bounded operators M_0, M_1 and an initial value $U_0 \in H$. This class of differential equations is known as *differential algebraic equations* since the operator M_0 is allowed to have a non-trivial kernel. Thus, (10.2) is a coupled problem of a differential equation (on $(\ker M_0)^{\perp}$) and an algebraic equation (on $\ker M_0$). We begin by treating these equations in the finite-dimensional case; that is, $H = \mathbb{C}^n$ and $M_0, M_1 \in \mathbb{C}^{n\times n}$ for some $n \in \mathbb{N}$.

10.1 The Finite-Dimensional Case

Throughout this section let $n \in \mathbb{N}$ and $M_0, M_1 \in \mathbb{C}^{n\times n}$.

Definition We define the *spectrum of the matrix pair* (M_0, M_1) by

$$\sigma(M_0, M_1) := \{z \in \mathbb{C};\ \det(zM_0 + M_1) = 0\},$$

and the *resolvent set of the matrix pair* (M_0, M_1) by

$$\rho(M_0, M_1) := \mathbb{C} \setminus \sigma(M_0, M_1).$$

Remark 10.1.1

(a) It is immediate that $\sigma(M_0, M_1)$ is closed since the mapping $z \mapsto \det(zM_0+M_1)$ is continuous.
(b) Note in particular that the spectrum (the set of eigenvalues) of a matrix A corresponds in this setting to the spectrum of the matrix pair $(1, -A)$.

In contrast to the case of the spectrum of one matrix, it may happen that $\sigma(M_0, M_1) = \mathbb{C}$ (for example we can choose $M_0 = 0$ and M_1 singular). More precisely, we have the following result.

Lemma 10.1.2 *The set $\sigma(M_0, M_1)$ is either finite or equals the whole complex plane $\mathbb{C}$. If $\sigma(M_0, M_1)$ is finite then* $\operatorname{card}(\sigma(M_0, M_1)) \leqslant n$.

Proof The function $z \mapsto \det(zM_0 + M_1)$ is a polynomial of order less than or equal to n. If it is constantly zero, then $\sigma(M_0, M_1) = \mathbb{C}$ and otherwise $\operatorname{card}(\sigma(M_0, M_1)) \leqslant n$. □

Definition The matrix pair (M_0, M_1) is called *regular* if $\sigma(M_0, M_1) \neq \mathbb{C}$.

The main problem in solving an initial value problem of the form (10.2) is that one cannot expect a solution for each initial value $U_0 \in \mathbb{C}^n$ as the following simple example shows.

Example 10.1.3 Let $M_0 = \begin{pmatrix} 1 & 1 \\ 0 & 0 \end{pmatrix}$, $M_1 = \begin{pmatrix} 1 & 0 \\ 0 & 1 \end{pmatrix}$ and let $U_0 \in \mathbb{C}^2$. We assume that there exists a solution $U : \mathbb{R}_{\geqslant 0} \to \mathbb{C}^2$ satisfying (10.2); that is,

$$\begin{aligned} U_1'(t) + U_2'(t) + U_1(t) &= 0 \quad (t > 0), \\ U_2(t) &= 0 \quad (t > 0), \\ U(0+) &= U_0. \end{aligned}$$

The second and third equation yield that the second coordinate of U_0 has to be zero. Then, for $U_0 = (x, 0) \in \mathbb{C}^2$ the unique solution of the above problem is given by

$$U(t) = \big(U_1(t), U_2(t)\big) = (x\mathrm{e}^{-t}, 0) \quad (t \geqslant 0).$$

Definition We call an initial value $U_0 \in \mathbb{C}^n$ *consistent* for (10.2) if there exists $\nu > 0$ and $U \in C(\mathbb{R}_{\geqslant 0}; \mathbb{C}^n) \cap L_{2,\nu}(\mathbb{R}_{\geqslant 0}; \mathbb{C}^n)$ such that (10.2) holds. We denote the set of all consistent initial values for (10.2) by

$$\mathrm{IV}(M_0, M_1) := \left\{U_0 \in \mathbb{C}^n \,;\, U_0 \text{ consistent}\right\}.$$

Remark 10.1.4 It is obvious that $\mathrm{IV}(M_0, M_1)$ is a subspace of $\mathbb{C}^n$. In particular, $0 \in \mathrm{IV}(M_0, M_1)$.

It is now our goal to determine the space $\mathrm{IV}(M_0, M_1)$. One possibility for doing so uses the so-called *quasi-Weierstraß normal form*.

Proposition 10.1.5 (Quasi-Weierstraß Normal Form) *Assume that (M_0, M_1) is regular. Then there exist invertible matrices $P, Q \in \mathbb{C}^{n\times n}$ such that*

$$PM_0Q = \begin{pmatrix} 1 & 0 \\ 0 & N \end{pmatrix}, \quad PM_1Q = \begin{pmatrix} C & 0 \\ 0 & 1 \end{pmatrix},$$

where $C \in \mathbb{C}^{k\times k}$ and $N \in \mathbb{C}^{(n-k)\times(n-k)}$ for some $k \in \{0, \ldots, n\}$. Moreover, the matrix N is nilpotent; that is, there exists $\ell \in \mathbb{N}$ such that $N^\ell = 0$.

Proof Since (M_0, M_1) is regular we find $\lambda \in \mathbb{C}$ such that $\lambda M_0 + M_1$ is invertible. We set $P_1 := (\lambda M_0 + M_1)^{-1}$ and obtain

$$\begin{aligned} M_{0,1} &:= P_1 M_0 = (\lambda M_0 + M_1)^{-1} M_0, \\ M_{1,1} &:= P_1 M_1 = (\lambda M_0 + M_1)^{-1} M_1 = 1 - \lambda M_{0,1}. \end{aligned}$$

Let now $P_2 \in \mathbb{C}^{n\times n}$ such that

$$M_{0,2} := P_2 M_{0,1} P_2^{-1} = \begin{pmatrix} J & 0 \\ 0 & \widetilde{N} \end{pmatrix}$$

for some invertible matrix $J \in \mathbb{C}^{k\times k}$ and a nilpotent matrix $\widetilde{N} \in \mathbb{C}^{(n-k)\times(n-k)}$ (use the Jordan normal form of $M_{0,1}$ here). Then

$$M_{1,2} := P_2 M_{1,1} P_2^{-1} = \begin{pmatrix} 1-\lambda J & 0 \\ 0 & 1-\lambda\widetilde{N} \end{pmatrix}.$$

Now, by the nilpotency of $\widetilde{N}$, the matrix $(1-\lambda\widetilde{N})$ is invertible by the Neumann series. We set

$$P_3 := \begin{pmatrix} J^{-1} & 0 \\ 0 & (1-\lambda\widetilde{N})^{-1} \end{pmatrix}$$

and obtain

$$P_3 M_{0,2} = \begin{pmatrix} 1 & 0 \\ 0 & (1-\lambda\widetilde{N})^{-1}\widetilde{N} \end{pmatrix}, \quad \text{and} \quad P_3 M_{1,2} = \begin{pmatrix} J^{-1}-\lambda & 0 \\ 0 & 1 \end{pmatrix}.$$

Note that $(1-\lambda\widetilde{N})^{-1}\widetilde{N}$ is nilpotent, since the matrices commute and $\widetilde{N}$ is nilpotent. Thus, the assertion follows with $N := (1-\lambda\widetilde{N})^{-1}\widetilde{N}$, $C := J^{-1}-\lambda$, $P = P_3P_2P_1$, and $Q = P_2^{-1}$. □

It is clear that the matrices P, Q, C and N in the previous proposition are not uniquely determined by M_0 and M_1. However, the size of N and C as well as the degree of nilpotency of N are determined by M_0 and M_1 as the following proposition shows.

Proposition 10.1.6 *Let $P, Q \in \mathbb{C}^{n\times n}$ be invertible such that*

$$PM_0Q = \begin{pmatrix} 1 & 0 \\ 0 & N \end{pmatrix}, \quad PM_1Q = \begin{pmatrix} C & 0 \\ 0 & 1 \end{pmatrix},$$

where $C \in \mathbb{C}^{k\times k}$, $N \in \mathbb{C}^{(n-k)\times(n-k)}$ for some $k \in \{0, \ldots, n\}$, and N is nilpotent. Then (M_0, M_1) is regular and

(a) *k is the degree of the polynomial $z \mapsto \det(zM_0 + M_1)$.*
(b) *$N^\ell = 0$ if and only if*

$$\sup_{|z|\geqslant r} \left\| z^{-\ell+1}(zM_0 + M_1)^{-1} \right\| < \infty$$

for one (or equivalently all) $r > 0$ such that $B(0, r) \supseteq \sigma(M_0, M_1)$.

Proof First, note that

$$\det(zM_0 + M_1) = \frac{1}{\det P \, \det Q} \det \begin{pmatrix} z + C & 0 \\ 0 & zN + 1 \end{pmatrix} = \frac{1}{\det P \, \det Q} \det(z + C)$$

for all $(z \in \mathbb{C})$. Hence, (M_0, M_1) is regular and

$$k = \deg \det((\cdot) + C) = \deg \det((\cdot)M_0 + M_1),$$

which shows (a). Moreover, we have $\rho(M_0, M_1) = \rho(-C)$ and

$$(zM_0 + M_1)^{-1} = Q \begin{pmatrix} (z + C)^{-1} & 0 \\ 0 & (zN + 1)^{-1} \end{pmatrix} P \quad (z \in \rho(M_0, M_1)),$$

and hence, for $r > 0$ with $B(0, r) \supseteq \sigma(M_0, M_1)$ we have

$$\left\| (zM_0 + M_1)^{-1} \right\| \leqslant K_1 \left\| (zN + 1)^{-1} \right\| \quad (|z| \geqslant r)$$

for some $K_1 \geqslant 0$, since $\sup_{|z| \geqslant r} \left\| (z + C)^{-1} \right\| < \infty$. Now let $\ell \in \mathbb{N}$ such that $N^\ell = 0$. Then

$$\left\| (zN + 1)^{-1} \right\| = \left\| \sum_{k=0}^{\ell - 1} (-1)^k z^k N^k \right\| \leqslant K_2 \, |z|^{\ell - 1} \quad (|z| \geqslant r)$$

for some constant $K_2 \geqslant 0$ and thus,

$$\left\| (zM_0 + M_1)^{-1} \right\| \leqslant K_1 K_2 \, |z|^{\ell - 1} \quad (|z| \geqslant r).$$

Assume on the other hand that

$$\sup_{|z| \geqslant r} \left\| z^{-\ell + 1} (zM_0 + M_1)^{-1} \right\| < \infty$$

for some $\ell \in \mathbb{N}$ and $r > 0$ with $\sigma(M_0, M_1) \subseteq B(0, r)$. Then there exist $\widetilde{K}_1, \widetilde{K}_2 \geqslant 0$ such that

$$\left\| (zN + 1)^{-1} \right\| \leqslant \left\| \begin{pmatrix} (z + C)^{-1} & 0 \\ 0 & (zN + 1)^{-1} \end{pmatrix} \right\| \leqslant \widetilde{K}_1 \left\| (zM_0 + M_1)^{-1} \right\| \leqslant \widetilde{K}_2 \, |z|^{\ell - 1}$$

for all $z \in \mathbb{C}$ with $|z| \geqslant r$. Now, let $p \in \mathbb{N}$ be minimal such that $N^p = 0$. We show that $p \leqslant \ell$ by contradiction. Assume $p > \ell$. Then we compute

$$\begin{aligned} 0 &= \lim_{n\to\infty} \frac{1}{n^\ell}(nN+1)^{-1}N^{p-\ell-1} = \lim_{n\to\infty} \sum_{k=0}^{p-1} (-1)^k n^{k-\ell} N^{k+p-\ell-1} \\ &= \lim_{n\to\infty} \sum_{k=0}^{\ell-1} (-1)^k n^{k-\ell} N^{k+p-\ell-1} + (-1)^\ell N^{p-1} \\ &= (-1)^\ell N^{p-1}, \end{aligned}$$

which contradicts the minimality of p. □

Theorem 10.1.7 *Let (M_0, M_1) be regular and $P, Q \in \mathbb{C}^{n\times n}$ be chosen according to Proposition 10.1.5. Let $k = \deg\det((\cdot)M_0 + M_1)$. Then*

$$\mathrm{IV}(M_0, M_1) = \left\{U_0 \in \mathbb{C}^n\,;\ Q^{-1}U_0 \in \mathbb{C}^k \times \{0\}\right\}.$$

Moreover, for each $U_0 \in \mathrm{IV}(M_0, M_1)$ the solution U of (10.2) is unique and satisfies $U \in C(\mathbb{R}_{\geqslant 0}; \mathbb{C}^n) \cap C^1(\mathbb{R}_{>0}; \mathbb{C}^n)$ as well as

$$\begin{aligned} M_0U'(t) + M_1U(t) &= 0 \quad (t > 0), \\ U(0+) &= U_0. \end{aligned}$$

Proof Let $C \in \mathbb{C}^{k\times k}$ and $N \in \mathbb{C}^{(n-k)\times(n-k)}$ be nilpotent as in Proposition 10.1.5. Obviously U is a solution of (10.2) if and only if $V := Q^{-1}U$ both is continuous on $\mathbb{R}_{\geqslant 0}$ and solves

$$\begin{aligned} \left(\partial_{t,\nu}\begin{pmatrix}1 & 0\\ 0 & N\end{pmatrix} + \begin{pmatrix}C & 0\\ 0 & 1\end{pmatrix}\right)V &= 0 \quad \text{on } (0,\infty), \\ V(0+) &= Q^{-1}U_0 =: V_0. \end{aligned} \tag{10.3}$$

Clearly, if $Q^{-1}U_0 = (x, 0) \in \mathbb{C}^k \times \{0\}$ then V given by $V(t) := (\mathrm{e}^{-tC}x, 0)$ for $t \geqslant 0$ is a solution of (10.3) for $\nu > 0$ large enough. On the other hand, if V given by $V(t) = (V_1(t), V_2(t)) \in \mathbb{C}^k \times \mathbb{C}^{n-k}$ $(t \geqslant 0)$ is a solution of (10.3) then we have

$$\partial_{t,\nu}NV_2 + V_2 = 0 \quad \text{on } (0,\infty)\,.$$

Since N is nilpotent, there exists $\ell \in \mathbb{N}$ with $N^\ell = 0$. Hence,

$$N^{\ell-1}V_2(t) = -N^{\ell-1}\partial_{t,\nu}NV_2(t) = \partial_{t,\nu}N^\ell V_2(t) = 0 \quad (t > 0),$$

which in turn implies $\partial_{t,\nu} N^{\ell-1} V_2 = 0$ on $(0,\infty)$. Using again the differential equation, we infer $N^{\ell-2} V_2(t) = 0$ for $t > 0$. Inductively, we deduce $V_2(t) = 0$ for $t > 0$ and by continuity $V_2(0+) = 0$, which yields $V_0 = Q^{-1} U_0 \in \mathbb{C}^k \times \{0\}$. The uniqueness follows from Proposition 10.2.7 below. □

10.2 The Infinite-Dimensional Case

Let now $M_0, M_1 \in L(H)$. Again, it is our aim to determine the space of consistent initial values for the problem

$$\begin{cases} \left(\partial_{t,\nu} M_0 + M_1\right) U = 0 \quad \text{on } (0,\infty), \\ U(0+) = U_0. \end{cases} \tag{10.4}$$

Here, consistent initial values are defined as in the finite-dimensional setting:

Definition We call an initial value $U_0 \in H$ *consistent* for (10.4) if there exist $\nu > 0$ and $U \in C(\mathbb{R}_{\geqslant 0}; H) \cap L_{2,\nu}(\mathbb{R}_{\geqslant 0}; H)$ such that (10.4) holds. We denote the set of all consistent initial values for (10.4) by

$$\mathrm{IV}(M_0, M_1) := \{U_0 \in H\,;\ U_0 \text{ consistent}\}.$$

Before we try to determine $\mathrm{IV}(M_0, M_1)$ we prove a regularity result for solutions of (10.4).

Proposition 10.2.1 *Let $\nu > 0$, $U_0 \in H$ and $U \in C(\mathbb{R}_{\geqslant 0}; H) \cap L_{2,\nu}(\mathbb{R}_{\geqslant 0}; H)$ be a solution of (10.4). Then $M_0(U - \mathbb{1}_{[0,\infty)} U_0) \in H_\nu^1(\mathbb{R}; H)$ and*

$$\partial_{t,\nu} M_0 \left(U - \mathbb{1}_{[0,\infty)} U_0\right) + M_1 U = 0.$$

Proof We extend U to $\mathbb{R}$ by 0. First, observe that $M_0(U - \mathbb{1}_{[0,\infty)} U_0) \colon \mathbb{R} \to H$ is continuous, since U is continuous and $U(0+) = U_0$. By Lemma 9.4.2 (with $A = 0$), we obtain

$$U - \mathbb{1}_{[0,\infty)} U_0 \in \operatorname{dom}\left(\overline{\partial_{t,\nu} M_0 + M_1}\right) \text{ and } \left(\overline{\partial_{t,\nu} M_0 + M_1}\right)(U - \mathbb{1}_{[0,\infty)} U_0) = -M_1 U_0 \mathbb{1}_{[0,\infty)}.$$

Since $\partial_{t,\nu}$ is closed and M_0 is bounded, $\partial_{t,\nu} M_0$ is closed as well. Since M_1 is bounded, therefore also $\partial_{t,\nu} M_0 + M_1$ is closed. Thus, $U - \mathbb{1}_{[0,\infty)} U_0 \in \operatorname{dom}(\partial_{t,\nu} M_0 + M_1) = \operatorname{dom}(\partial_{t,\nu} M_0)$ and therefore $M_0(U - \mathbb{1}_{[0,\infty)} U_0) \in \operatorname{dom}(\partial_{t,\nu})$, and

$$\partial_{t,\nu} M_0 (U - \mathbb{1}_{[0,\infty)} U_0) + M_1 U = 0.$$

□

We now come back to the space $\mathrm{IV}(M_0, M_1)$. Since we are now dealing with an infinite-dimensional setting, we cannot use normal forms to determine $\mathrm{IV}(M_0, M_1)$ without dramatically restricting the class of operators. Thus, we follow a different approach using so-called Wong sequences.

Definition We set

$$\mathrm{IV}_0 := H$$

and for $k \in \mathbb{N}_0$ we set

$$\mathrm{IV}_{k+1} := M_1^{-1}[M_0[\mathrm{IV}_k]].$$

The sequence $(\mathrm{IV}_k)_{k\in\mathbb{N}_0}$ is called the *Wong sequence* associated with (M_0, M_1).

Remark 10.2.2 By induction, we infer $\mathrm{IV}_{k+1} \subseteq \mathrm{IV}_k$ for each $k \in \mathbb{N}_0$.

As in the matrix case, we denote by

$$\rho(M_0, M_1) := \left\{z \in \mathbb{C};\ (zM_0 + M_1)^{-1} \in L(H)\right\}$$

the *resolvent set of* (M_0, M_1).

Lemma 10.2.3 *Let $k \in \mathbb{N}_0$. Then:*

(a) $M_1(zM_0 + M_1)^{-1}M_0 = M_0(zM_0 + M_1)^{-1}M_1$ *for each* $z \in \rho(M_0, M_1)$.
(b) $(zM_0 + M_1)^{-1}M_0[\mathrm{IV}_k] \subseteq \mathrm{IV}_{k+1}$ *for each* $z \in \rho(M_0, M_1)$.
(c) *If* $x \in \mathrm{IV}_k$ *we find* $x_1, \ldots, x_{k+1} \in H$ *such that for each* $z \in \rho(M_0, M_1) \setminus \{0\}$

$$(zM_0 + M_1)^{-1}M_0x = \frac{1}{z}x + \sum_{\ell=1}^{k} \frac{1}{z^{\ell+1}}x_\ell + \frac{1}{z^{k+1}}(zM_0 + M_1)^{-1}x_{k+1}.$$

(d) *If* $\rho(M_0, M_1) \neq \varnothing$ *then* $M_1^{-1}[M_0[\overline{\mathrm{IV}_k}]] \in \overline{\mathrm{IV}_{k+1}}$.

Proof The proofs of the statements (a) to (c) are left as Exercise 10.6. We now prove (d). If $k = 0$ there is nothing to show. So assume that the statement holds for some $k \in \mathbb{N}_0$ and let $x \in M_1^{-1}\left[M_0\left[\overline{\mathrm{IV}_{k+1}}\right]\right]$. Since $\overline{\mathrm{IV}_{k+1}} \subseteq \overline{\mathrm{IV}_k}$, we infer $x \in M_1^{-1}\left[M_0\left[\overline{\mathrm{IV}_k}\right]\right] \subseteq \overline{\mathrm{IV}_{k+1}}$ by induction hypothesis. Hence, we find a sequence $(w_n)_{n\in\mathbb{N}}$ in IV_{k+1} with $w_n \to x$. Let now $z \in \rho(M_0, M_1)$. Then, by (b), we have $(zM_0 + M_1)^{-1}M_0w_n \in \mathrm{IV}_{k+2}$ for each $n \in \mathbb{N}$ and hence, $(zM_0 + M_1)^{-1}M_0x \in \overline{\mathrm{IV}_{k+2}}$. Moreover, since $M_1x \in M_0\left[\overline{\mathrm{IV}_{k+1}}\right]$, we find a sequence $(y_n)_{n\in\mathbb{N}}$ in IV_{k+1} with $M_0y_n \to M_1x$. Setting now

$$x_n := (zM_0 + M_1)^{-1}zM_0x + (zM_0 + M_1)^{-1}M_0y_n \in \overline{\mathrm{IV}_{k+2}}$$

(where, again, we have used (b)) for $n \in \mathbb{N}$, we derive

$$\begin{aligned} x_n &= (zM_0 + M_1)^{-1} zM_0 x + (zM_0 + M_1)^{-1} M_0 y_n \\ &= x - (zM_0 + M_1)^{-1} (M_1 x - M_0 y_n) \to x \end{aligned}$$

as $n \to \infty$ and thus, $x \in \overline{\mathrm{IV}_{k+2}}$. □

The importance of the Wong sequence becomes apparent if we consider solutions of (10.4).

Lemma 10.2.4 *Assume that* $\rho(M_0, M_1) \neq \varnothing$. *Let* $\nu > 0$ *and* $U \in L_{2,\nu}(\mathbb{R}_{\geqslant 0}; H) \cap C(\mathbb{R}_{\geqslant 0}; H)$ *be a solution of (10.4). Then* $U(t) \in \bigcap_{k\in\mathbb{N}_0} \overline{\mathrm{IV}_k}$ *for each* $t \geqslant 0$.

Proof We prove the claim, $U(t) \in \overline{\mathrm{IV}_k}$ for all $t \geqslant 0$ and $k \in \mathbb{N}_0$, by induction. For $k = 0$ there is nothing to show. Assume now that $U(t) \in \overline{\mathrm{IV}_k}$ for each $t \geqslant 0$ and some $k \in \mathbb{N}_0$. By Proposition 10.2.1 we know that

$$\partial_{t,\nu} M_0 (U - \mathbb{1}_{[0,\infty)} U_0) + M_1 U = 0$$

and thus, in particular,

$$M_0 U(t) - M_0 U_0 + \int_0^t M_1 U(s)\, \mathrm{d}s = 0 \quad (t \geqslant 0).$$

Let now $t \geqslant 0$ and $h > 0$. Then we infer

$$M_0 U(t+h) - M_0 U(t) + M_1 \int_t^{t+h} U(s)\, \mathrm{d}s = 0$$

and hence,

$$\int_t^{t+h} U(s)\, \mathrm{d}s \in M_1^{-1}\big[M_0[\overline{\mathrm{IV}_k}]\big] \subseteq \overline{\mathrm{IV}_{k+1}}$$

by Lemma 10.2.3(d). Since U is continuous, the fundamental theorem of calculus implies $U(t) \in \overline{\mathrm{IV}_{k+1}}$, which yields the assertion. □

In particular, the space of consistent initial values has to be a subspace of $\bigcap_{k\in\mathbb{N}_0} \overline{\mathrm{IV}_k}$. We now impose an additional constraint on the operator pair (M_0, M_1), which is equivalent to being regular in the finite-dimensional setting (cf. Proposition 10.1.6).

Definition We call the operator pair (M_0, M_1) *regular* if there exists $\nu_0 \geqslant 0$ such that

(a) $\mathbb{C}_{\mathrm{Re}>\nu_0} \subseteq \rho(M_0, M_1)$, and
(b) there exist $C \geqslant 0$ and $\ell \in \mathbb{N}$ such that for all $z \in \mathbb{C}_{\mathrm{Re}>\nu_0}$ we have $\left\|(zM_0 + M_1)^{-1}\right\| \leqslant C\,|z|^{\ell-1}$.

Moreover, we call the smallest $\ell \in \mathbb{N}$ satisfying (b) the *index of* (M_0, M_1), which is denoted by $\operatorname{ind}(M_0, M_1)$.

Remark 10.2.5 Note that for matrices M_0 and M_1 the index equals the degree of nilpotency of N in the quasi-Weierstraß normal form by Proposition 10.1.6.

From now on, we will require that (M_0, M_1) is regular. First, we prove an important result on the Wong sequence in this case.

Proposition 10.2.6 *Let (M_0, M_1) be regular, $k \in \mathbb{N}_0$, and $k \geqslant \operatorname{ind}(M_0, M_1)$. Then*

$$\overline{\mathrm{IV}_k} = \overline{\mathrm{IV}_{\operatorname{ind}(M_0,M_1)}}.$$

Proof We show that $\overline{\mathrm{IV}_k} = \overline{\mathrm{IV}_{k+1}}$ for each $k \geqslant \operatorname{ind}(M_0, M_1)$. Since the inclusion "$\supseteq$" holds trivially, it suffices to show $\mathrm{IV}_k \subseteq \overline{\mathrm{IV}_{k+1}}$. For doing so, let $k \geqslant \operatorname{ind}(M_0, M_1)$ and $x \in \mathrm{IV}_k$. By Lemma 10.2.3(c) we find $x_1, \ldots, x_{k+1} \in H$ such that

$$(zM_0 + M_1)^{-1} M_0 x = \frac{1}{z} x + \sum_{\ell=1}^{k} \frac{1}{z^{\ell+1}} x_\ell + \frac{1}{z^{k+1}} (zM_0 + M_1)^{-1} x_{k+1}$$

for each $z \in \mathbb{C}_{\mathrm{Re}>\nu_0}$. Since $k \geqslant \operatorname{ind}(M_0, M_1)$, we derive

$$z(zM_0 + M_1)^{-1} M_0 x \to x \quad (\mathrm{Re}\, z \to \infty),$$

and since the elements on the left-hand side belong to IV_{k+1}, by Lemma 10.2.3(b), the assertion immediately follows. □

We now prove that in case of a regular operator pair (M_0, M_1) the solution of (10.4) for a consistent initial value U_0 is uniquely determined.

Proposition 10.2.7 *Let (M_0, M_1) be regular, $U_0 \in \mathrm{IV}(M_0, M_1)$, and $\nu > 0$ such that a solution $U \in C(\mathbb{R}_{\geqslant 0}; H) \cap L_{2,\nu}(\mathbb{R}_{\geqslant 0}; H)$ of (10.4) exists. Then this solution is unique. In particular*

$$(\mathcal{L}_\rho U)(t) = \frac{1}{\sqrt{2\pi}} \big((\mathrm{i}t + \rho)M_0 + M_1\big)^{-1} M_0 U_0 \quad (a.e.\ t \in \mathbb{R})$$

for each $\rho > \max\{\nu, \nu_0\}$.

Proof By Proposition 10.2.1 we have $M_0(U - \mathbb{1}_{[0,\infty)} U_0) \in H^1_\nu(\mathbb{R}; H)$ and

$$\partial_{t,\nu} M_0(U - \mathbb{1}_{[0,\infty)} U_0) + M_1 U = 0.$$

Applying the Fourier–Laplace transformation, $\mathcal{L}_\rho$, for $\rho > \max\{\nu, \nu_0\}$ we deduce

$$(\mathrm{i}t + \rho)M_0\Big(\mathcal{L}_\rho U(t) - \frac{1}{\sqrt{2\pi}} \frac{1}{\mathrm{i}t + \rho} U_0\Big) + M_1 \mathcal{L}_\rho U(t) = 0 \quad (\text{a.e. } t \in \mathbb{R})$$

which in turn yields

$$\mathcal{L}_\rho U(t) = \frac{1}{\sqrt{2\pi}}\big((\mathrm{i}t+\rho)M_0+M_1\big)^{-1}M_0U_0 \quad (\text{a.e. } t\in\mathbb{R})$$

and, in particular, proves the uniqueness of the solution. □

Remark 10.2.8 Let U be a solution of (10.4) for a consistent initial value U_0. Then the formula in Proposition 10.2.7 shows that $U\in\bigcap_{\rho>\nu_0}L_{2,\rho}(\mathbb{R};H)$ and hence, we also have $M_0(U-\mathbb{1}_{[0,\infty)}U_0)\in\bigcap_{\rho>\nu_0}H^1_\rho(\mathbb{R};H)$. If $\nu_0>0$ then we even obtain $U\in L_{2,\nu_0}(\mathbb{R};H)$ since $\sup_{\rho>\nu_0}\|U\|_{L_{2,\rho}(\mathbb{R};H)}=\sup_{\rho>\nu_0}\left\|\mathcal{L}_\rho U\right\|_{L_2(\mathbb{R};H)}<\infty$ (cp. Lemma 8.1.1), and therefore also $M_0(U-\mathbb{1}_{[0,\infty)}U_0)\in H^1_{\nu_0}(\mathbb{R};H)$.

One interesting consequence of the latter proposition is the following.

Corollary 10.2.9 *Let (M_0,M_1) be regular. Then the operator $M_0\colon \mathrm{IV}(M_0,M_1)\to H$ is injective.*

Proof Let $U_0\in\mathrm{IV}(M_0,M_1)$ with $M_0U_0=0$. By Proposition 10.2.7, the solution U of (10.4) with $U(0+)=U_0$ satisfies

$$\mathcal{L}_\rho U(t)=\frac{1}{\sqrt{2\pi}}\big((\mathrm{i}t+\rho)M_0+M_1\big)^{-1}M_0U_0=0$$

and hence, $U=0$, which in turn implies $U_0=U(0+)=0$. □

We now want to determine the space $\mathrm{IV}(M_0,M_1)$ in terms of the Wong sequence.

Proposition 10.2.10 *Let (M_0,M_1) be regular. Then*

$$\mathrm{IV}_{\mathrm{ind}(M_0,M_1)}\subseteq\mathrm{IV}(M_0,M_1)\subseteq\overline{\mathrm{IV}_{\mathrm{ind}(M_0,M_1)}}.$$

Proof The second inclusion follows from Lemma 10.2.4 and Proposition 10.2.6. Let now $U_0\in\mathrm{IV}_{\mathrm{ind}(M_0,M_1)}$ and set

$$V(z):=\frac{1}{\sqrt{2\pi}}(zM_0+M_1)^{-1}M_0U_0 \quad (z\in\mathbb{C}_{\mathrm{Re}>\nu_0}).$$

Let $k:=\mathrm{ind}(M_0,M_1)$. By Lemma 10.2.3(c) we find $x_1,\ldots,x_{k+1}\in H$ such that

$$V(z)=\frac{1}{\sqrt{2\pi}}\left(\frac{1}{z}U_0+\sum_{\ell=1}^{k}\frac{1}{z^{\ell+1}}x_\ell+\frac{1}{z^{k+1}}(zM_0+M_1)^{-1}x_{k+1}\right) \quad (z\in\mathbb{C}_{\mathrm{Re}>\nu_0}).$$

In particular, we read off that $V\in\mathcal{H}_2(\mathbb{C}_{\mathrm{Re}>\nu};H)$ for all $\nu>\nu_0$. Now, let $\nu>\nu_0$. By the Theorem of Paley–Wiener (more precisely by Corollary 8.1.3) there exists $U\in L_{2,\nu}(\mathbb{R}_{\geqslant 0};H)$ such that

$$\big(\mathcal{L}_\rho U\big)(t)=V(\mathrm{i}t+\rho) \quad (\text{a.e. } t\in\mathbb{R},\ \rho>\nu).$$

Moreover,

$$zV(z) - \frac{1}{\sqrt{2\pi}}U_0 = \frac{1}{\sqrt{2\pi}}\left(\sum_{\ell=1}^{k}\frac{1}{z^\ell}x_\ell + \frac{1}{z^k}(zM_0+M_1)^{-1}x_{k+1}\right) \quad (z \in \mathbb{C}_{\mathrm{Re}>\nu})$$

and hence $\left(z \mapsto zV(z) - \frac{1}{\sqrt{2\pi}}U_0\right) \in \mathcal{H}_2(\mathbb{C}_{\mathrm{Re}>\nu}; H)$ as well. Since

$$\begin{aligned}\left(\mathcal{L}_\rho \partial_{t,\rho}(U - \mathbb{1}_{[0,\infty)}U_0)\right)(t) &= (\mathrm{i}t+\rho)\left(\mathcal{L}_\rho U\right)(t) - \frac{1}{\sqrt{2\pi}}U_0 \\ &= (\mathrm{i}t+\rho)V(\mathrm{i}t+\rho) - \frac{1}{\sqrt{2\pi}}U_0 \quad (\text{a.e. } t \in \mathbb{R}, \rho > \nu),\end{aligned}$$

we infer $U - \mathbb{1}_{[0,\infty)}U_0 \in H_\nu^1(\mathbb{R}; H)$ and, thus, $U - \mathbb{1}_{[0,\infty)}U_0$ is continuous by Theorem 4.1.2. Hence, $U \in C(\mathbb{R}_{\geqslant 0}; H)$ and since $\operatorname{spt} U \subseteq \mathbb{R}_{\geqslant 0}$ we derive $U(0+) = U_0$. Finally, by the definition of V,

$$M_0\left(zV(z) - \frac{1}{\sqrt{2\pi}}U_0\right) = -\frac{1}{\sqrt{2\pi}}M_1(zM_0+M_1)^{-1}M_0U_0 = -M_1V(z)$$

for all $z \in \mathbb{C}_{\mathrm{Re}>\nu}$. Hence,

$$\partial_{t,\nu}M_0(U - \mathbb{1}_{[0,\infty)}U_0) + M_1U = 0,$$

from which we see that U solves (10.4). □

Finally, we treat the case when $\mathrm{IV}(M_0, M_1)$ is closed.

Theorem 10.2.11 *Let (M_0, M_1) be regular and $\mathrm{IV}(M_0, M_1)$ closed. Then the operator $S\colon \mathrm{IV}(M_0, M_1) \to C(\mathbb{R}_{\geqslant 0}; H)$, which assigns to each initial state, $U_0 \in \mathrm{IV}(M_0, M_1)$, its corresponding solution, $U \in C(\mathbb{R}_{\geqslant 0}; H)$, of (10.4) is bounded in the sense that*

$$S_n\colon \mathrm{IV}(M_0, M_1) \to C([0,n]; H), \quad U_0 \mapsto SU_0|_{[0,n]}$$

is bounded for each $n \in \mathbb{N}$.

Proof By Proposition 10.2.10 we infer that $\mathrm{IV}(M_0, M_1) = \overline{\mathrm{IV}_k}$ with $k := \operatorname{ind}(M_0, M_1)$. Let $\nu > \nu_0 \geqslant 0$. By Proposition 10.2.7 and Corollary 8.1.3, there exists $C \geqslant 0$ such that

$$\begin{aligned}\sqrt{2\pi}\left\|\partial_{t,\nu}^{-k}SU_0\right\|_{L_{2,\nu}(\mathbb{R}_{\geq 0}; H)} &= \left\|\left(z \mapsto z^{-k}(zM_0+M_1)^{-1}M_0U_0\right)\right\|_{\mathcal{H}_2(\mathbb{C}_{\mathrm{Re}>\nu}; H)} \\ &\leqslant C\sqrt{\frac{\pi}{\nu}}\,\|M_0U_0\|_H\end{aligned}$$

for each $U_0 \in \mathrm{IV}(M_0, M_1)$, where we have used the regularity of (M_0, M_1) and

$$\left\|(z \mapsto z^{-1} M_0 U_0)\right\|_{\mathcal{H}_2(\mathbb{C}_{\mathrm{Re}>\nu}; H)} = \sqrt{\frac{\pi}{\nu}}\, \|M_0 U_0\|_H .$$

In particular, $S\colon \mathrm{IV}(M_0, M_1) \to H^{-1}(\partial_{t,\nu}^k)$ is bounded. Since $L_{2,\nu_0}(\mathbb{R}_{\geqslant 0}; H) \hookrightarrow H^{-1}(\partial_{t,\nu}^k)$ continuously, we infer that $S\colon \mathrm{IV}(M_0, M_1) \to L_{2,\nu_0}(\mathbb{R}_{\geqslant 0}; H)$ is bounded by the closed graph theorem. Hence, also

$$S_n\colon \mathrm{IV}(M_0, M_1) \to L_2([0, n]; H), \quad U_0 \mapsto SU_0|_{[0,n]}$$

is bounded for each $n \in \mathbb{N}$ and since $C([0, n]; H) \hookrightarrow L_2([0, n]; H)$ continuously, we infer that S_n is bounded with values in $C([0, n]; H)$ again by the closed graph theorem. □

Remark 10.2.12 The variant of the closed graph theorem used in the proof above is the following: Let X, Y be Banach spaces and Z a Hausdorff topological vector space (e.g. a Banach space) such that $Y \hookrightarrow Z$ continuously. Let $T\colon X \to Z$ be linear and continuous with $T[X] \subseteq Y$. Then $T \in L(X, Y)$. Indeed, by the closed graph theorem it suffices to show that $T : X \to Y$ is closed. For doing so, let $(x_n)_n$ be a sequence in X with $x_n \to x$ and $Tx_n \to y$ for some $x \in X$, $y \in Y$. Then $Tx_n \to Tx$ in Z by the continuity of T and $Tx_n \to y$ in Z by the continuous embedding. Hence, $y = Tx$ and thus, T is closed.

10.3 Comments

The theory of differential algebraic equations in finite dimensions is a very active field. The main motivation for studying these equations comes from the modelling of electrical circuits and from control theory (see e.g. [28] and Exercise 10.5). The main reference for the statements presented in the first part of this chapter is the book by Kunkel and Mehrmann [57]. Of course, also in the finite-dimensional case Wong sequences can be used to determine the consistent initial values, see Exercise 10.1. For instance, in [13] the connection between Wong sequences and the quasi-Weierstraß normal form for matrix pairs is studied. Of course, the theory is not restricted to linear and homogeneous problems. Indeed, in the non-homogeneous case it turns out that the set of consistent initial values also depends on the given right-hand side.

The theory of differential algebraic equations in infinite dimensions is less well studied than the finite-dimensional case. We refer to [114], where the theory of C_0-semigroups is used to deal with such equations. Moreover, we refer to [97, 98], where sequences of projectors are used to decouple the system. Moreover, there exist several references in the Russian literature, where the equations are called Sobolev type equations (see e.g. [111]). The results on infinite-dimensional problems presented here are based on [121, 124, 125]. In [124] the focus was on systems with index 0 with an emphasis on exponential stability and dichotomy.

We also add the following remark concerning the result in Theorem 10.2.11. By Corollary 10.2.9 we know that $M_0\colon \operatorname{IV}(M_0, M_1) \to H$ is injective. If $\operatorname{IV}(M_0, M_1)$ is closed, it follows that the operator $C\colon \operatorname{dom}(C) \subseteq \operatorname{IV}(M_0, M_1) \to \operatorname{IV}(M_0, M_1)$ given by

$$\operatorname{dom}(C) := \{U_0 \in \operatorname{IV}(M_0, M_1)\, ;\ M_1 U_0 \in M_0 [\operatorname{IV}(M_0, M_1)]\},$$

$$CU_0 := M_0^{-1} M_1 U_0 \quad (U_0 \in \operatorname{dom}(C))$$

is well-defined and closed. Using this operator, C, Theorem 10.2.11 states that if $\operatorname{IV}(M_0, M_1)$ is closed then $-C$ generates a C_0-semigroup on $\operatorname{IV}(M_0, M_1)$. The precise statement can be found in [121, Theorem 5.7]. Moreover, C is bounded if $\operatorname{IV}_{\operatorname{ind}(M_0,M_1)}$ is closed (cf. Exercise 10.7).

Exercises

Exercise 10.1 Let $M_0, M_1 \in \mathbb{C}^{n\times n}$ such that (M_0, M_1) is regular and define the Wong sequence $(\operatorname{IV}_j)_{j\in\mathbb{N}_0}$ associated with (M_0, M_1). Moreover, let $P, Q \in \mathbb{C}^{n\times n}$, $C \in \mathbb{C}^{k\times k}$, and $N \in \mathbb{C}^{(n-k)\times(n-k)}$ be as in the quasi-Weierstraß normal form for (M_0, M_1) with N nilpotent (cf. Proposition 10.1.5). We decompose a vector $x \in \mathbb{C}^n$ into $\check{x} \in \mathbb{C}^k$ and $\widehat{x} \in \mathbb{C}^{n-k}$ such that $x = (\check{x}, \widehat{x})$. Prove that

$$x \in \operatorname{IV}_j \Leftrightarrow \widehat{Q^{-1}x} \in \operatorname{ran} N^j \quad (j \in \mathbb{N}_0).$$

Moreover, show that for each $z \in \rho(M_0, M_1)$ we have

$$\operatorname{IV}_j = \operatorname{ran}\left((zM_0 + M_1)^{-1} M_0\right)^j \quad (j \in \mathbb{N}_0).$$

Exercise 10.2 Let $E \in \mathbb{C}^{n\times n}$. We set $k := \operatorname{ind}(E, 1)$, where 1 denotes the identity matrix in $\mathbb{C}^{n\times n}$. A matrix $X \in \mathbb{C}^{n\times n}$ is called a *Drazin inverse of E* if the following properties hold:

- $EX = XE$,
- $XEX = X$,
- $XE^{k+1} = E^k$.

Prove that each matrix $E \in \mathbb{C}^{n\times n}$ has a unique Drazin inverse.
Hint: For the existence consider the quasi-Weierstraß form for $(E, 1)$.

Exercise 10.3 Let $M_0, M_1 \in \mathbb{C}^{n\times n}$ with (M_0, M_1) regular and $M_0 M_1 = M_1 M_0$. Denote by M_0^{D} the Drazin inverse of M_0 (see Exercise 10.2). Prove:

(a) $M_0^{\mathrm{D}} M_1 = M_1 M_0^{\mathrm{D}}$,
(b) $\operatorname{ran} M_0^{\mathrm{D}} M_0 = \mathrm{IV}(M_0, M_1)$,
(c) For all $U_0 \in \mathrm{IV}(M_0, M_1)$ the solution U of (10.2) is given by

$$U(t) = \mathrm{e}^{-tM_0^{\mathrm{D}} M_1} U_0 \quad (t \geqslant 0).$$

Exercise 10.4 Let $M_0, M_1 \in \mathbb{C}^{n\times n}$ with (M_0, M_1) regular. Prove that there exist two matrices $E, A \in \mathbb{C}^{n\times n}$ with (E, A) regular and $EA = AE$ such that

- $\mathrm{IV}(E, A) = \mathrm{IV}(M_0, M_1)$,
- U solves the initial value problem (10.2) for the matrices M_0, M_1 if and only if U solves the initial value problem (10.2) for the matrices E, A with the same initial value $U_0 \in \mathrm{IV}(M_0, M_1)$.

Exercise 10.5 We consider the following electrical circuit (see Fig. 10.1) with a resistor with resistance $R > 0$, an inductor with inductance $L > 0$ and a capacitor with capacitance $C > 0$. We denote the respective voltage drops by v_R, v_L and v_C. Moreover, the current is denoted by i. The constitutive relations for resistor, inductor and capacitor are given by

$$\begin{aligned} Ri &= v_R, \\ Li' &= v_L, \\ Cv_C' &= i, \end{aligned}$$

respectively. Moreover, by Kirchhoff's second law we have

$$v_R + v_C + v_L = 0.$$

Write these equations as a differential algebraic equation and compute the index and the space of consistent initial values. Moreover, compute the solution for each consistent initial value for $R = 2$ and $C = L = 1$.

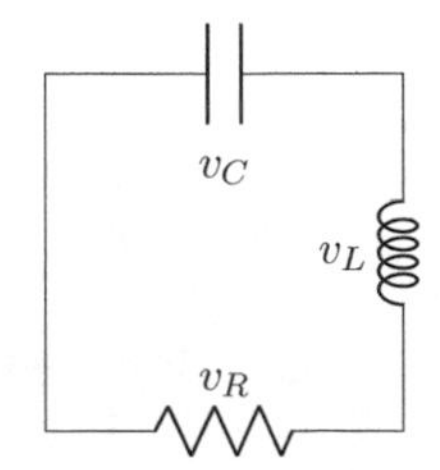

Fig. 10.1 Electrical circuit

Exercise 10.6 Prove the assertions (a) to (c) in Lemma 10.2.3.

Exercise 10.7 Let $M_0, M_1 \in L(H)$.

(a) Assume that $\rho(M_0, M_1) \neq \varnothing$. Prove that for each $k \in \mathbb{N}$ the space IV_k is closed if and only if $M_0\,[\mathrm{IV}_{k-1}]$ is closed.
(b) Assume that (M_0, M_1) is regular with $\operatorname{ind}(M_0, M_1) \geqslant 1$. Prove that if $\mathrm{IV}_{\operatorname{ind}(M_0,M_1)}$ is closed then the operator

$$M_0|_{\mathrm{IV}_{\operatorname{ind}(M_0,M_1)}} : \mathrm{IV}_{\operatorname{ind}(M_0,M_1)} \to M_0\left[\mathrm{IV}_{\operatorname{ind}(M_0,M_1)-1}\right]$$

is an isomorphism.

References

13. T. Berger, A. Ilchmann, S. Trenn, The quasi-Weierstrass form for regular matrix pencils. Linear Algebra Appl. **436**(10), 4052–4069 (2012)
28. L. Dai, *Singular Control Systems*, vol. 118 (Springer, Berlin, 1989)
57. P. Kunkel, V. Mehrmann, *Differential-Algebraic Equations*. EMS Textbooks in Mathematics. Analysis and Numerical Solution (European Mathematical Society (EMS), Zürich, 2006)
97. T. Reis, Consistent initialization and perturbation analysis for abstract differential-algebraic equations. Math. Control Signals Syst. **19**(3), 255–281 (2007)
98. T. Reis, C. Tischendorf, Frequency domain methods and decoupling of linear infinite dimensional differential algebraic systems. J. Evol. Equ. **5**(3), 357–385 (2005)
111. G.A. Sviridyuk, V.E. Fedorov, *Linear Sobolev Type Equations and Degenerate Semigroups of Operators*. Inverse and Ill-posed Problems Series (VSP, Utrecht, 2003)
114. B. Thaller, S. Thaller, Factorization of degenerate Cauchy problems: the linear case. J. Oper. Theory **36**(1), 121–146 (1996)
121. S. Trostorff, Semigroups associated with differential-algebraic equations. In: *Semi-groups of Operators – Theory and Applications*. Selected papers based on the presentations at the conference, SOTA 2018, Kazimierz Dolny, Poland, September 30–October 5, 2018. In honour of Jan Kisyński's 85th birthday (Springer, Cham, 2020), pp. 79–94
124. S. Trostorff, M. Waurick, On differential-algebraic equations in infinite dimensions. J. Differ. Equ. **266**(1), 526–561 (2019)
125. S. Trostorff, M. Waurick, On higher index differential-algebraic-equations in infinite dimensions, in *The Diversity and Beauty of Applied Operator Theory*, vol. 268, ed. by P.S. Albrecht Böttcher, D. Potts, D. Wenzel. Operator Theory: Advances and Applications (Birkhäuser, Basel, 2018), pp. 477–486

Chapter 11
Exponential Stability of Evolutionary Equations

In this chapter we study the exponential stability of evolutionary equations. Roughly speaking, exponential stability of a well-posed evolutionary equation

$$\left(\partial_{t,\nu} M(\partial_{t,\nu}) + A\right) U = F$$

means that exponentially decaying right-hand sides F lead to exponentially decaying solutions U. The main problem in defining the notion of exponential decay for a solution of an evolutionary equation is the lack of continuity with respect to time, so a pointwise definition would not make sense in this framework. Instead, we will use our exponentially weighted spaces $L_{2,\nu}(\mathbb{R}; H)$, but this time for negative ν, and define the exponential stability by the invariance of these spaces under the solution operator associated with the evolutionary equation under consideration.

11.1 The Notion of Exponential Stability

Throughout this section, let H be a Hilbert space, $M\colon \operatorname{dom}(M) \subseteq \mathbb{C} \to L(H)$ a material law and $A\colon \operatorname{dom}(A) \subseteq H \to H$ a skew-selfadjoint operator. Moreover, we assume that there exist $\nu_0 > \mathrm{s_b}(M)$ and $c > 0$ such that

$$\operatorname{Re} zM(z) \geqslant c \quad (z \in \mathbb{C}_{\operatorname{Re} \geqslant \nu_0}).$$

By Picard's theorem (Theorem 6.2.1) we know that for $\nu \geqslant \nu_0$ the operator

$$S_\nu := \overline{\left(\partial_{t,\nu} M(\partial_{t,\nu}) + A\right)}^{-1} \in L(L_{2,\nu}(\mathbb{R}; H))$$

is causal and independent of the particular choice of ν. We now define the notion of exponential stability.

C. Seifert et al., *Evolutionary Equations*, Operator Theory: Advances and Applications 287, https://doi.org/10.1007/978-3-030-89397-2_11

Definition We call the solution operators $(S_\nu)_{\nu\geqslant\nu_0}$ *exponentially stable with decay rate* $\rho_0 > 0$ if for all $\rho \in [0, \rho_0)$ and $\nu \geqslant \nu_0$ we have

$$S_\nu F \in L_{2,-\rho}(\mathbb{R}; H) \quad (F \in L_{2,\nu}(\mathbb{R}; H) \cap L_{2,-\rho}(\mathbb{R}; H)).$$

Remark 11.1.1 We emphasise that the definition of exponential stability does not mean that the evolutionary equation is just solvable for some negative weights. Indeed, if we consider $H = \mathbb{C}$, $A = 0$ and $M(z) = 1$ for $z \in \mathbb{C}$ we obtain that the corresponding evolutionary equation

$$\partial_{t,\nu} U = F \tag{11.1}$$

is well-posed for each $\nu \neq 0$. However, we also place a demand for causality on our solution operator. Thus, we only have to consider parameters $\nu > 0$. We obtain the solution U by

$$U(t) = \int_{-\infty}^{t} F(s)\,\mathrm{d}s.$$

As it turns out, the problem (11.1) is not exponentially stable. Indeed, for $F := \mathbb{1}_{[0,1]} \in \bigcap_{\nu\in\mathbb{R}} L_{2,\nu}(\mathbb{R})$ the solution U is given by

$$U(t) = \begin{cases} 0 & \text{if } t < 0, \\ t & \text{if } 0 \leqslant t \leqslant 1, \\ 1 & \text{if } t > 1, \end{cases}$$

which does not belong to the space $L_{2,-\rho}(\mathbb{R})$ for any $\rho > 0$.

We first show that the aforementioned notion of exponential stability also yields a pointwise exponential decay of solutions if we assume more regularity for our source term F.

Proposition 11.1.2 *Let* $(S_\nu)_{\nu\geqslant\nu_0}$ *be exponentially stable with decay rate* $\rho_0 > 0$, $\nu \geqslant \nu_0$, $\rho \in [0, \rho_0)$ *and* $F \in \operatorname{dom}(\partial_{t,\nu}) \cap \operatorname{dom}(\partial_{t,-\rho})$. *Then* $U := S_\nu F$ *is continuous and satisfies*

$$U(t)\mathrm{e}^{\rho t} \to 0 \quad (t \to \infty).$$

Proof We first note that $\partial_{t,\nu} F = \partial_{t,-\rho} F$ by Exercise 11.1. Moreover, since S_ν is a material law operator (i.e., $S_\nu = S(\partial_{t,\nu})$ for some material law S; see Remark 6.3.4) we have

$$S_\nu \partial_{t,\nu} \subseteq \partial_{t,\nu} S_\nu.$$

Thus, in particular, we have

$$S_\nu \partial_{t,\nu} F = \partial_{t,\nu} S_\nu F = \partial_{t,\nu} U;$$

that is, $U \in \operatorname{dom}(\partial_{t,\nu})$. Moreover, since $\partial_{t,\nu} F = \partial_{t,-\rho} F \in L_{2,-\rho}(\mathbb{R}; H)$, we infer also $U, \partial_{t,\nu} U \in L_{2,-\rho}(\mathbb{R}; H)$ by exponential stability. By Exercise 11.1 this yields $U \in \operatorname{dom}(\partial_{t,-\rho})$ with $\partial_{t,-\rho} U = \partial_{t,\nu} U$. The assertion now follows from the Sobolev embedding theorem (Theorem 4.1.2 and Corollary 4.1.3). □

11.2 A Criterion for Exponential Stability of Parabolic-Type Equations

In this section we will prove a useful criterion for exponential stability of a certain class of evolutionary equations. The easiest example we have in mind is the heat equation with homogeneous Dirichlet boundary conditions, which can be written as an evolutionary equation of the form (cf. Theorem 6.2.4)

$$\left(\partial_{t,\nu} \begin{pmatrix} 1 & 0 \\ 0 & 0 \end{pmatrix} + \begin{pmatrix} 0 & 0 \\ 0 & a^{-1} \end{pmatrix} + \begin{pmatrix} 0 & \operatorname{div} \\ \operatorname{grad}_0 & 0 \end{pmatrix} \right) \begin{pmatrix} \theta \\ q \end{pmatrix} = \begin{pmatrix} Q \\ 0 \end{pmatrix}$$

in $L_{2,\nu}(\mathbb{R}; H)$, where $H = L_2(\Omega) \oplus L_2(\Omega)^d$ with $\Omega \subseteq \mathbb{R}^d$ open, and $a \in L(L_2(\Omega)^d)$ with

$$\operatorname{Re} a \geqslant c$$

for some $c > 0$ which models the heat conductivity, and $\nu > 0$.

Theorem 11.2.1 *Let H_0, H_1 be Hilbert spaces and $C\colon \operatorname{dom}(C) \subseteq H_0 \to H_1$ a densely defined closed linear operator which is boundedly invertible. Moreover, let $M_0 \in L(H_0)$ be selfadjoint with*

$$M_0 \geqslant c_0$$

for some $c_0 > 0$ and $M_1\colon \operatorname{dom}(M_1) \subseteq \mathbb{C} \to L(H_1)$ be a material law satisfying $\mathrm{s_b}(M_1) < -\rho_1$ for some $\rho_1 > 0$ and

$$\exists c_1 > 0\, \forall z \in \mathbb{C}_{\mathrm{Re} > -\rho_1} : \operatorname{Re} M_1(z) \geqslant c_1.$$

Then

$$S_\nu := \overline{\left(\partial_{t,\nu} \begin{pmatrix} M_0 & 0 \\ 0 & 0 \end{pmatrix} + \begin{pmatrix} 0 & 0 \\ 0 & M_1(\partial_{t,\nu}) \end{pmatrix} + \begin{pmatrix} 0 & -C^* \\ C & 0 \end{pmatrix} \right)}^{-1} \in L\big(L_{2,\nu}(\mathbb{R}; H_0 \oplus H_1)\big)$$

for each $\nu > 0$. *Moreover, for all* $\nu_0 > 0$ *the family* $(S_\nu)_{\nu \geqslant \nu_0}$ *is exponentially stable with decay rate* $\rho_0 := \min\left\{\rho_1, c_1/\left(\|M_1\|^2_{\infty,\mathbb{C}_{\mathrm{Re}>-\rho_1}} \|M_0\| \left\|C^{-1}\right\|^2\right)\right\}$.

In order to prove this theorem we need a preparatory result.

Lemma 11.2.2 *Assume the hypotheses of Theorem 11.2.1. Then for each* $z \in \mathbb{C}_{\mathrm{Re}>-\rho_0}$ *the operator*

$$T(z) := \begin{pmatrix} zM_0 & 0 \\ 0 & M_1(z) \end{pmatrix} + \begin{pmatrix} 0 & -C^* \\ C & 0 \end{pmatrix} : \mathrm{dom}(C) \times \mathrm{dom}(C^*) \subseteq H_0 \oplus H_1 \to H_0 \oplus H_1$$

is boundedly invertible. Moreover,

$$\sup_{z \in \mathbb{C}_{\mathrm{Re}\geqslant -\rho}} \left\| T(z)^{-1} \right\| < \infty$$

for each $\rho < \rho_0$.

Proof Let $z \in \mathbb{C}_{\mathrm{Re}\geqslant -\rho}$ for some $\rho < \rho_0$. We note that $M_1(z)$ is boundedly invertible with $\left\|M_1(z)^{-1}\right\| \leqslant 1/c_1$ (see Proposition 6.2.3(b)) and $(C^*)^{-1} = (C^{-1})^* \in L(H_0, H_1)$ (see Lemmas 2.2.2 and 2.2.9). The beginning of the proof deals with a reformulation of $T(z)$. For this, let $u, f \in H_0$, $v, g \in H_1$. Then, by definition, $(u, v) \in \mathrm{dom}(T(z)) = \mathrm{dom}(C) \times \mathrm{dom}(C^*)$ and $T(z)(u, v) = (f, g)$ if and only if $v \in \mathrm{dom}(C^*)$ and $u \in \mathrm{dom}(C)$ together with

$$\begin{aligned} zM_0u - C^*v &= f \\ Cu + M_1(z)v &= g. \end{aligned}$$

Since both C^* and $M_1(z)$ are continuously invertible, we obtain equivalently $u \in \mathrm{dom}(C)$ together with

$$\begin{aligned} z(C^*)^{-1}M_0u - v &= (C^*)^{-1}f \\ M_1(z)^{-1}Cu + v &= M_1(z)^{-1}g. \end{aligned}$$

Adding the latter two equations and retaining the first equation, we obtain the following equivalent system subject to the condition $u \in \mathrm{dom}(C)$

$$\begin{aligned} v &= z(C^*)^{-1}(zM_0u - f) \in \mathrm{dom}(C^*), \\ (z(C^*)^{-1}M_0C^{-1} + M_1(z)^{-1})Cu &= M_1(z)^{-1}g + (C^*)^{-1}f. \end{aligned}$$

We now inspect the operator $S(z) := z(C^{-1})^* M_0 C^{-1} + M_1(z)^{-1} \in L(H_1)$. By Proposition 6.2.3 for $x \in H_1$ we estimate

$$\begin{aligned}
\operatorname{Re}\langle x, S(z)x\rangle &= \operatorname{Re}\left\langle C^{-1}x, zM_0C^{-1}x\right\rangle + \operatorname{Re}\left\langle x, M_1(z)^{-1}x\right\rangle \\
&\geqslant -\rho \|M_0\| \left\|C^{-1}\right\|^2 \|x\|^2 + \frac{c_1}{\|M_1(z)\|^2}\|x\|^2 \\
&\geqslant \underbrace{\Big(\frac{c_1}{\|M_1\|^2_{\infty,\mathbb{C}_{\mathrm{Re}>-\rho_1}}} - \rho \|M_0\| \left\|C^{-1}\right\|^2\Big)}_{=:\mu} \|x\|^2 .
\end{aligned}$$

Since $\rho < \rho_0$ and by the definition of ρ_0 we infer that $\mu > 0$. Hence, $S(z)$ is boundedly invertible with

$$\left\|S(z)^{-1}\right\| \leqslant \frac{1}{\mu}.$$

We now set

$$\begin{aligned}
u &:= C^{-1}S(z)^{-1}\big((C^*)^{-1}f + M_1(z)^{-1}g\big) \in \operatorname{dom}(C), \\
v &:= (C^*)^{-1}(zM_0u - f) \in \operatorname{dom}(C^*).
\end{aligned}$$

By the first part of the proof we have that (u, v) is the unique solution of $T(z)(u, v) = (f, g)$. Moreover, we can estimate

$$\begin{aligned}
\|u\| &\leqslant \left\|C^{-1}\right\| \frac{1}{\mu}\Big(\left\|(C^*)^{-1}\right\| \|f\| + \frac{1}{c_1}\|g\|\Big), \text{ and} \\
\|v\| &\leqslant \frac{1}{c_1}(\|g\| + \|Cu\|) \leqslant \frac{1}{c_1}\Big(\|g\| + \frac{1}{\mu}\big(\left\|(C^*)^{-1}\right\| \|f\| + \frac{1}{c_1}\|g\|\big)\Big),
\end{aligned}$$

which proves that $T(z)$ is boundedly invertible with

$$\sup_{z \in \mathbb{C}_{\mathrm{Re} \geqslant -\rho}} \left\|T(z)^{-1}\right\| < \infty.$$

□

Proof of Theorem 11.2.1 Let $H := H_0 \oplus H_1$. We set

$$M(z) := \begin{pmatrix} M_0 & 0 \\ 0 & z^{-1}M_1(z) \end{pmatrix} \quad (z \in \operatorname{dom}(M_1) \setminus \{0\}).$$

Let $\nu > 0$. Then

$$\forall z \in \mathbb{C}_{\mathrm{Re} \geqslant \nu} : \mathrm{Re}\, z M(z) \geqslant \min\{\nu c_0, c_1\}$$

and hence, the first assertion of the theorem follows from Theorem 6.2.1.

Next, we focus on exponential stability. For $\nu > 0$, we have that

$$S_\nu = T(\partial_{t,\nu})^{-1},$$

where T is defined in Lemma 11.2.2. Moreover, by Lemma 11.2.2, the mapping $T^{-1} \colon \mathbb{C}_{\mathrm{Re} > -\rho_0} \to L(H)$ with $T^{-1}(z) = T(z)^{-1}$ defines a material law with $\mathrm{s_b}\left(T^{-1}\right) = -\rho_0$ (the holomorphy of T is obvious and hence, T^{-1} is also holomorphic). Thus, we may apply Theorem 5.3.6 to obtain (note that $T^{-1}(\partial_{t,\nu}) = T(\partial_{t,\nu})^{-1}$)

$$S_\nu(f) = T(\partial_{t,\nu})^{-1} f = T(\partial_{t,\rho})^{-1} f \in L_{2,\rho}(\mathbb{R}; H)$$

for each $f \in L_{2,\nu}(\mathbb{R}; H) \cap L_{2,\rho}(\mathbb{R}; H)$ with $\rho > -\rho_0$, which shows exponential stability. □

11.3 Three Exponentially Stable Models for Heat Conduction

The Classical Heat Equation

We recall the classical heat equation (cf. Theorem 6.2.4) on an open subset $\Omega \subseteq \mathbb{R}^d$ consisting of two equations, the heat flux balance

$$\partial_t \theta + \operatorname{div} q = f$$

and Fourier's law

$$q = -a \operatorname{grad} \theta,$$

where f is a given source term and $a \in L(L_2(\Omega)^d)$ is an operator modelling the heat conductivity of the underlying medium. We will impose Dirichlet boundary conditions which will be incorporated in our equation by replacing the operator grad by grad_0 in Fourier's law (cf. Sect. 6.1).

In order to apply Theorem 11.2.1 we need that grad_0 is boundedly invertible in some sense. This can be shown using *Poincaré's inequality*.

Proposition 11.3.1 (Poincaré Inequality) *Let $\Omega \subseteq \mathbb{R}^d$ be open and contained in a slab; that is, there exist $e \in \mathbb{R}^d$ with $\|e\| = 1$ and $a, b \in \mathbb{R}$, $a < b$ such that*

$$\Omega \subseteq \left\{ x \in \mathbb{R}^d \, ; \, a < \langle e, x \rangle < b \right\}.$$

Then for each $u \in \operatorname{dom}(\operatorname{grad}_0)$ we have

$$\|u\|_{L_2(\Omega)} \leqslant (b-a) \left\| \operatorname{grad}_0 u \right\|_{L_2(\Omega)^d} .$$

Proof Without loss of generality, let $e = (1, 0, \ldots, 0)$. Recall that, by definition, $C_c^\infty(\Omega)$ is a core for grad_0. Thus, it suffices to prove the assertion for functions in $C_c^\infty(\Omega)$. Let $\varphi \in C_c^\infty(\Omega)$. We identify φ with its extension by 0 to the whole of $\mathbb{R}^d$. By the fundamental theorem of calculus, we may compute

$$\varphi(x) = \int_a^{x_1} \partial_1 \varphi(s, x_2, \ldots, x_d) \, \mathrm{d}s \quad (x \in \Omega).$$

Hence, by the Cauchy–Schwarz inequality and Tonelli's theorem

$$\begin{aligned}
\int_\Omega |\varphi(x)|^2 \, \mathrm{d}x &= \int_\Omega \left| \int_a^{x_1} \partial_1 \varphi(s, x_2, \ldots, x_d) \, \mathrm{d}s \right|^2 \mathrm{d}x \\
&\leqslant \int_\Omega (b-a) \int_a^b |\partial_1 \varphi(s, x_2, \ldots, x_d)|^2 \, \mathrm{d}s \, \mathrm{d}x = (b-a)^2 \int_\Omega |\partial_1 \varphi(x)|^2 \, \mathrm{d}x \\
&\leqslant (b-a)^2 \left\| \operatorname{grad}_0 \varphi \right\|_{L_2(\Omega)^d}^2 ,
\end{aligned}$$

which shows the assertion. □

Corollary 11.3.2 *Under the assumptions of Proposition 11.3.1 the operator* grad_0 *is one-to-one and* $\operatorname{ran}(\operatorname{grad}_0)$ *is closed.*

Proof The injectivity follows immediately from Poincaré's inequality. To prove the closedness of $\operatorname{ran}(\operatorname{grad}_0)$, let $(u_k)_{k\in\mathbb{N}}$ in $\operatorname{dom}(\operatorname{grad}_0)$ with $\operatorname{grad}_0 u_k \to v$ in $L_2(\Omega)^d$ for some $v \in L_2(\Omega)^d$. By Poincaré's inequality, we infer that $(u_k)_{k\in\mathbb{N}}$ is a Cauchy-sequence in $L_2(\Omega)$ and hence convergent to some $u \in L_2(\Omega)$. By the closedness of grad_0 we obtain $u \in \operatorname{dom}(\operatorname{grad}_0)$ and $v = \operatorname{grad}_0 u \in \operatorname{ran}(\operatorname{grad}_0)$. □

We need another auxiliary result which is interesting in its own right.

Lemma 11.3.3 *Let H be a Hilbert space and $V \subseteq H$ a closed subspace. We denote by*

$$\iota_V \colon V \to H, \quad x \mapsto x$$

the canonical embedding of V into H. Then $\iota_V \iota_V^* \colon H \to H$ is the orthogonal projection on V and $\iota_V^* \iota_V \colon V \to V$ is the identity on V.

Proof The proof is left as Exercise 11.2. □

We now come to the exponential stability of the heat equation. First, we need to formulate both the heat flux balance and Fourier's law as a suitable evolutionary equation. For doing so, we assume that $\Omega \subseteq \mathbb{R}^d$ is open and contained in a slab. Then $\operatorname{ran}(\operatorname{grad}_0)$ is closed by Corollary 11.3.2. It is clear that we can write Fourier's law as

$$q = -a \operatorname{grad}_0 \theta = -a\iota_{\operatorname{ran}(\operatorname{grad}_0)}\iota^*_{\operatorname{ran}(\operatorname{grad}_0)} \operatorname{grad}_0 \theta.$$

Hence, defining $\widetilde{q} := \iota^*_{\operatorname{ran}(\operatorname{grad}_0)} q$ and $\widetilde{a} := \iota^*_{\operatorname{ran}(\operatorname{grad}_0)} a \iota_{\operatorname{ran}(\operatorname{grad}_0)} \in L(\operatorname{ran}(\operatorname{grad}_0))$, we arrive at

$$\widetilde{q} = -\widetilde{a}\iota^*_{\operatorname{ran}(\operatorname{grad}_0)} \operatorname{grad}_0 \theta.$$

Moreover, since $\operatorname{ran}(\operatorname{grad}_0)^\perp = \ker(\operatorname{div})$, we derive from the heat flux balance

$$f = \partial_t \theta + \operatorname{div} q = \partial_t \theta + \operatorname{div} \iota_{\operatorname{ran}(\operatorname{grad}_0)} \widetilde{q}$$

and hence, assuming that $\widetilde{a}$ is invertible, we may write both equations with the unknowns $(\theta, \widetilde{q})$ as an evolutionary equation in $L_{2,\nu}(\mathbb{R}; H)$ for $\nu > 0$, where $H := L_2(\Omega) \oplus \operatorname{ran}(\operatorname{grad}_0)$. This yields

$$\left(\partial_{t,\nu} \begin{pmatrix} 1 & 0 \\ 0 & 0 \end{pmatrix} + \begin{pmatrix} 0 & 0 \\ 0 & \widetilde{a}^{-1} \end{pmatrix} + \begin{pmatrix} 0 & \operatorname{div} \iota_{\operatorname{ran}(\operatorname{grad}_0)} \\ \iota^*_{\operatorname{ran}(\operatorname{grad}_0)} \operatorname{grad}_0 & 0 \end{pmatrix}\right) \begin{pmatrix} \theta \\ \widetilde{q} \end{pmatrix} = \begin{pmatrix} f \\ 0 \end{pmatrix}. \tag{11.2}$$

For notational convenience, we set

$$C := \iota^*_{\operatorname{ran}(\operatorname{grad}_0)} \operatorname{grad}_0 \colon \operatorname{dom}(\operatorname{grad}_0) \subseteq L_2(\Omega) \to \operatorname{ran}(\operatorname{grad}_0). \tag{11.3}$$

Lemma 11.3.4 *Let $\Omega \subseteq \mathbb{R}^d$ be open and contained in a slab and C as above. Then C is densely defined, closed and boundedly invertible. Moreover*

$$C^* = -\operatorname{div} \iota_{\operatorname{ran}(\operatorname{grad}_0)}.$$

Proof The proof is left as Exercise 11.3. □

Proposition 11.3.5 *Let $\Omega \subseteq \mathbb{R}^d$ be open and contained in a slab, $a \in L(L_2(\Omega)^d)$, and $c_1 > 0$ such that*

$$\operatorname{Re} a \geqslant c_1.$$

Then $\widetilde{a} := \iota^*_{\mathrm{ran}(\mathrm{grad}_0)} a \iota_{\mathrm{ran}(\mathrm{grad}_0)}$ is boundedly invertible and the solution operators associated with (11.2) are exponentially stable.

Proof For $x \in \mathrm{ran}(\mathrm{grad}_0)$ we have

$$\begin{aligned}\mathrm{Re}\,\langle x, \widetilde{a}x\rangle_{\mathrm{ran}(\mathrm{grad}_0)} &= \mathrm{Re}\left\langle \iota_{\mathrm{ran}(\mathrm{grad}_0)} x, a\iota_{\mathrm{ran}(\mathrm{grad}_0)} x\right\rangle_{L_2(\Omega)^d} \\ &\geqslant c_1 \left\|\iota_{\mathrm{ran}(\mathrm{grad}_0)} x\right\|^2_{L_2(\Omega)^d} = c_1 \left\|x\right\|^2_{\mathrm{ran}(\mathrm{grad}_0)},\end{aligned}$$

and thus, $\widetilde{a}$ is boundedly invertible. Hence, (11.2) is an evolutionary equation of the form considered in Theorem 11.2.1 with $M_0 := 1$, $M_1(z) := \widetilde{a}^{-1}$ for $z \in \mathbb{C}$ and C given by (11.3). Since $\mathrm{Re}\,\widetilde{a}^{-1} \geqslant \frac{c_1}{\|\widetilde{a}\|^2}$, Theorem 11.2.1 is applicable and we derive the exponential stability. □

The Heat Equation with Additional Delay

Again we consider the heat equation, but now we replace Fourier's law by

$$q = -a_1 \,\mathrm{grad}_0\, \theta - a_2 \tau_{-h}\, \mathrm{grad}_0\, \theta$$

for some operators $a_1, a_2 \in L(L_2(\Omega)^d)$ and $h > 0$. As above, we assume that $\Omega \subseteq \mathbb{R}^d$ is open and contained in a slab. We may introduce $\widetilde{q} := \iota^*_{\mathrm{ran}(\mathrm{grad}_0)} q$ and $\widetilde{a}_j := \iota^*_{\mathrm{ran}(\mathrm{grad}_0)} a_j \iota_{\mathrm{ran}(\mathrm{grad}_0)} \in L(L_2(\Omega)^d)$ for $j \in \{1, 2\}$. Moreover, we assume that there exists $c > 0$ such that

$$\mathrm{Re}\, a_1 \geqslant c.$$

By Lemma 7.3.1 there exists $\nu_0 > 0$ such that the operator $\widetilde{a}_1 + \widetilde{a}_2\tau_{-h}$ is boundedly invertible in $L_{2,\nu}(\mathbb{R}; \mathrm{ran}(\mathrm{grad}_0))$ and its inverse is uniformly strictly positive definite for each $\nu \geqslant \nu_0$. Hence, we may write the heat equation with additional delay as an evolutionary equation of the form

$$\left(\partial_{t,\nu}\begin{pmatrix}1 & 0\\ 0 & 0\end{pmatrix} + \begin{pmatrix}0 & 0\\ 0 & (\widetilde{a}_1 + \widetilde{a}_2\tau_{-h})^{-1}\end{pmatrix} + \begin{pmatrix}0 & -C^*\\ C & 0\end{pmatrix}\right)\begin{pmatrix}\theta\\ \widetilde{q}\end{pmatrix} = \begin{pmatrix}f\\ 0\end{pmatrix} \tag{11.4}$$

with C given by (11.3).

Proposition 11.3.6 *Let $\Omega \subseteq \mathbb{R}^d$ be open and contained in a slab, $h > 0$, $a_1, a_2 \in L(L_2(\Omega)^d)$, and $c > 0$ such that*

$$\mathrm{Re}\, a_1 \geqslant c$$

and $\|a_2\| < c$. Then the solution operators $(S_\nu)_{\nu \geqslant \nu_0}$ associated with (11.4) are exponentially stable.

Proof Note that $\|\widetilde{a}_2\| \leqslant \|a_2\| < c$. We choose

$$0 < \rho_1 < \frac{1}{h} \log \frac{c}{\|\widetilde{a}_2\|}.$$

Then we estimate for $z \in \mathbb{C}_{\mathrm{Re}>-\rho_1}$

$$\mathrm{Re}\left\langle x, (\widetilde{a}_1 + \widetilde{a}_2 \mathrm{e}^{-zh})x \right\rangle_{\mathrm{ran}(\mathrm{grad}_0)} \geqslant (c - \|\widetilde{a}_2\| \, \mathrm{e}^{\rho_1 h}) \, \|x\|^2_{\mathrm{ran}(\mathrm{grad}_0)}.$$

By the choice of ρ_1, we infer $\widetilde{c} := (c - \|\widetilde{a}_2\| \, \mathrm{e}^{\rho_1 h}) > 0$. Hence,

$$M_1(z) := \left(\widetilde{a}_1 + \widetilde{a}_2 \mathrm{e}^{-hz}\right)^{-1} \quad (z \in \mathbb{C}_{\mathrm{Re}>-\rho_1})$$

is well-defined and satisfies

$$\mathrm{Re}\, M_1(z) \geqslant c_1 \quad (z \in \mathbb{C}_{\mathrm{Re}>-\rho_1})$$

for some $c_1 > 0$ by Proposition 6.2.3. Thus, Theorem 11.2.1 is applicable and yields the exponential stability of (11.4). □

A Dual Phase Lag Model

In this last variant of heat conduction, we replace Fourier's law by

$$(1 + s_q \partial_t)q = (1 + s_\theta \partial_t) \, \mathrm{grad}_0 \, \theta,$$

where $s_q, s_\theta > 0$ are the so-called "phases" (cf. Sect. 7.4, where a different type of dual phase lag model is studied). The latter equation can be reformulated as

$$(1 + s_q \partial_{t,\nu})(1 + s_\theta \partial_{t,\nu})^{-1} q = \mathrm{grad}_0 \, \theta$$

for $\nu > 0$. Assuming that $\Omega \subseteq \mathbb{R}^d$ is open and contained in a slab, and defining $\widetilde{q} := \iota^*_{\mathrm{ran}(\mathrm{grad}_0)} q$, the dual phase lag model may be written as

$$\left(\partial_{t,\nu} \begin{pmatrix} 1 & 0 \\ 0 & 0 \end{pmatrix} + \begin{pmatrix} 0 & 0 \\ 0 & (1 + s_q \partial_{t,\nu})(1 + s_\theta \partial_{t,\nu})^{-1} \end{pmatrix} + \begin{pmatrix} 0 & -C^* \\ C & 0 \end{pmatrix}\right) \begin{pmatrix} \theta \\ \widetilde{q} \end{pmatrix} = \begin{pmatrix} f \\ 0 \end{pmatrix} \tag{11.5}$$

with C given by (11.3).

Proposition 11.3.7 *Let $\Omega \subseteq \mathbb{R}^d$ be open and contained in a slab, $\nu_0 > 0$. Moreover, let $s_\theta > s_q > 0$. Then the solution operators $(S_\nu)_{\nu \geqslant \nu_0}$ associated with (11.5) are exponentially stable.*

Proof Again, we note that (11.5) is of the form considered in Theorem 11.2.1 with $M_0 := 1$ and

$$M_1(z) := \frac{1+s_q z}{1+s_\theta z} \quad (z \in \mathbb{C} \setminus \{-s_\theta^{-1}\}).$$

Setting $\mu := \frac{s_q}{s_\theta} < 1$ we compute

$$\operatorname{Re} M_1(z) = \operatorname{Re}\left(\mu + \frac{(1-\mu)}{1+s_\theta z}\right) = \mu + (1-\mu)\frac{1+s_\theta \operatorname{Re} z}{|1+s_\theta z|^2} \geqslant \mu \quad (z \in \mathbb{C}_{\operatorname{Re}>-s_\theta^{-1}}).$$

Thus, Theorem 11.2.1 is applicable and hence, the claim follows. □

11.4 Exponential Stability for Hyperbolic-Type Equations

Important examples of exponentially stable equations do not fit in the class of parabolic-like equations studied in Sect. 11.2. As a motivating example we consider the damped wave equation, which can be written as a second-order equation of the form

$$\partial_{t,\nu}^2 M_0 u + \partial_{t,\nu} M_1 u - \operatorname{div} \operatorname{grad}_0 u = f, \tag{11.6}$$

where $M_0, M_1 \in L(L_2(\Omega))$, M_0 is selfadjoint and $M_0, \operatorname{Re} M_1 \geqslant c > 0$, with $\Omega \subseteq \mathbb{R}^d$ modelling the underlying medium. It is well-known that this equation is exponentially stable if Ω is bounded. However, if we write this equation as an evolutionary problem in the canonical way; that is, we introduce $v := \partial_{t,\nu} u$ and $q := -\operatorname{grad}_0 u$ as new unknowns, we end up with an equation of the form

$$\left(\partial_{t,\nu}\begin{pmatrix} M_0 & 0 \\ 0 & 1\end{pmatrix} + \begin{pmatrix} M_1 & 0 \\ 0 & 0\end{pmatrix} + \begin{pmatrix} 0 & \operatorname{div} \\ \operatorname{grad}_0 & 0\end{pmatrix}\right)\begin{pmatrix} v \\ q\end{pmatrix} = \begin{pmatrix} f \\ 0\end{pmatrix}, \tag{11.7}$$

which is not of the form discussed in Sect. 11.2. However, another formulation of (11.6) as an evolutionary equation allows to show exponential stability in a similar way as for parabolic-type equations. More precisely, we aim for a formulation, such that the second block operator matrix in (11.7) has non-vanishing diagonal entries. This leads to a damping effect for both unknowns.

We start to provide a general reformulation scheme of second-order equations as suitable evolutionary equations and afterwards discuss the exponential stability of those.

An Alternative Reformulation for Hyperbolic-Type Equations

Throughout we assume that $C\colon \operatorname{dom}(C) \subseteq H_0 \to H_1$ is a densely defined closed linear operator between two Hilbert spaces H_0 and H_1, which is additionally assumed to be boundedly invertible. Furthermore, let $M\colon \operatorname{dom}(M) \subseteq \mathbb{C} \to L(H_0)$ be a material law of the form

$$M(z) = M_0(z) + z^{-1} M_1(z) \quad (z \in \operatorname{dom}(M)),$$

where $M_0, M_1\colon \operatorname{dom}(M) \subseteq \mathbb{C} \to L(H)$ are material laws themselves. We consider second-order problems of the form

$$\left(\partial_{t,\nu}^2 M(\partial_{t,\nu}) + C^*C\right) u = f, \tag{11.8}$$

for a given right-hand side $f \in L_{2,\nu}(\mathbb{R}; H_0)$ and aim for conditions on M to ensure the exponential stability in a suitable sense.

Example 11.4.1 The wave equation (11.6) on a bounded domain $\Omega \subseteq \mathbb{R}^n$ is indeed of the form (11.8). We set $C := \iota_{\operatorname{ran}(\operatorname{grad}_0)}^* \operatorname{grad}_0\colon \operatorname{dom}(\operatorname{grad}_0) \subseteq L_2(\Omega) \to \operatorname{ran}(\operatorname{grad}_0)$, which is boundedly invertible by Poincaré's inequality (see Proposition 11.3.1 and Lemma 11.3.4) and

$$M(z) = M_0 + z^{-1} M_1 \quad (z \in \mathbb{C} \setminus \{0\})$$

for $M_0, M_1 \in L(L_2(\Omega))$.

We now introduce two new unknowns to rewrite (11.8) as an evolutionary equation. For this let $d > 0$ and set $v_d := \partial_{t,\nu} u + du$ and $q := -Cu$. Then we formally get

$$\partial_{t,\nu} q = -C\partial_{t,\nu} u = -C(v_d - du) = -Cv_d + dCu = -Cv_d - dq$$

and

$$\begin{aligned}
\partial_{t,\nu} M(\partial_{t,\nu}) v_d &= \partial_{t,\nu}^2 M(\partial_{t,\nu}) u + d\partial_{t,\nu} M(\partial_{t,\nu}) u \\
&= f - C^*Cu + d\partial_{t,\nu} M_0(\partial_{t,\nu}) u + dM_1(\partial_{t,\nu}) u \\
&= f + C^*q + dM_0(\partial_{t,\nu})(v_d - du) + dM_1(\partial_{t,\nu}) u \\
&= f + C^*q + dM_0(\partial_{t,\nu}) v_d - d\left(M_1(\partial_{t,\nu}) - dM_0(\partial_{t,\nu})\right) C^{-1} q.
\end{aligned}$$

Thus, the new unknowns, v_d and q, satisfy an evolutionary equation of the form

$$\begin{aligned}
&\left(\partial_{t,\nu} \begin{pmatrix} M(\partial_{t,\nu}) & 0 \\ 0 & 1 \end{pmatrix} + d \begin{pmatrix} -M_0(\partial_{t,\nu}) & \left(M_1(\partial_{t,\nu}) - dM_0(\partial_{t,\nu})\right) C^{-1} \\ 0 & 1 \end{pmatrix}\right. \\
&\qquad\qquad \left. + \begin{pmatrix} 0 & -C^* \\ C & 0 \end{pmatrix}\right) \begin{pmatrix} v_d \\ q \end{pmatrix} = \begin{pmatrix} f \\ 0 \end{pmatrix},
\end{aligned} \tag{11.9}$$

with a new material law $M_d\colon \operatorname{dom}(M) \subseteq \mathbb{C} \to L(H_0 \oplus H_1)$ given by

$$M_d(z) := \begin{pmatrix} M(z) & 0 \\ 0 & 1 \end{pmatrix} + z^{-1} d \begin{pmatrix} -M_0(z) & (M_1(z) - dM_0(z))\, C^{-1} \\ 0 & 1 \end{pmatrix}.$$

Remark 11.4.2 We remark that the above formal computation can be done rigorously (both forward and backwards), so that indeed (11.8) and (11.9) are equivalent problems in the sense that the solutions u and (v_d, q) are linked via

$$v_d = \partial_{t,\nu} u + du, \quad q = -Cu.$$

11.5 A Criterion for Exponential Stability of Hyperbolic-Type Equations

In this section we provide sufficient conditions on the material law M in order to obtain a well-posed and exponentially stable problem (11.9) for a suitable $d > 0$. So, we assume the same assumptions to be in effect as in the previous section.

Remark 11.5.1 Assume that (11.9) is exponentially stable with decay rate $\rho_0 > 0$; that is, $v_d \in L_{2,-\rho}(\mathbb{R}; H_0)$, $q \in L_{2,-\rho}(\mathbb{R}; H_1)$ if $f \in L_{2,-\rho}(\mathbb{R}; H_0) \cap L_{2,\nu}(\mathbb{R}; H_0)$ for all $\rho \in [0, \rho_0)$ and $\nu > 0$ large enough. Then $u, \partial_{t,\nu} u \in L_{2,-\rho}(\mathbb{R}; H_0)$ as well. Indeed, since

$$u = -C^{-1} q \in L_{2,-\rho}(\mathbb{R}; H_0),$$

we derive

$$\partial_{t,\nu} u = v_d - du \in L_{2,-\rho}(\mathbb{R}; H_0).$$

Employing Exercise 11.1, we even infer $u \in \operatorname{dom}(\partial_{t,-\rho})$ and hence, $u \in C_{-\rho}(\mathbb{R}; H_0)$ by Sobolev's embedding theorem (see Theorem 4.1.2). Thus, we also obtain the exponential stability of (11.8) in this case.

In order to prove the exponential stability of (11.9), we have to show how a positive definiteness assumption on M allows for positive definiteness of M_d for some $d > 0$. We start with the following observation.

Lemma 11.5.2 *Let $z \in \operatorname{dom}(M)$, $c > 0$. Assume*

$$\operatorname{Re} \langle u, zM(z)u \rangle_{H_0} \geqslant c \, \|u\|_{H_0}^2 \quad (u \in H_0).$$

Then for $d > 0$ and $(v, q) \in H_0 \oplus H_1$ it follows that

$$\operatorname{Re} \langle (v, q), zM_d(z)(v, q) \rangle_{H_0 \oplus H_1} \geqslant \min \left\{ c - dK(d), \frac{3}{4} d + \operatorname{Re} z \right\} \|(v, q)\|_{H_0 \oplus H_1}^2 ,$$

where $K(d) := m_0 + (dm_0 + m_1)^2 \left\| C^{-1} \right\|^2$ and $m_j := \left\| M_j \right\|_\infty$ for $j \in \{0, 1\}$.

Proof Let $v \in H_0$ and $q \in H_1$. Then we estimate

$$
\begin{aligned}
&\operatorname{Re} \langle (v, q), zM_d(z)(v, q)\rangle_{H_0 \oplus H_1} \\
&= \operatorname{Re} \left\langle v, zM(z)v - dM_0(z)v + d(M_1(z) - dM_0(z))C^{-1}q \right\rangle_{H_0} + \operatorname{Re} \langle q, zq + dq\rangle_{H_1} \\
&\geqslant (c - dm_0) \|v\|_{H_0}^2 - d(m_1 + dm_0) \left\| C^{-1} \right\| \|q\|_{H_1} \|v\|_{H_0} + (\operatorname{Re} z + d) \|q\|_{H_1}^2 \\
&\geqslant \left(c - dm_0 - \frac{1}{4\varepsilon} d^2 (m_1 + dm_0)^2 \left\| C^{-1} \right\|^2 \right) \|v\|_{H_0}^2 + (\operatorname{Re} z + d - \varepsilon) \|q\|_{H_1}^2 ,
\end{aligned}
$$

for each $\varepsilon > 0$, where we have used the Peter–Paul inequality. Choosing $\varepsilon = \frac{d}{4}$, we obtain the assertion. □

This estimate allows us to derive the positive definiteness of M_d for a suitable choice of $d > 0$.

Proposition 11.5.3 *Let $c > 0$ and assume that*

$$
\operatorname{Re} \langle u, zM(z)u\rangle_{H_0} \geqslant c \|u\|_{H_0}^2 \quad (u \in H_0,\ z \in \operatorname{dom}(M)).
$$

Then there exist $\widetilde{c}, d, \rho_0 > 0$ such that

$$
\operatorname{Re} \langle (v, q), zM_d(z)(v, q)\rangle_{H_0 \oplus H_1} \geqslant \widetilde{c} \|(v, q)\|_{H_0 \oplus H_1}^2
$$

for all $z \in \operatorname{dom}(M) \cap \mathbb{C}_{\operatorname{Re} > -\rho_0}$ and $(v, q) \in H_0 \oplus H_1$.

Proof We note that $dK(d) \to 0$ as $d \to 0$, where $K(d)$ is given as in Lemma 11.5.2. Hence, we find $d > 0$ such that $dK(d) < c$. Choosing $\rho_0 < \frac{3}{4}d$ and using Lemma 11.5.2, we estimate for each $z \in \operatorname{dom}(M) \cap \mathbb{C}_{\operatorname{Re} > -\rho_0}$ and $(v, q) \in H_0 \oplus H_1$

$$
\operatorname{Re} \langle (v, q), zM_d(z)(v, q)\rangle_{H_0 \oplus H_1} \geqslant \widetilde{c} \|(v, q)\|_{H_0 \oplus H_1}^2 ,
$$

where $\widetilde{c} := \min \left\{ c - dK(d), \frac{3}{4}d - \rho_0 \right\} > 0$ showing the assertion. □

We are now in the position to state the main result for exponential stability of hyperbolic-type equations.

Theorem 11.5.4 *Let $C\colon \operatorname{dom}(C) \subseteq H_0 \to H_1$ be a densely defined closed linear and boundedly invertible operator between two Hilbert spaces H_0 and H_1. Furthermore, let $M\colon \operatorname{dom}(M) \subseteq \mathbb{C} \to L(H_0)$ be a material law of the form*

$$
M(z) = M_0(z) + z^{-1} M_1(z) \quad (z \in \operatorname{dom}(M)),
$$

where $M_0, M_1 \colon \operatorname{dom}(M) \subseteq \mathbb{C} \to L(H)$ *are bounded analytic functions. Assume that there exist* $c, \nu_0 > 0$ *such that* $\mathbb{C}_{\mathrm{Re}>-\nu_0} \setminus \operatorname{dom}(M)$ *is discrete and*

$$\mathrm{Re}\, \langle u, zM(z)u \rangle_{H_0} \geqslant c\, \|u\|_{H_0}^2$$

for each $u \in H_0, z \in \operatorname{dom}(M)$. *Then there exists some* $d > 0$ *such that problem (11.9) is well-posed and exponentially stable.*

Proof We first note that by Proposition 11.5.3 there exist $\rho_0, d, \widetilde{c} > 0$ such that

$$\mathrm{Re}\, \langle (v, q), zM_d(z)(v, q) \rangle_{H_0 \oplus H_1} \geqslant \widetilde{c}\, \|(v, q)\|_{H_0 \oplus H_1}^2$$

for all $z \in \operatorname{dom}(M) \cap \mathbb{C}_{\mathrm{Re}>-\rho_0}$ and $(v, q) \in H_0 \oplus H_1$. Since M is a material law, so is M_d and thus, well-posedness of (11.9) follows from Picard's theorem (see Theorem 6.2.1). Since

$$\begin{pmatrix} 0 & -C^* \\ C & 0 \end{pmatrix}$$

is skew-selfadjoint, the above estimate yields that $zM_d(z) + \begin{pmatrix} 0 & -C^* \\ C & 0 \end{pmatrix}$ is boundedly invertible for each $z \in \operatorname{dom}(M) \cap \mathbb{C}_{\mathrm{Re}>-\rho_0}$ with

$$\sup_{z \in \operatorname{dom}(M) \cap \mathbb{C}_{\mathrm{Re}>-\rho_0}} \|T_d(z)\| \leqslant \frac{1}{\widetilde{c}},$$

where

$$T_d(z) := \left(zM_d(z) + \begin{pmatrix} 0 & -C^* \\ C & 0 \end{pmatrix} \right)^{-1}.$$

Setting $\mu := \min\{\nu_0, \rho_0\}$, we infer that T_d is defined on the whole $\mathbb{C}_{\mathrm{Re}>-\mu}$ despite a discrete set. Since T_d is holomorphic and bounded, Riemann's theorem on removable singularities implies that T_d can be extended to a holomorphic and bounded function on $\mathbb{C}_{\mathrm{Re}>-\mu}$. We denote this extension again by T_d. In particular, T_d is a material law with $s_b(T_d) \leqslant -\mu$. Let now $\rho \in [0, \mu)$ and $(f, g) \in L_{2,\nu}(\mathbb{R}; H_0 \oplus H_1) \cap L_{2,-\rho}(\mathbb{R}; H_0 \oplus H_1)$, where $\nu > 0$ is large enough to ensure well-posedness. By Theorem 5.3.6 we derive

$$T_d(\partial_{t,\nu})(f, g) = T_d(\partial_{t,-\rho})(f, g) \in L_{2,-\rho}(\mathbb{R}; H_0 \oplus H_1)$$

and since $T_d(\partial_{t,\nu})(f, g)$ is nothing but the solution of (11.9) with the right-hand side replaced by (f, g), exponential stability follows. □

Definition We call the equation

$$\left(\partial_{t,\nu}^2 M(\partial_{t,\nu}) + C^*C\right) u = f$$

exponentially stable if there exists some $d > 0$ such that the equation

$$\left(\partial_{t,\nu} M_d(\partial_{t,\nu}) + \begin{pmatrix} 0 & -C^* \\ C & 0 \end{pmatrix}\right) v = g$$

is exponentially stable.

11.6 Examples of Exponentially Stable Hyperbolic Problems

We will illustrate our findings by providing two concrete examples. Firstly, we discuss the damped wave equation in an abstract form and, secondly, we consider the dual phase lag model, as it was introduced in Sect. 7.4.

The Damped Wave Equation
We start by formulating an immediate corollary of our main stability theorem.

Corollary 11.6.1 *Let $C\colon \operatorname{dom}(C) \subseteq H_0 \to H_1$ be a densely defined closed linear and boundedly invertible operator between two Hilbert spaces H_0 and H_1 and let $M_0, M_1 \in L(H_0)$ such that M_0 is selfadjoint and $M_0 \geqslant 0$, $\operatorname{Re} M_1 \geqslant c > 0$. Then the second order problem*

$$\left(\partial_{t,\nu}^2 M_0 + \partial_{t,\nu} M_1 + C^*C\right) u = f$$

is exponentially stable.

Proof We have to prove that the material law

$$M(z) := M_0 + z^{-1} M_1 \quad (z \in \mathbb{C} \setminus \{0\})$$

satisfies the assumptions of Theorem 11.5.4. For $\operatorname{Re} z \geqslant 0$ we have

$$\operatorname{Re} \langle u, zM(z)u\rangle_{H_0} \geqslant c\, \|u\|_{H_0}^2 \quad (u \in H_0),$$

since $\operatorname{Re} zM_0 \geqslant 0$. Moreover, for $\operatorname{Re} z \in [-\rho_0, 0]$ with $\rho_0 < \frac{c}{\|M_0\|}$ (we set $\frac{c}{0} := \infty$) we have that

$$\operatorname{Re} \langle u, zM(z)u\rangle_{H_0} \geqslant (-\rho_0 \|M_0\| + c)\, \|u\|_{H_0}^2 \quad (u \in H_0).$$

Since $\mathbb{C}_{\operatorname{Re} > -\rho_0} \setminus \operatorname{dom}(M) = \{0\}$, we can apply Theorem 11.5.4. □

We now come to a concrete realisation of the operator C. Let $\Omega \subseteq \mathbb{R}^d$ be open and contained in a slab. According to Corollary 11.3.2 the space $\operatorname{ran}(\operatorname{grad}_0)$ is closed and by Lemma 11.3.4 the operator

$$C := \iota^*_{\operatorname{ran}(\operatorname{grad}_0)} \operatorname{grad}_0 \colon \operatorname{dom}(\operatorname{grad}_0) \subseteq L_2(\Omega) \to \operatorname{ran}(\operatorname{grad}_0)$$

is densely defined, closed and boundedly invertible, and its adjoint is given by

$$C^* = -\operatorname{div} \iota_{\operatorname{ran}(\operatorname{grad}_0)}.$$

Thus, we have that

$$C^*C = -\operatorname{div} \iota_{\operatorname{ran}(\operatorname{grad}_0)} \iota^*_{\operatorname{ran}(\operatorname{grad}_0)} \operatorname{grad}_0 = -\operatorname{div} \operatorname{grad}_0 .$$

Let now $M_0, M_1 \in L(L_2(\Omega))$ with M_0 selfadjoint and $M_0 \geqslant 0$, $\operatorname{Re} M_1 \geqslant c > 0$. By Corollary 11.6.1 the equation

$$\left(\partial_{t,\nu}^2 M_0 + \partial_{t,\nu} M_1 - \operatorname{div} \operatorname{grad}_0\right) u = f \tag{11.10}$$

is exponentially stable.

Remark 11.6.2 We emphasise that this result yields the classical exponential stability for the damped wave equation; i.e., the situation where $M_0 = 1$. However, Corollary 11.6.1 is also applicable in the situation where $M_0 = \mathbb{1}_{\Omega_0}$ for some $\Omega_0 \subseteq \Omega$ and $\operatorname{Re} M_1 \geqslant c$. In this case, Eq. (11.10) is a coupled system of the damped wave equation inside Ω_0 and of the heat equation outside Ω_0.

Dual Phase Lag Heat Conduction

We recall the setting of Sect. 7.4, where we have discussed the equations of dual phase lag heat conduction on an open and bounded subset $\Omega \subseteq \mathbb{R}^d$ within the framework of evolutionary equations. The equations under consideration consist of the heat flux balance

$$\partial_{t,\nu}\theta + \operatorname{div} q = Q,$$

and a modified Fourier's law

$$(1 + s_q \partial_{t,\nu} + \frac{1}{2} s_q^2 \partial_{t,\nu}^2) q = -(1 + s_\theta \partial_{t,\nu}) \operatorname{grad} \theta, \tag{11.11}$$

where $s_q \in \mathbb{R}$, $s_\theta > 0$ are given. Note that $(1 + s_\theta \partial_{t,\nu})$ is boundedly invertible for $\nu > -\frac{1}{s_\theta}$ and hence, (11.11) yields

$$-\operatorname{grad} \theta = \partial_{t,\nu}(\partial_{t,\nu}^{-1} + s_q + \frac{1}{2} s_q^2 \partial_{t,\nu})(1 + s_\theta \partial_{t,\nu})^{-1} q.$$

Applying the operator $\partial_{t,\nu}(\partial_{t,\nu}^{-1}+s_q+\frac{1}{2}s_q^2\partial_{t,\nu})(1+s_\theta\partial_{t,\nu})^{-1}$ to the heat flux balance equation (and assuming that $Q \in \operatorname{dom}(\partial_{t,\nu})$) we obtain the following second order problem

$$\partial_{t,\nu}^2(\partial_{t,\nu}^{-1} + s_q + \frac{1}{2}s_q^2\partial_{t,\nu})(1 + s_\theta\partial_{t,\nu})^{-1}\theta - \operatorname{div}\operatorname{grad}\theta = \widetilde{Q}, \tag{11.12}$$

for a suitable source term $\widetilde{Q}$. Assuming Dirichlet boundary conditions for θ, the equation takes the form

$$\left(\partial_{t,\nu}^2 M(\partial_{t,\nu}) + C^*C\right)\theta = \widetilde{Q},$$

with $C := \iota_{\operatorname{ran}(\operatorname{grad}_0)}^* \operatorname{grad}_0 \colon \operatorname{dom}(\operatorname{grad}_0) \subseteq L_2(\Omega) \to \operatorname{ran}(\operatorname{grad}_0)$ and

$$M(z) = \frac{z^{-1} + s_q + \frac{1}{2}s_q^2 z}{1 + s_\theta z} \qquad \left(z \in \mathbb{C}\setminus\left\{0, -\frac{1}{s_\theta}\right\}\right).$$

Note that

$$M(z) = \frac{s_q + \frac{1}{2}s_q^2 z}{1 + s_\theta z} + z^{-1}\frac{1}{1 + s_\theta z}$$

and hence, M is indeed of the form considered in Sect. 11.5 with

$$M_0(z) = \frac{s_q + \frac{1}{2}s_q^2 z}{1 + s_\theta z}, \qquad M_1(z) = \frac{1}{1 + s_\theta z},$$

which are both bounded if we restrict the domain of M to a right half-plane $\mathbb{C}_{\operatorname{Re}>-\frac{1}{s_\theta}+\varepsilon}$ for some $\varepsilon > 0$.

Proposition 11.6.3 *If* $0 < \frac{s_q}{s_\theta} < 2$ *then the dual phase lag model (11.12) is exponentially stable.*

Proof We apply Theorem 11.5.4. For this we need to show that there exists $c > 0$ such that

$$\operatorname{Re}\langle u, zM(z)u\rangle_{L_2(\Omega)} \geqslant c\,\|u\|_{L_2(\Omega)}^2$$

for each $u \in L_2(\Omega)$ and $z \in \mathbb{C}_{\operatorname{Re}>-\nu_0} \cap \operatorname{dom}(M)$ for some $0 < \nu_0 < \frac{1}{s_\theta}$. Indeed, this is sufficient for exponential stability, since $\mathbb{C}_{\operatorname{Re}>-\nu_0}\setminus\operatorname{dom}(M) = \{0\}$ is discrete and $C = \iota_{\operatorname{ran}(\operatorname{grad}_0)}^* \operatorname{grad}_0$ is boundedly invertible. Similar to the proof of Lemma 7.4.3 we set $\sigma := \frac{s_q}{s_\theta}$ and obtain

$$zM(z) = \frac{1}{2}s_q z\sigma + \sigma\left(1 - \frac{1}{2}\sigma\right) + \frac{1 - \sigma(1 - \frac{1}{2}\sigma)}{1 + s_\theta z}$$

for each $z \in \operatorname{dom}(M)$. Since $0 < \sigma < 2$ we obtain $0 < \sigma\left(1 - \frac{1}{2}\sigma\right) \leqslant \frac{1}{2}$ and hence,

$$\begin{aligned}\operatorname{Re} zM(z) &= \frac{1}{2}s_q \operatorname{Re} z\sigma + \sigma\left(1 - \frac{1}{2}\sigma\right) + \frac{\left(1 - \sigma(1 - \frac{1}{2}\sigma)\right)(1 + s_\theta \operatorname{Re} z)}{|1 + s_\theta z|^2} \\ &\geqslant -\frac{1}{2}s_q \nu_0 \sigma + \sigma(1 - \frac{1}{2}\sigma) =: c_{\nu_0}\end{aligned}$$

for each $z \in \mathbb{C}_{\operatorname{Re} > -\nu_0} \cap \operatorname{dom}(M)$ with $0 < \nu_0 < \frac{1}{s_\theta}$. Choosing now $0 < \nu_0 < \min\{\frac{1}{s_\theta}, \frac{2-\sigma}{s_q}\}$, we obtain $c_{\nu_0} > 0$ and thus, Theorem 11.5.4 is applicable which yields the assertion. □

11.7 Comments

The results of this chapter are based on the results obtained in [116, Section 2]. There, Laplace transform techniques are used to characterise the exponential stability of evolutionary equations in a slightly more general setting. In particular, further criteria for exponential stability of parabolic- and hyperbolic-type equations are given, which also allow for the treatment of integro-differential equations.

In general whether or not a given partial differential equation is (exponentially) stable is both an important and classical question in the area of equations depending on time. The understanding of this question for instance contributes to the study of equilibria of non-linear equations. In the linear case, in particular in the framework of C_0-semigroups, stability has been studied intensively resulting in an abundance of criteria. Due to strong continuity of the semigroup and, thus, of the considered solutions (exponential) stability is defined via pointwise estimates. As an example criterion we mention Datko's theorem [29] (see also [6, Theorem 5.1.2]), which states that a C_0-semigroup is exponentially stable if and only if the solution operator associated with the equation

$$\left(\partial_{t,\nu} + A\right) U = F$$

leaves $L_p(\mathbb{R}_{\geqslant 0}; H)$ invariant for some (or equivalently all) $p \in [1, \infty)$. As it turns out, the latter is equivalent to the invariance of $L_{2,-\rho}(\mathbb{R}; H)$ for some $\rho > 0$ and thus, our notion of exponential stability coincides with the usual one used in the theory of C_0-semigroups. Another important theorem on the exponential stability of C_0-semigroups on Hilbert spaces is the Theorem of Gearhart–Prüß [96] (see also [38, Chapter 5, Theorem 1.11]), where the exponential stability of a C_0-semigroup is characterised in terms of the resolvent of its generator.

The wave equation without damping is not exponentially stable. In fact one can even show that energy is preserved during the evolution. Hence, it is a natural question whether it is possible to introduce suitable 'dampers' (i.e., lower order coefficients) leading to an exponentially stable equation. The criterion in Corollary 11.6.1 shows that if the damper M_1 is 'global' in the sense that it is induced by a multiplication operator $a(\mathrm{m})$ for a strictly positive function a, the resulting damped wave equation is exponentially stable.

A less general, more detailed analysis of the actual wave equation shows that it is possible to obtain an exponentially stable damped wave equation if the damper is only local or introduced via boundary conditions. Indeed, in [9] the authors proved exponential stability of the damped equation if the damping area $[a > 0] := \{x \in \Omega\,;\, a(x) > 0\}$ satisfies the geometric optics condition. This is, for instance, the case if $[a > 0]$ contains a neighbourhood of the boundary $\partial\Omega$.

Besides exponential stability, which is the only type of stability studied so far within the current framework of evolutionary equations, different kinds of asymptotic behaviours were addressed and characterised for C_0-semigroups. We just mention the celebrated Arendt–Batty–Lyubich–Vu theorem [4, 61] on strong stability of C_0-semigroups or the Theorem of Borichev–Tomilov [15] on the polynomial stability of C_0-semigroups on Hilbert spaces.

Exercises

Exercise 11.1 Let H be a Hilbert space, $\nu, \rho \in \mathbb{R}$ and $u \in L_{1,\mathrm{loc}}(\mathbb{R}; H)$. Prove the following statements:

(a) If $u \in \mathrm{dom}(\partial_{t,\nu}) \cap \mathrm{dom}(\partial_{t,\rho})$ then $\partial_{t,\nu}u = \partial_{t,\rho}u$.
(b) If $u \in \mathrm{dom}(\partial_{t,\nu})$ such that $u, \partial_{t,\nu}u \in L_{2,\rho}(\mathbb{R}; H)$ then $u \in \mathrm{dom}(\partial_{t,\rho})$.

Exercise 11.2 Prove Lemma 11.3.3.

Exercise 11.3 Let H_0, H_1 be Hilbert spaces and $A\colon \mathrm{dom}(A) \subseteq H_0 \to H_1$ a densely defined closed linear operator. Moreover, we assume that A has closed range. Show that the adjoint of the operator $\iota_{\mathrm{ran}(A)}^* A\colon \mathrm{dom}(A) \subseteq H_0 \to \mathrm{ran}(A)$ is given by $A^*\iota_{\mathrm{ran}(A)}$. If additionally A is one-to-one, show that $\iota_{\mathrm{ran}(A)}^* A$ is boundedly invertible.

Exercise 11.4 Let $\Omega \subseteq \mathbb{R}^d$ be open and contained in a slab. We consider the heat conduction with a memory term given by the equations

$$\begin{aligned} \partial_{t,\nu}\theta + \operatorname{div} q &= f, \\ q &= -(1 - k*)\operatorname{grad}_0 \theta, \end{aligned} \tag{11.13}$$

where $k \in L_{1,-\rho_1}(\mathbb{R}_{\geqslant 0}; \mathbb{R})$ for some $\rho_1 > 0$ with

$$\int_0^\infty |k(t)| \, \mathrm{d}t < 1.$$

Write (11.13) as a suitable evolutionary equation and prove that this equation is exponentially stable.

Exercise 11.5 Let $A \in \mathbb{C}^{n\times n}$ for some $n \in \mathbb{N}$ and consider the evolutionary equation

$$(\partial_{t,\nu} + A)U = F.$$

Prove that the solution operators associated with this problem are exponentially stable if and only if A has only eigenvalues with strictly positive real part.

Exercise 11.6 Let $\Omega \subseteq \mathbb{R}^d$ be open.

(a) Let $\varphi \in C_c^\infty(\Omega)^d$. Prove *Korn's inequality*

$$\|\mathrm{Grad}\,\varphi\|^2_{L_2(\Omega)^{d\times d}_{\mathrm{sym}}} \geqslant \frac{1}{2}\sum_{j=1}^d \|\mathrm{grad}\,\varphi_j\|^2_{L_2(\Omega)^d}.$$

(b) Use Korn's inequality to prove that for $u \in L_2(\Omega)^d$ we have

$$u \in \mathrm{dom}(\mathrm{Grad}_0) \quad \iff \quad \forall j \in \{1,\ldots,d\} : u_j \in \mathrm{dom}(\mathrm{grad}_0).$$

Moreover, show that in either case

$$\frac{1}{2}\sum_{j=1}^d \|\mathrm{grad}_0\,u_j\|^2_{L_2(\Omega)^d} \leqslant \|\mathrm{Grad}_0\,u\|^2_{L_2(\Omega)^{d\times d}_{\mathrm{sym}}} \leqslant \sum_{j=1}^d \|\mathrm{grad}_0\,u_j\|^2_{L_2(\Omega)^d}.$$

(c) Let now Ω be contained in a slab. Prove that Grad_0 is one-to-one and has closed range.

Exercise 11.7 Let $\Omega \subseteq \mathbb{R}^d$ be open and $a \in L(L_2(\Omega)^d)$ with $\mathrm{Re}\,a \geqslant c > 0$.

(a) Let $\nu > 0$ and $f \in L_{2,\nu}(\mathbb{R}; L_2(\Omega))$. Moreover, assume that Ω is contained in a slab and define $\widetilde{a} := \iota^*_{\mathrm{ran}(\mathrm{grad}_0)} a \iota_{\mathrm{ran}(\mathrm{grad}_0)}$. Let $\theta \in L_{2,\nu}(\mathbb{R}; L_2(\Omega))$, $q \in L_{2,\nu}(\mathbb{R}; L_2(\Omega)^d)$ satisfy

$$\left(\partial_{t,\nu}\begin{pmatrix}1 & 0\\ 0 & 0\end{pmatrix} + \begin{pmatrix}0 & 0\\ 0 & a^{-1}\end{pmatrix} + \begin{pmatrix}0 & \mathrm{div}\\ \mathrm{grad}_0 & 0\end{pmatrix}\right)\begin{pmatrix}\theta\\ q\end{pmatrix} = \begin{pmatrix}f\\ 0\end{pmatrix}$$

and $\widetilde{\theta} \in L_{2,\nu}(\mathbb{R}; L_2(\Omega))$, $\widetilde{q} \in L_{2,\nu}(\mathbb{R}; \operatorname{ran}(\operatorname{grad}_0))$ satisfy

$$\left(\partial_{t,\nu}\begin{pmatrix}1 & 0\\ 0 & 0\end{pmatrix} + \begin{pmatrix}0 & 0\\ 0 & \widetilde{a}^{-1}\end{pmatrix} + \begin{pmatrix}0 & \operatorname{div} \iota_{\operatorname{ran}(\operatorname{grad}_0)}\\ \iota^*_{\operatorname{ran}(\operatorname{grad}_0)} \operatorname{grad}_0 & 0\end{pmatrix}\right)\begin{pmatrix}\widetilde{\theta}\\ \widetilde{q}\end{pmatrix} = \begin{pmatrix}f\\ 0\end{pmatrix}.$$

Show that $(\theta, \iota^*_{\operatorname{ran}(\operatorname{grad}_0)} q) = (\widetilde{\theta}, \widetilde{q})$.

(b) Let Ω be bounded and consider the evolutionary equation

$$\left(\partial_{t,\nu}\begin{pmatrix}1 & 0\\ 0 & 0\end{pmatrix} + \begin{pmatrix}0 & 0\\ 0 & a^{-1}\end{pmatrix} + \begin{pmatrix}0 & \operatorname{div}_0\\ \operatorname{grad} & 0\end{pmatrix}\right)\begin{pmatrix}\theta\\ q\end{pmatrix} = \begin{pmatrix}f\\ 0\end{pmatrix}.$$

Show that the associated solution operators are not exponentially stable.

References

4. W. Arendt, C.J.K. Batty, Tauberian theorems and stability of one-parameter semigroups. Trans. Am. Math. Soc. **306**(2), 837–852 (1988)
6. W. Arendt et al., *Vector-Valued Laplace Transforms and Cauchy Problems*, 2nd edn. (Birkhäuser, Basel, 2011)
9. C. Bardos, G. Lebeau, J. Rauch, Sharp sufficient conditions for the observation, control, and stabilization of waves from the boundary. SIAM J. Control Optim. **30**(5), 1024–1065 (1992)
15. A. Borichev, Y. Tomilov, Optimal polynomial decay of functions and operator semigroups. Math. Ann. **347**(2), 455–478 (2010)
29. R. Datko, Uniform asymptotic stability of evolutionary processes in a Banach space. SIAM J. Math. Anal. **3**, 428–445 (1972)
38. K.-J. Engel, R. Nagel, *One-Parameter Semigroups for Linear Evolution Equations*, vol. 194. Graduate Texts in Mathematics. With contributions by S. Brendle, M. Campiti, T. Hahn, G. Metafune, G. Nickel, D. Pallara, C. Perazzoli, A. Rhandi, S. Romanelli, R. Schnaubelt (Springer, New York, 2000)
61. Y.I. Lyubich, Q.P. Vũ, Asymptotic stability of linear differential equations in Banach spaces. Studia Math. **88**(1), 37–42 (1988)
96. J. Prüss, On the spectrum of C_0-semigroups. Trans. Am. Math. Soc. **284**(2), 847–857 (1984)
116. S. Trostorff, Exponential Stability and Initial Value Problems for Evolutionary Equations. Habilitation Thesis. TU Dresden, 2018

Chapter 12
Boundary Value Problems and Boundary Value Spaces

This chapter is devoted to the study of inhomogeneous boundary value problems. For this, we shall reformulate the boundary value problem again into a form which fits within the general framework of evolutionary equations. In order to have an idea of the type of boundary values which make sense to study, we start off with a section that deals with the boundary values of functions in the domain of the gradient operator defined on a half-space in $\mathbb{R}^d$ (for $d = 1$ we have $L_2(\mathbb{R}^{d-1}) = \mathbb{K}$).

12.1 The Boundary Values of $H^1(\mathbb{R}^{d-1} \times \mathbb{R}_{>0})$

In this section we let $\Omega := \mathbb{R}^{d-1} \times \mathbb{R}_{>0}$ and $f \in H^1(\Omega)$; our aim is to make sense of the function $\mathbb{R}^{d-1} \ni \check{x} \mapsto f(\check{x}, 0)$. Note that this makes no sense if we only assume $f \in L_2(\Omega)$ since $\mathbb{R}^{d-1} \times \{0\} = \partial\Omega$ is a set of (d-dimensional) Lebesgue-measure zero. However, if we assume f to be weakly differentiable, something more can be said and the boundary values can be defined by means of a continuous extension of the so-called trace map. In order to properly formulate this, we need the following density result.

Theorem 12.1.1 *The set* $\mathcal{D} := \left\{\phi \colon \Omega \to \mathbb{K}\,;\ \exists \psi \in C_c^\infty(\mathbb{R}^d) \colon \psi|_\Omega = \phi\right\}$ *is dense in the space* $H^1(\Omega)$.

We will need a density result for $H^1(\mathbb{R}^d)$ first.

Lemma 12.1.2 $C_c^\infty(\mathbb{R}^d)$ *is dense in* $H^1(\mathbb{R}^d)$.

Proof Let $f \in H^1(\mathbb{R}^d)$. We first show that f can be approximated by functions with compact support. For this let $\phi \in C_c^\infty(\mathbb{R}^d)$ with the properties $0 \leqslant \phi \leqslant 1$, $\phi = 1$ on $B(0, 1/2)$ and $\phi = 0$ on $\mathbb{R} \setminus B(0, 1)$. For all $k \in \mathbb{N}$ we put $\phi_k := \phi(\cdot/k)$ and $f_k := \phi_k f \in L_2(\mathbb{R}^d)$. Then f_k has support contained in $B[0, k]$. The dominated convergence theorem implies that $f_k \to f$ in $L_2(\mathbb{R}^d)$ as $k \to \infty$. Next,

C. Seifert et al., *Evolutionary Equations*, Operator Theory: Advances and Applications 287, https://doi.org/10.1007/978-3-030-89397-2_12

let $\psi \in C_c^\infty(\mathbb{R}^d)^d$ and compute for all $k \in \mathbb{N}$

$$\begin{aligned}
-\langle f_k, \operatorname{div}\psi\rangle &= -\langle \phi_k f, \operatorname{div}\psi\rangle = -\langle f, \phi_k \operatorname{div}\psi\rangle = -\langle f, \operatorname{div}(\phi_k\psi) - (\operatorname{grad}\phi_k)\cdot\psi\rangle \\
&= -\langle f, \operatorname{div}(\phi_k\psi)\rangle + \langle f \operatorname{grad}\phi_k, \psi\rangle \\
&= \left\langle (\operatorname{grad} f)\phi_k + \frac{1}{k} f(\operatorname{grad}\phi)(\cdot/k), \psi\right\rangle,
\end{aligned}$$

which shows that $f_k \in \operatorname{dom}(\operatorname{grad}) = H^1(\mathbb{R}^d)$ and

$$\operatorname{grad} f_k = (\operatorname{grad} f)\phi_k + \frac{1}{k} f\,(\operatorname{grad}\phi)\,(\cdot/k).$$

From this expression of $\operatorname{grad} f_k$ we observe $\operatorname{grad} f_k \to \operatorname{grad} f$ in $L_2(\mathbb{R}^d)^d$ by dominated convergence. Hence, $f_k \to f$ in $\operatorname{dom}(\operatorname{grad}) = H^1(\mathbb{R}^d)$.

To conclude the proof of this lemma it suffices to revisit Exercise 3.2. For this, let $(\psi_k)_k$ in $C_c^\infty(\mathbb{R}^d)$ be a δ-sequence. Then, by Exercise 3.2, we infer $\psi_k * f \to f$ in $L_2(\mathbb{R}^d)$ as $k \to \infty$ and hence, by Exercise 12.1, it follows also that $\operatorname{grad}(\psi_k * f) = \psi_k * \operatorname{grad} f \to \operatorname{grad} f$ (note the component-wise definition of the convolution). A combination of the first part of this proof together with an estimate for the support of the convolution (see again Exercise 3.2) yields the assertion. □

Proof of Theorem 12.1.1 Let $f \in H^1(\Omega)$. The approximation of f by functions in $\mathcal{D}$ is done in two steps. First, we shift f in the negative e_d-direction to avoid the boundary, and then we convolve the shifted f to obtain smooth approximants in $\mathcal{D}$.

Let $\widetilde{f} \in L_2(\mathbb{R}^d)$ be the extension of f by zero. Put $e_d := (\delta_{jd})_{j\in\{1,\ldots d\}}$, the d-th unit vector. Then for all $\tau > 0$ we have $\Omega + \tau e_d \subseteq \Omega$ and, thus by Exercise 12.2, we deduce $f_\tau := \widetilde{f}(\cdot + \tau e_d)|_\Omega \to f$ in $H^1(\Omega)$ as $\tau \to 0$. Thus, it suffices to approximate f_τ for $\tau > 0$.

Let $\tau > 0$ and let $(\psi_k)_k$ in $C_c^\infty(\mathbb{R}^d)$ be a δ-sequence. Then $\psi_k * \widetilde{f}(\cdot + \tau e_d) \in H^1(\mathbb{R}^d)$, by Exercise 12.1. Define $f_{k,\tau} := \big(\psi_k * \widetilde{f}(\cdot + \tau e_d)\big)|_\Omega$. Then we obtain that $f_{k,\tau} \to f_\tau$ in $H^1(\Omega)$ as $k \to \infty$. Indeed, the only thing left to prove is that $\operatorname{grad} f_{k,\tau} \to \operatorname{grad} f_\tau$ in $L_2(\Omega)^d$ as $k \to \infty$. For this, we denote by g the extension of $\operatorname{grad} f$ by 0. Since $g \in L_2(\mathbb{R}^d)^d$ it suffices to show that $\operatorname{grad} f_{k,\tau} = \psi_k * g_\tau$ on Ω for all large enough $k \in \mathbb{N}$, where $g_\tau = g(\cdot + \tau e_d)$. Let $k > \frac{1}{\tau}$. Then for all $x \in \Omega$ and $y \in \operatorname{spt}\psi_k \subseteq [-1/k, 1/k]^d$ we infer $x - y + \tau e_d \in \Omega$. In particular, $f(\cdot - y + \tau e_d) \in H^1(\Omega)$ and $\operatorname{grad} f(\cdot - y + \tau e_d) = g(\cdot - y + \tau e_d)$. Take $\eta \in C_c^\infty(\Omega)^d$ and compute

$$\begin{aligned}
-\langle f_{k,\tau}, \operatorname{div}\eta\rangle_{L_2(\Omega)} &= -\int_\Omega \int_{\mathbb{R}^d} \psi_k(x-y)\widetilde{f}(y + \tau e_d)^*\,\mathrm{d}y \operatorname{div}\eta(x)\,\mathrm{d}x \\
&= -\int_\Omega \int_{\mathbb{R}^d} \psi_k(y)\widetilde{f}(x - y + \tau e_d)^*\,\mathrm{d}y \operatorname{div}\eta(x)\,\mathrm{d}x \\
&= -\int_\Omega \int_{[-1/k,1/k]^d} \psi_k(y) f(x - y + \tau e_d)^*\,\mathrm{d}y \operatorname{div}\eta(x)\,\mathrm{d}x
\end{aligned}$$

$$
\begin{aligned}
&= -\int_{[-1/k,1/k]^d} \psi_k(y) \,\langle f(\cdot - y + \tau e_d), \operatorname{div}\eta\rangle_{L_2(\Omega)} \,\mathrm{d}y \\
&= \int_{[-1/k,1/k]^d} \psi_k(y) \,\langle g(\cdot - y + \tau e_d), \eta\rangle_{L_2(\Omega)^d} \,\mathrm{d}y \\
&= \langle \psi_k * g_\tau, \eta\rangle_{L_2(\Omega)^d} .
\end{aligned}
$$

As $\psi_k * \widetilde{f}(\cdot + \tau e_d) \in H^1(\mathbb{R}^d)$, we conclude the proof using Lemma 12.1.2. □

With these preparations at hand, we can define the boundary trace of $H^1(\Omega)$.

Theorem 12.1.3 *The operator*

$$
\begin{aligned}
\gamma \colon \mathcal{D} \subseteq H^1(\Omega) &\to L_2(\mathbb{R}^{d-1}) \\
f &\mapsto \left(\mathbb{R}^{d-1} \ni \check{x} \mapsto f(\check{x}, 0)\right)
\end{aligned}
$$

is continuous, densely defined and, thus, admits a unique continuous extension to $H^1(\Omega)$ again denoted by γ. Moreover, we have

$$
\|\gamma f\|_{L_2(\mathbb{R}^{d-1})} \leqslant \left(2\, \|f\|_{L_2(\Omega)} \,\|\operatorname{grad} f\|_{L_2(\Omega)^d}\right)^{\frac{1}{2}} \leqslant \|f\|_{H^1(\Omega)} \quad (f \in H^1(\Omega)).
$$

Proof Note that γ is densely defined by Theorem 12.1.1. Let $f \in C_c^\infty(\mathbb{R}^d)$ and $\check{x} \in \mathbb{R}^{d-1}$. Let $R > 0$ be such that $\operatorname{spt} f \subseteq B(0, R)$. Then

$$
\begin{aligned}
\int_{\mathbb{R}^{d-1}} \left|f(\check{x}, 0)\right|^2 \,\mathrm{d}\check{x} &= -\int_{\mathbb{R}^{d-1}} \int_0^R \partial_d \left|f(\check{x}, \widehat{x})\right|^2 \,\mathrm{d}\widehat{x}\,\mathrm{d}\check{x} \\
&= -\int_\Omega \left(f(x)^* \partial_d f(x) + \partial_d f^*(x) f(x)\right) \mathrm{d}x \\
&\leqslant 2\, \|f\|_{L_2(\Omega)} \,\|\operatorname{grad} f\|_{L_2(\Omega)^d} .
\end{aligned}
$$

The remaining inequality follows from $2ab \leqslant a^2 + b^2$ for all $a, b \in \mathbb{R}$. □

Except for one spatial dimension, where the boundary trace can be obtained by point evaluation, the boundary trace γ does not map onto the whole of $L_2(\mathbb{R}^{d-1})$. Hence, in order to define the space of all possible boundary values for a function in H^1 one uses a quotient construction: we set

$$
H^{1/2}(\mathbb{R}^{d-1}) := \left\{\gamma f\,;\; f \in H^1(\Omega)\right\}
$$

and endow $H^{1/2}(\mathbb{R}^{d-1})$ with the norm

$$
\|\gamma f\|_{H^{1/2}(\mathbb{R}^{d-1})} := \inf\left\{\|g\|_{H^1(\Omega)}\,;\; g \in H^1(\Omega), \gamma g = \gamma f\right\}.
$$

It is not difficult to see that $H^{1/2}(\mathbb{R}^{d-1})$ is unitarily equivalent to $(\ker\gamma)^{\perp}$, where the orthogonal complement is computed with respect to the scalar product in $H^1(\Omega)$. Thus, $H^{1/2}(\mathbb{R}^{d-1})$ is a Hilbert space.

Remark 12.1.4 The norm defined on the space $H^{1/2}(\mathbb{R}^{d-1})$ given above is not the standard norm defined on this space. Indeed, following [72, Section 2.3.8] the usual norm is given by

$$\left(\|u\|^2_{L_2(\mathbb{R}^{d-1})} + \int_{\mathbb{R}^{d-1}}\int_{\mathbb{R}^{d-1}} \frac{|u(x)-u(y)|^2}{|x-y|^d}\,\mathrm{d}x\,\mathrm{d}y\right)^{1/2}$$

for $u \in H^{1/2}(\mathbb{R}^{d-1})$. However, this norm turns out to be equivalent to the norm given above, see e.g. [115, Section 4].

As the notation of this space suggests, it can also be defined as an interpolation space between $H^1(\mathbb{R}^{d-1})$ and $L_2(\mathbb{R}^{d-1})$, see [60, Theorem 15.1].

12.2 The Boundary Values of $H(\mathrm{div}, \mathbb{R}^{d-1} \times \mathbb{R}_{>0})$

Let $\Omega := \mathbb{R}^{d-1} \times \mathbb{R}_{>0}$. There is also a space of corresponding boundary traces for the divergence operator. Similarly to the boundary values for the domain of the gradient operator, $H^1(\Omega)$, the construction of the boundary trace for $H(\mathrm{div})$-vector fields rests on a density result. The proof can be done along the lines of Theorem 12.1.1 and will be addressed in Exercise 12.3.

Theorem 12.2.1 *$\mathcal{D}^d$ is dense in $H(\mathrm{div}, \Omega)$, where $\mathcal{D}$ is defined as in Theorem 12.1.1.*

Equipped with this result, we can describe all possible boundary values of $H(\mathrm{div}, \Omega)$. It will turn out that vector fields in $H(\mathrm{div}, \Omega)$ have a well-defined *normal* trace, which for $\Omega = \mathbb{R}^{d-1} \times \mathbb{R}_{>0}$ is just the negative of the last coordinate of the vector field.

Theorem 12.2.2 *The operator*

$$\begin{aligned} \gamma_{\mathrm{n}} \colon \mathcal{D}^d \subseteq H(\mathrm{div}, \Omega) &\to \left(H^{1/2}(\mathbb{R}^{d-1})\right)' =: H^{-1/2}(\mathbb{R}^{d-1}) \\ q &\mapsto \left(\mathbb{R}^{d-1} \ni \check{x} \mapsto -q_d(\check{x}, 0)\right), \end{aligned}$$

is densely defined, continuous with norm bounded by 1 *and has dense range. Thus γ_{n} admits a unique extension to $H(\mathrm{div}, \Omega)$ again denoted by γ_{n}. Here, $-q_d$ is the negative of the d-th component of q pointing into the outward normal direction of Ω and $-q_d$ is identified with the linear functional*

$$H^{1/2}(\mathbb{R}^d) \ni \gamma f \mapsto \langle -q_d(\cdot, 0), \gamma f\rangle_{L_2(\mathbb{R}^{d-1})}.$$

Moreover, for all $f \in \operatorname{dom}(\operatorname{grad})$ *and* $q \in \operatorname{dom}(\operatorname{div})$ *we have*

$$\langle \operatorname{div} q, f\rangle + \langle q, \operatorname{grad} f\rangle = (\gamma_{\mathrm{n}} q)(\gamma f). \tag{12.1}$$

Proof Let $f \in \mathcal{D}$ and $q \in \mathcal{D}^d$. Then integration by parts yields

$$\begin{aligned}\langle \operatorname{div} q, f\rangle + \langle q, \operatorname{grad} f\rangle &= \int_\Omega \operatorname{div}(q^* f) = \int_{\mathbb{R}^{d-1}} \big\langle q^*(\check{x}, 0) f(\check{x}, 0), -e_d\big\rangle \,\mathrm{d}\check{x} \\ &= -\int_{\mathbb{R}^{d-1}} \gamma q_d^* \gamma f = \langle \gamma_{\mathrm{n}} q, \gamma f\rangle_{L_2(\mathbb{R}^{d-1})} = (\gamma_{\mathrm{n}} q)(\gamma f).\end{aligned}$$

Hence,

$$\left|\langle \gamma_{\mathrm{n}} q, \gamma f\rangle_{L_2(\mathbb{R}^{d-1})}\right| \leqslant \|q\|_{H(\operatorname{div})} \|f\|_{H^1}.$$

Since $\mathcal{D}$ is dense in $H^1(\Omega)$, the inequality remains true for all $f \in H^1(\Omega)$. Thus,

$$\left|\langle \gamma_{\mathrm{n}} q, \gamma f\rangle_{L_2(\mathbb{R}^{d-1})}\right| \leqslant \|q\|_{H(\operatorname{div})} \|f\|_{H^1} \quad (f \subset H^1(\Omega)).$$

Computing the infimum over all $g \in H^1(\Omega)$ with $\gamma g = \gamma f$, we deduce

$$\left|\langle \gamma_{\mathrm{n}} q, \gamma f\rangle_{L_2(\mathbb{R}^{d-1})}\right| \leqslant \|q\|_{H(\operatorname{div})} \|\gamma f\|_{H^{1/2}(\mathbb{R}^{d-1})} \quad (f \in H^1(\Omega)).$$

Therefore $\gamma_{\mathrm{n}} q \in H^{-1/2}(\mathbb{R}^{d-1})$ and $\|\gamma_{\mathrm{n}} q\|_{H^{-1/2}} \leqslant \|q\|_{H(\operatorname{div})}$, which shows continuity of γ_{n}. It is left to show that γ_{n} has dense range. For this, take $\gamma f \in H^{1/2}(\mathbb{R}^{d-1})$ for some $f \in H^1(\Omega)$ such that

$$\langle \gamma_{\mathrm{n}} g, \gamma f\rangle_{L_2(\mathbb{R}^{d-1})} = 0$$

for all $g \in \mathcal{D}^d$. Next, take $\widetilde{g} \in C_{\mathrm{c}}^\infty(\mathbb{R}^{d-1})$ and $\psi \in C_{\mathrm{c}}^\infty(\mathbb{R})$ with $\psi(0) = 1$. Then we set $g \colon \Omega \ni (\check{x}, \widehat{x}) \mapsto -e_d \widetilde{g}(\check{x})\psi(\widehat{x}) \in \mathcal{D}^d$ and note that $\gamma_{\mathrm{n}} g = \widetilde{g}$. Hence

$$\langle \gamma f, \widetilde{g}\rangle_{L_2(\mathbb{R}^{d-1})} = 0 \quad (\widetilde{g} \in C_{\mathrm{c}}^\infty(\mathbb{R}^{d-1})).$$

Thus, $\gamma f = 0$, which implies that the range of γ_{n} is dense, as $H^{-1/2}(\mathbb{R}^{d-1})$ is a Hilbert space. The remaining formula (12.1) follows by continuously extending both the left- and right-hand side of the integration by parts formula from the beginning of the proof. Note that for this, we have used both Theorems 12.1.1 and 12.2.1. □

Corollary 12.2.3 *Let $f \in H^1(\Omega)$, $q \in H(\mathrm{div}, \Omega)$. Then $f \in \mathrm{dom}(\mathrm{grad}_0)$ if and only if $\gamma f = 0$, and $q \in \mathrm{dom}(\mathrm{div}_0)$ if and only if $\gamma_{\mathrm{n}} q = 0$.*

Proof We only show the statement for q. The proof for f is analogous. If $q \in \mathrm{dom}(\mathrm{div}_0)$, then there exists a sequence $(\psi_n)_n$ in $C_c^\infty(\Omega)^d$ such that $\psi_n \to q$ in $H(\mathrm{div}, \Omega)$ as $n \to \infty$. Thus, by continuity of γ_{n}, we infer $0 = \gamma_{\mathrm{n}}\psi_n \to \gamma_{\mathrm{n}} q$. Assume on the other hand that $q \in \mathrm{dom}(\mathrm{div})$ with $\gamma_{\mathrm{n}} q = 0$. Using (12.1), we obtain for all $f \in \mathrm{dom}(\mathrm{grad})$

$$\langle \mathrm{div}\, q, f\rangle + \langle q, \mathrm{grad}\, f\rangle = 0.$$

This equality implies that $q \in \mathrm{dom}(\mathrm{grad}^*) = \mathrm{dom}(\mathrm{div}_0)$, which shows the remaining assertion. □

The remaining part of this section is devoted to showing that the continuous extension of γ_{n} maps onto $H^{-1/2}(\mathbb{R}^{d-1})$. For this we require the following observation, which will also be needed later on.

Proposition 12.2.4 *Let $U \subseteq \mathbb{R}^d$ be open. Then*

$$H_0(\mathrm{div}, U)^{\perp_{H(\mathrm{div},U)}} = \left\{q \in H(\mathrm{div}, U)\,;\ \mathrm{div}\, q \in H^1(U), q = \mathrm{grad}\,\mathrm{div}\, q\right\}.$$

Proof Let $q \in H(\mathrm{div}, U)$. Then $q \in H_0(\mathrm{div}, U)^{\perp_{H(\mathrm{div},U)}}$ if and only if for all $r \in H_0(\mathrm{div}, U)$ we have

$$\begin{aligned} 0 = \langle r, q\rangle_{H(\mathrm{div},U)} &= \langle r, q\rangle_{L_2(U)^d} + \langle \mathrm{div}\, r, \mathrm{div}\, q\rangle_{L_2(U)} \\ &= \langle r, q\rangle_{L_2(U)^d} + \langle \mathrm{div}_0\, r, \mathrm{div}\, q\rangle_{L_2(U)}\,. \end{aligned}$$

The latter, in turn, is equivalent to $\mathrm{div}\, q \in \mathrm{dom}(\mathrm{div}_0^*) = \mathrm{dom}(\mathrm{grad}) = H^1(U)$ and $-\mathrm{grad}\,\mathrm{div}\, q = \mathrm{div}_0^*\,\mathrm{div}\, q = -q$. □

Theorem 12.2.5 *γ_{n} maps onto $H^{-1/2}(\mathbb{R}^{d-1})$. In particular, we have*

$$\|q\|_{H(\mathrm{div},\Omega)} \leqslant \|\gamma_{\mathrm{n}} q\|_{H^{-1/2}(\mathbb{R}^{d-1})}$$

for all $q \in H_0(\mathrm{div}, \Omega)^{\perp_{H(\mathrm{div},\Omega)}}$.

Proof By Theorem 12.2.2 it suffices to show that γ_{n} has closed range. For this, it suffices to show that there exists $c > 0$ such that

$$\|q\|_{H(\mathrm{div},\Omega)} \leqslant c\, \|\gamma_{\mathrm{n}} q\|_{H^{-1/2}(\mathbb{R}^{d-1})}$$

for all $q \in \ker(\gamma_{\mathrm{n}})^{\perp_{H(\mathrm{div},\Omega)}}$. By Corollary 12.2.3, we obtain $\ker(\gamma_{\mathrm{n}}) = H_0(\mathrm{div}, \Omega)$. Hence, by Proposition 12.2.4, we deduce that $q \in \ker(\gamma_{\mathrm{n}})^{\perp_{H(\mathrm{div},\Omega)}}$ if and only if $q \in \mathrm{dom}(\mathrm{grad}\,\mathrm{div})$ and $q = \mathrm{grad}\,\mathrm{div}\, q$. So, assume that $q \in \mathrm{dom}(\mathrm{grad}\,\mathrm{div})$ with

$q = \operatorname{grad}\operatorname{div} q$. Then (12.1) applied to $q \in \operatorname{dom}(\operatorname{div})$ and $f = \operatorname{div} q \in \operatorname{dom}(\operatorname{grad})$ yields

$$\begin{aligned}(\gamma_{\mathrm{n}} q)(\gamma \operatorname{div} q) &= \langle \operatorname{div} q, \operatorname{div} q\rangle + \langle q, \operatorname{grad}\operatorname{div} q\rangle = \langle \operatorname{div} q, \operatorname{div} q\rangle + \langle q, q\rangle \\ &= \|q\|^2_{H(\operatorname{div},\Omega)},\end{aligned}$$

where we used $\operatorname{grad}\operatorname{div} q = q$. Hence

$$\begin{aligned}\|q\|^2_{H(\operatorname{div},\Omega)} &\leqslant \|\gamma \operatorname{div} q\|_{H^{1/2}} \|\gamma_{\mathrm{n}} q\|_{H^{-1/2}} \leqslant \|\operatorname{div} q\|_{H^1(\Omega)} \|\gamma_{\mathrm{n}} q\|_{H^{-1/2}} \\ &= \|q\|_{H(\operatorname{div},\Omega)} \|\gamma_{\mathrm{n}} q\|_{H^{-1/2}}\end{aligned}$$

where we again used that $\operatorname{grad}\operatorname{div} q = q$. This yields the assertion. □

12.3 Inhomogeneous Boundary Value Problems

Let $\Omega := \mathbb{R}^{d-1} \times \mathbb{R}_{>0}$. With the notion of traces we now have a tool at hand that allows us to formulate inhomogeneous boundary value problems. Here we focus on the scalar wave type equation for given Neumann data $\widetilde{g} \in H^{-1/2}(\mathbb{R}^{d-1})$. We shall address other boundary value problems in the exercises. Let $M\colon \operatorname{dom}(M) \subseteq \mathbb{C} \to L\big(L_2(\Omega) \times L_2(\Omega)^d\big)$ be a material law with $\mathrm{s_b}(M) < \nu_0$ for some $\nu_0 \in \mathbb{R}$. We assume that M satisfies the positive definiteness condition in Theorem 6.2.1; that is, we assume there exists $c > 0$ such that for all $z \in \mathbb{C}_{\operatorname{Re}\geqslant \nu_0}$ we have $\operatorname{Re} zM(z) \geqslant c$. For $\nu \geqslant \nu_0$ we want to solve

$$\begin{cases}\left(\partial_{t,\nu} M(\partial_{t,\nu}) + \begin{pmatrix} 0 & \operatorname{div} \\ \operatorname{grad} & 0\end{pmatrix}\right)\begin{pmatrix} v \\ q\end{pmatrix} = \begin{pmatrix} 0 \\ 0\end{pmatrix} & \text{on } \Omega, \\ \gamma_{\mathrm{n}} q(t,\cdot) = \widetilde{g} & \text{on } \partial\Omega \text{ for all } t > 0.\end{cases}$$

Let us reformulate this problem. Let $\phi \in C^\infty(\mathbb{R})$ such that $0 \leqslant \phi \leqslant 1$ with $\phi = 1$ on $[0, \infty)$ and $\phi = 0$ on $(-\infty, -1]$. We define the function

$$g := \big(t \mapsto \phi(t)\widetilde{g} \in H^{-1/2}(\mathbb{R}^{d-1})\big) \in \bigcap_{\nu>0} L_{2,\nu}(\mathbb{R}; H^{-1/2}(\mathbb{R}^{d-1}))$$

and consider

$$\begin{cases}\left(\partial_{t,\nu} M(\partial_{t,\nu}) + \begin{pmatrix} 0 & \operatorname{div} \\ \operatorname{grad} & 0\end{pmatrix}\right)\begin{pmatrix} v \\ q\end{pmatrix} = \begin{pmatrix} 0 \\ 0\end{pmatrix} & \text{on } \Omega, \\ \gamma_{\mathrm{n}} q(t) = g(t) & \text{for all } t > 0\end{cases} \tag{12.2}$$

instead.

Theorem 12.3.1 *Let* $\nu \geqslant \max\{\nu_0, 0\}$, $\nu \neq 0$. *Then (12.2) admits a unique solution* $(v, q) \in H^1_\nu\Big(\mathbb{R}; \operatorname{dom}\Big(\begin{pmatrix} 0 & \operatorname{div} \\ \operatorname{grad} & 0 \end{pmatrix}\Big)\Big)$.

Proof We start with the existence part. By Theorem 12.2.5, we find $\widetilde{G} \in H(\operatorname{div}, \Omega)$ such that $\gamma_{\mathrm{n}}\widetilde{G} = \widetilde{g}$; set $G := \phi(\cdot)\widetilde{G} \in H^3_\nu(\mathbb{R}; H(\operatorname{div}, \Omega))$. Consider the following evolutionary equation

$$\overline{\left(\partial_{t,\nu} M(\partial_{t,\nu}) + \begin{pmatrix} 0 & \operatorname{div}_0 \\ \operatorname{grad} & 0 \end{pmatrix}\right)} \begin{pmatrix} u \\ r \end{pmatrix} = \partial_{t,\nu} M(\partial_{t,\nu}) \begin{pmatrix} 0 \\ -G \end{pmatrix} + \begin{pmatrix} -\operatorname{div} G \\ 0 \end{pmatrix}.$$

Note that the right-hand side is in $H^2_\nu(\mathbb{R}; L_2(\Omega) \times L_2(\Omega)^d)$. By Theorem 6.2.1, we obtain

$$\begin{aligned} \begin{pmatrix} u \\ r \end{pmatrix} &= \overline{\left(\partial_{t,\nu} M(\partial_{t,\nu}) + \begin{pmatrix} 0 & \operatorname{div}_0 \\ \operatorname{grad} & 0 \end{pmatrix}\right)}^{-1} \left(\partial_{t,\nu} M(\partial_{t,\nu}) \begin{pmatrix} 0 \\ -G \end{pmatrix} + \begin{pmatrix} -\operatorname{div} G \\ 0 \end{pmatrix}\right) \\ &\in H^1_\nu(\mathbb{R}; L_2(\Omega) \times L_2(\Omega)^d) \cap L_{2,\nu}\Big(\mathbb{R}; \operatorname{dom}\Big(\begin{pmatrix} 0 & \operatorname{div} \\ \operatorname{grad} & 0 \end{pmatrix}\Big)\Big). \end{aligned}$$

Indeed, since the solution operator commutes with $\partial_{t,\nu}$ and the right-hand side lies in H^2_ν, it even follows that $\begin{pmatrix} u \\ r \end{pmatrix} \in H^2_\nu(\mathbb{R}; L_2(\Omega) \times L_2(\Omega)^d)$. From the equality

$$\left(\partial_{t,\nu} M(\partial_{t,\nu}) + \begin{pmatrix} 0 & \operatorname{div}_0 \\ \operatorname{grad} & 0 \end{pmatrix}\right) \begin{pmatrix} u \\ r \end{pmatrix} = \partial_{t,\nu} M(\partial_{t,\nu}) \begin{pmatrix} 0 \\ -G \end{pmatrix} + \begin{pmatrix} -\operatorname{div} G \\ 0 \end{pmatrix}$$

it follows that

$$\left(\begin{pmatrix} 0 & \operatorname{div}_0 \\ \operatorname{grad} & 0 \end{pmatrix}\right) \begin{pmatrix} u \\ r \end{pmatrix} \in H^1_\nu(\mathbb{R}; L_2(\Omega) \times L_2(\Omega)^d).$$

Hence,

$$\begin{aligned} \begin{pmatrix} u \\ r \end{pmatrix} &\in \left(1 + \begin{pmatrix} 0 & \operatorname{div}_0 \\ \operatorname{grad} & 0 \end{pmatrix}\right)^{-1} [H^1_\nu(\mathbb{R}; L_2(\Omega) \times L_2(\Omega)^d] \\ &\subseteq H^1_\nu\Big(\mathbb{R}; \operatorname{dom}\Big(\begin{pmatrix} 0 & \operatorname{div}_0 \\ \operatorname{grad} & 0 \end{pmatrix}\Big)\Big), \end{aligned}$$

where the resolvent is well-defined since $\begin{pmatrix} 0 & \operatorname{div}_0 \\ \operatorname{grad} & 0 \end{pmatrix}$ is skew-selfadjoint. Also, we deduce that

$$\left(\partial_{t,\nu} M(\partial_{t,\nu}) + \begin{pmatrix} 0 & \operatorname{div} \\ \operatorname{grad} & 0 \end{pmatrix}\right) \begin{pmatrix} u \\ r+G \end{pmatrix} = \begin{pmatrix} 0 \\ 0 \end{pmatrix}.$$

Since $r \in H_\nu^1(\mathbb{R}; \operatorname{dom}(\operatorname{div}_0))$, by Corollary 12.2.3 and Theorem 4.1.2 we obtain

$$\gamma_{\mathrm{n}}\left((r+G)(t)\right) = \gamma_{\mathrm{n}} G(t) = g(t) \quad (t \in \mathbb{R}).$$

Hence, $(u, r+G)$ solves (12.2).

Next we address the uniqueness result. For this we note that a straightforward computation shows

$$\begin{pmatrix} v \\ q-G \end{pmatrix} = \left(\overline{\partial_{t,\nu} M(\partial_{t,\nu}) + \begin{pmatrix} 0 & \operatorname{div}_0 \\ \operatorname{grad} & 0 \end{pmatrix}}\right)^{-1} \left(\partial_{t,\nu} M(\partial_{t,\nu}) \begin{pmatrix} 0 \\ -G \end{pmatrix} + \begin{pmatrix} -\operatorname{div} G \\ 0 \end{pmatrix}\right),$$

which coincides with the formula for $(u, r+G)$. □

The upshot of the rationale exemplified in the proof is that inhomogeneous boundary value problems can be reduced to an evolutionary equation of the standard form with non-vanishing right-hand side. The treatment of inhomogeneous Dirichlet data works along similar lines.

12.4 Abstract Boundary Data Spaces

Of course inhomogeneous boundary value problems can be addressed for other domains Ω than the half-space $\mathbb{R}^{d-1} \times \mathbb{R}_{>0}$. Classically, some more specific properties need to be imposed on the description of the boundary $\partial\Omega$. In this section, however, we deviate from the classical perspective in as much as we like to consider *arbitrary* open sets $\Omega \subseteq \mathbb{R}^d$. For this we introduce

$$\operatorname{BD}(\operatorname{div}) = \{q \in H(\operatorname{div}, \Omega)\,;\ \operatorname{div} q \in \operatorname{dom}(\operatorname{grad}), \operatorname{grad}\operatorname{div} q = q\},$$

$$\operatorname{BD}(\operatorname{grad}) = \left\{u \in H^1(\Omega)\,;\ \operatorname{grad} u \in \operatorname{dom}(\operatorname{div}), \operatorname{div}\operatorname{grad} u = u\right\}.$$

By Proposition 12.2.4 and Exercise 6.7, these spaces are closed subspaces of $H(\operatorname{div}, \Omega)$ and $H^1(\Omega)$, respectively, and therefore Hilbert spaces. Indeed,

$$\operatorname{BD}(\operatorname{div}) = H_0(\operatorname{div}, \Omega)^{\perp_{H(\operatorname{div},\Omega)}}$$

and

$$\mathrm{BD}(\mathrm{grad}) = H_0^1(\Omega)^{\perp_{H^1(\Omega)}}.$$

Now, we are in a position to solve inhomogeneous boundary value problems, where the trace mappings γ and γ_{n} are replaced by the canonical orthogonal projections $\pi_{\mathrm{BD(grad)}}$ and $\pi_{\mathrm{BD(div)}}$ respectively; see Exercise 12.4. We devote the rest of this section to describe the relationship between the classical trace spaces introduced before and the BD-spaces. In the perspective outlined here, there is not much of a difference between Neumann boundary values and Dirichlet boundary values. The next result is an incarnation of this.

Proposition 12.4.1 *We have*

$$\mathrm{grad}[\mathrm{BD}(\mathrm{grad})] \subseteq \mathrm{BD}(\mathrm{div}) \quad \textit{and} \quad \mathrm{div}[\mathrm{BD}(\mathrm{div})] \subseteq \mathrm{BD}(\mathrm{grad}).$$

Moreover, the mappings

$$\begin{aligned} \mathrm{grad}_{\mathrm{BD}} \colon \mathrm{BD}(\mathrm{grad}) &\to \mathrm{BD}(\mathrm{div}), \\ u &\mapsto \mathrm{grad}\, u \end{aligned}$$

and

$$\begin{aligned} \mathrm{div}_{\mathrm{BD}} \colon \mathrm{BD}(\mathrm{div}) &\to \mathrm{BD}(\mathrm{grad}), \\ q &\mapsto \mathrm{div}\, q \end{aligned}$$

are unitary, and $\mathrm{grad}_{\mathrm{BD}}^* = \mathrm{div}_{\mathrm{BD}}$.

Proof Let $\phi \in \mathrm{BD}(\mathrm{grad})$. Then $\mathrm{grad}\,\phi \in H(\mathrm{div}, \Omega)$ and $\mathrm{div}\,\mathrm{grad}\,\phi = \phi$. This implies $\mathrm{div}\,\mathrm{grad}\,\phi \in \mathrm{dom}(\mathrm{grad})$ and $\mathrm{grad}\,\mathrm{div}\,\mathrm{grad}\,\phi = \mathrm{grad}\,\phi$, which yields $\mathrm{grad}\,\phi \in \mathrm{BD}(\mathrm{div})$. Thus, $\mathrm{grad}_{\mathrm{BD}}$ is defined everywhere; interchanging the roles of grad and div, we obtain $\mathrm{div}_{\mathrm{BD}}$ is also defined everywhere. We infer $\mathrm{div}_{\mathrm{BD}}\,\mathrm{grad}_{\mathrm{BD}} = 1_{\mathrm{BD(grad)}}$ and $\mathrm{grad}_{\mathrm{BD}}\,\mathrm{div}_{\mathrm{BD}} = 1_{\mathrm{BD(div)}}$ and thus $\mathrm{grad}_{\mathrm{BD}}$ is bijective with $\mathrm{grad}_{\mathrm{BD}}^{-1} = \mathrm{div}_{\mathrm{BD}}$. It remains to show that $\mathrm{grad}_{\mathrm{BD}}$ preserves the norm. For this we compute

$$\begin{aligned} \left\langle \mathrm{grad}_{\mathrm{BD}}\,\phi, \mathrm{grad}_{\mathrm{BD}}\,\phi \right\rangle_{\mathrm{BD(div)}} &= \langle \mathrm{grad}\,\phi, \mathrm{grad}\,\phi \rangle_{H(\mathrm{div})} \\ &= \langle \mathrm{grad}\,\phi, \mathrm{grad}\,\phi \rangle_{L_2(\Omega)^d} + \langle \mathrm{div}\,\mathrm{grad}\,\phi, \mathrm{div}\,\mathrm{grad}\,\phi \rangle_{L_2(\Omega)} \\ &= \langle \mathrm{grad}\,\phi, \mathrm{grad}\,\phi \rangle_{L_2(\Omega)^d} + \langle \phi, \phi \rangle_{L_2(\Omega)} \\ &= \langle \phi, \phi \rangle_{\mathrm{dom(grad)}} = \langle \phi, \phi \rangle_{\mathrm{BD(grad)}}, \end{aligned}$$

which implies that $\mathrm{grad}_{\mathrm{BD}}$ is unitary. Hence, $\mathrm{div}_{\mathrm{BD}} = \mathrm{grad}_{\mathrm{BD}}^{-1} = \mathrm{grad}_{\mathrm{BD}}^*$. □

It is also possible to show an 'integration by parts' formula analogous to (12.1) for the abstract situation:

Proposition 12.4.2 *Let $u \in H^1(\Omega)$ and $q \in H(\mathrm{div}, \Omega)$. Then*

$$\begin{aligned}\langle \operatorname{div} q, u\rangle_{L_2(\Omega)} + \langle q, \operatorname{grad} u\rangle_{L_2(\Omega)^d} &= \left\langle \operatorname{div}_{\mathrm{BD}} \pi_{\mathrm{BD(div)}} q, \pi_{\mathrm{BD(grad)}} u\right\rangle_{\mathrm{BD(grad)}} \\ &= \left\langle \pi_{\mathrm{BD(div)}} q, \operatorname{grad}_{\mathrm{BD}} \pi_{\mathrm{BD(grad)}} u\right\rangle_{\mathrm{BD(div)}} .\end{aligned}$$

Proof We decompose $u = u_0 + u_1$ and $q = q_0 + q_1$ with $u_0 \in H_0^1(\Omega)$, $q_0 \in H_0(\mathrm{div}, \Omega)$, $u_1 = \pi_{\mathrm{BD(grad)}} u$ and $q_1 = \pi_{\mathrm{BD(div)}} q$. Then we obtain

$$\begin{aligned}&\langle \operatorname{div} q, u\rangle_{L_2(\Omega)} + \langle q, \operatorname{grad} u\rangle_{L_2(\Omega)^d} \\ &= \langle \operatorname{div}_0 q_0, u\rangle_{L_2(\Omega)} + \langle \operatorname{div} q_1, u\rangle_{L_2(\Omega)} + \langle q_0, \operatorname{grad} u\rangle_{L_2(\Omega)^d} + \langle q_1, \operatorname{grad} u\rangle_{L_2(\Omega)^d} \\ &= \langle q_0, -\operatorname{grad} u\rangle_{L_2(\Omega)^d} + \langle \operatorname{div} q_1, u\rangle_{L_2(\Omega)} + \langle q_0, \operatorname{grad} u\rangle_{L_2(\Omega)^d} + \langle q_1, \operatorname{grad} u\rangle_{L_2(\Omega)^d} \\ &= \langle \operatorname{div} q_1, u_0\rangle_{L_2(\Omega)} + \langle \operatorname{div} q_1, u_1\rangle_{L_2(\Omega)} + \langle q_1, \operatorname{grad} u_0\rangle_{L_2(\Omega)^d} + \langle q_1, \operatorname{grad} u_1\rangle_{L_2(\Omega)^d} \\ &= \left\langle q_1, -\operatorname{grad}_0 u_0\right\rangle_{L_2(\Omega)^d} + \langle \operatorname{div} q_1, u_1\rangle_{L_2(\Omega)} + \left\langle q_1, \operatorname{grad}_0 u_0\right\rangle_{L_2(\Omega)^d} + \langle q_1, \operatorname{grad} u_1\rangle_{L_2(\Omega)^d} \\ &= \langle \operatorname{div} q_1, u_1\rangle_{L_2(\Omega)} + \langle q_1, \operatorname{grad} u_1\rangle_{L_2(\Omega)^d} \\ &= \langle \operatorname{div} q_1, u_1\rangle_{L_2(\Omega)} + \langle \operatorname{grad} \operatorname{div} q_1, \operatorname{grad} u_1\rangle_{L_2(\Omega)^d} = \langle \operatorname{div} q_1, u_1\rangle_{\mathrm{BD(grad)}} .\end{aligned}$$

The remaining equality follows from $\operatorname{div}_{\mathrm{BD}}^* = \operatorname{grad}_{\mathrm{BD}}$ by Proposition 12.4.1. □

In view of Proposition 12.4.2 the proper replacement of γ_{n} appears to be $\operatorname{div}_{\mathrm{BD}} \pi_{\mathrm{BD(div)}}$ instead of just $\pi_{\mathrm{BD(div)}}$. Next, we show the equivalence of the trace spaces for the half-space and the abstract ones introduced in this section.

Theorem 12.4.3 *Let $\Omega := \mathbb{R}^{d-1} \times \mathbb{R}_{>0}$. Then $\gamma|_{\mathrm{BD(grad)}} \colon \mathrm{BD(grad)} \to H^{1/2}(\mathbb{R}^{d-1})$ and $\gamma_{\mathrm{n}}|_{\mathrm{BD(div)}} \colon \mathrm{BD(div)} \to H^{-1/2}(\mathbb{R}^{d-1})$ are unitary mappings.*

Proof We begin with γ_{n}. We have shown in Theorem 12.2.2 that $\gamma_{\mathrm{n}}|_{\mathrm{BD(div)}}$ is continuous and in Theorem 12.2.5 it has been shown that $(\gamma_{\mathrm{n}}|_{\mathrm{BD(div)}})^{-1}$ is continuous. Also the two norm inequalities have been established.

The injectivity of $\gamma|_{\mathrm{BD(grad)}}$ follows from $\ker \gamma = H_0^1(\Omega)$ by Corollary 12.2.3. All that remains simply relies upon recalling that $H^{1/2}(\mathbb{R}^{d-1})$ is isomorphic to $(\ker \gamma)^{\perp}$ with the orthogonal complement computed in $H^1(\Omega)$. □

12.5 Robin Boundary Conditions

The classical Robin boundary conditions involve both traces, the Dirichlet trace γ and the Neumann trace γ_{n}. To motivate things, let us again have a look at the case $\Omega = \mathbb{R}^{d-1} \times \mathbb{R}_{>0}$. We consider the boundary condition for given $q \in H(\mathrm{div}, \Omega)$

and $u \in H^1(\Omega)$

$$\gamma_{\mathrm{n}} q + \mathrm{i}\gamma u = 0,$$

in the sense that

$$(\gamma_{\mathrm{n}} q)(v) = \langle -\mathrm{i}\gamma u, v\rangle_{L_2(\mathbb{R}^{d-1})} \quad (v \in H^{1/2}(\mathbb{R}^{d-1})).$$

Note that this is an implicit regularity statement as $\gamma_{\mathrm{n}} q \in H^{-1/2}(\mathbb{R}^{d-1})$ is representable as an $L_2(\mathbb{R}^{d-1})$ function. The next result asserts that an evolutionary equation with a spatial operator of the type $\begin{pmatrix} 0 & \operatorname{div} \\ \operatorname{grad} & 0 \end{pmatrix}$ with the above Robin boundary condition fits into the setting rendered by Theorem 6.2.1. In other words:

Theorem 12.5.1 *Let $\Omega = \mathbb{R}^{d-1} \times \mathbb{R}_{>0}$. Then the operator $A \colon \operatorname{dom}(A) \subseteq L_2(\Omega)^{d+1} \to L_2(\Omega)^{d+1}$ with $A \subseteq \begin{pmatrix} 0 & \operatorname{div} \\ \operatorname{grad} & 0 \end{pmatrix}$ with domain*

$$\operatorname{dom}(A) = \left\{(u, q) \in H^1(\Omega) \times H(\operatorname{div}, \Omega)\,;\ \gamma_{\mathrm{n}} q + \mathrm{i}\gamma u = 0\right\}$$

is skew-selfadjoint.

Proof Let $(u, q), (v, r) \in H^1(\Omega) \times H(\operatorname{div}, \Omega)$. Then, by (12.1) we obtain

$$\left\langle \begin{pmatrix} 0 & \operatorname{div} \\ \operatorname{grad} & 0 \end{pmatrix}\begin{pmatrix} u \\ q \end{pmatrix}, \begin{pmatrix} v \\ r \end{pmatrix}\right\rangle + \left\langle \begin{pmatrix} u \\ q \end{pmatrix}, \begin{pmatrix} 0 & \operatorname{div} \\ \operatorname{grad} & 0 \end{pmatrix}\begin{pmatrix} v \\ r \end{pmatrix}\right\rangle$$
$$= \langle \operatorname{div} q, v\rangle + \langle \operatorname{grad} u, r\rangle + \langle u, \operatorname{div} r\rangle + \langle q, \operatorname{grad} v\rangle = (\gamma_{\mathrm{n}} q)(\gamma v) + ((\gamma_{\mathrm{n}} r)(\gamma u))^*$$

If, in addition, $(u, q) \in \operatorname{dom}(A)$, we obtain

$$\left\langle A\begin{pmatrix} u \\ q \end{pmatrix}, \begin{pmatrix} v \\ r \end{pmatrix}\right\rangle + \left\langle \begin{pmatrix} u \\ q \end{pmatrix}, \begin{pmatrix} 0 & \operatorname{div} \\ \operatorname{grad} & 0 \end{pmatrix}\begin{pmatrix} v \\ r \end{pmatrix}\right\rangle$$
$$= (\gamma_{\mathrm{n}} q)(\gamma v) + ((\gamma_{\mathrm{n}} r)(\gamma u))^* = \langle -\mathrm{i}\gamma u, \gamma v\rangle_{L_2(\mathbb{R}^{d-1})} + ((\gamma_{\mathrm{n}} r)(\gamma u))^*$$
$$= \langle \gamma u, \mathrm{i}\gamma v\rangle_{L_2(\mathbb{R}^{d-1})} + ((\gamma_{\mathrm{n}} r)(\gamma u))^* = ((\mathrm{i}\gamma v + \gamma_{\mathrm{n}} r)(\gamma u))^*.$$

Since for every $u \in \mathcal{D}$, we find $q \in \mathcal{D}^d$ such that $(u, q) \in \operatorname{dom}(A)$,

$$\gamma[\mathcal{D}] \subseteq \{\gamma u\,;\ \exists q \in H(\operatorname{div}, \Omega) \colon (u, q) \in \operatorname{dom}(A)\}.$$

Thus, the set on the right-hand side is dense in $H^{1/2}(\mathbb{R}^{d-1})$. This in turn implies that $(v, r) \in \operatorname{dom}(A^*)$ if and only if $\mathrm{i}\gamma v + \gamma_{\mathrm{n}} r = 0$, and in this case we have $A^*(v, r) = -A(v, r)$. This implies that A is skew-selfadjoint. □

Remark 12.5.2 The factor i in front of γu is chosen as a mere convenience in order to render the corresponding operator A in Theorem 12.5.1 skew-selfadjoint. It is also possible to choose $\beta \in L(H^{1/2}(\partial\Omega))$ with $-\operatorname{Re}\beta \geqslant 0$ instead of i. Then one obtains for all $U \in \operatorname{dom}(A)$ and $V \in \operatorname{dom}(A^*)$ the estimates $\operatorname{Re}\langle U, AU\rangle \geqslant 0$ and $\operatorname{Re}\langle V, A^*V\rangle \geqslant 0$. Appealing to Remark 6.3.3, it can be shown that the corresponding evolutionary equation

$$(\partial_{t,\nu}M(\partial_{t,\nu}) + A)U = F$$

for a suitable material law M as in Theorem 6.2.1 is well-posed.

Next, one could argue that in the case of arbitrary Ω, the condition

$$\mathrm{i}\pi_{\mathrm{BD(grad)}}u + \operatorname{div}_{\mathrm{BD}}\pi_{\mathrm{BD(div)}}q = 0 \tag{12.3}$$

amounts to a generalisation of the Robin boundary condition just considered. However, this is not true as the following proposition shows.

Proposition 12.5.3 *Let $u \in H^1(\Omega)$, and $q \in H(\operatorname{div}, \Omega)$. Moreover, we set $\kappa\colon \mathrm{BD(grad)} \to L_2(\mathbb{R}^{d-1})$ with $\kappa v = \gamma v$ for $v \in \mathrm{BD(grad)}$. Then $\gamma_{\mathrm{n}}q + \mathrm{i}\gamma u = 0$ if and only if*

$$\operatorname{div}_{\mathrm{BD}}\pi_{\mathrm{BD(div)}}q + \mathrm{i}\kappa^*\kappa\pi_{\mathrm{BD(grad)}}u = 0.$$

Proof We first observe that $\kappa\pi_{\mathrm{BD(grad)}}w = \gamma w$ for each $w \in H^1(\Omega)$.
Assume now that $\gamma_{\mathrm{n}}q + \mathrm{i}\gamma u = 0$ and let $v \in \mathrm{BD(grad)}$. Then we compute, using Proposition 12.4.2 and (12.1)

$$\begin{aligned}\left\langle \mathrm{i}\kappa^*\kappa\pi_{\mathrm{BD(grad)}}u, v\right\rangle_{\mathrm{BD(grad)}} &= \left\langle \mathrm{i}\kappa\pi_{\mathrm{BD(grad)}}u, \kappa v\right\rangle_{L_2(\mathbb{R}^{d-1})} = \langle \mathrm{i}\gamma u, \gamma v\rangle_{L_2(\mathbb{R}^{d-1})} \\ &= -(\gamma_{\mathrm{n}}q)(\gamma v) = \langle -\operatorname{div} q, v\rangle_{L_2(\Omega)} + \langle -q, \operatorname{grad} v\rangle_{L_2(\Omega)^d} \\ &= \left\langle -\operatorname{div}_{\mathrm{BD}}\pi_{\mathrm{BD(div)}}q, v\right\rangle_{\mathrm{BD(grad)}},\end{aligned}$$

which proves one of the asserted implications.

Assume that $\operatorname{div}_{\mathrm{BD}}\pi_{\mathrm{BD(div)}}q + \mathrm{i}\kappa^*\kappa\pi_{\mathrm{BD(grad)}}u = 0$ and let $v \in H^{1/2}(\mathbb{R}^{d-1})$. We take $w \in H^1(\Omega)$ with $\gamma w = v$ and compute

$$\begin{aligned}(\gamma_{\mathrm{n}}q)(v) &= \langle \operatorname{div} q, w\rangle_{L_2(\Omega)} + \langle q, \operatorname{grad} w\rangle_{L_2(\Omega)^d} \\ &= \left\langle \operatorname{div}_{\mathrm{BD}}\pi_{\mathrm{BD(div)}}q, \pi_{\mathrm{BD(grad)}}w\right\rangle_{\mathrm{BD(grad)}} \\ &= \left\langle -\mathrm{i}\kappa^*\kappa\pi_{\mathrm{BD(grad)}}u, \pi_{\mathrm{BD(grad)}}w\right\rangle_{\mathrm{BD(grad)}} \\ &= \left\langle -\mathrm{i}\kappa\pi_{\mathrm{BD(grad)}}u, \kappa\pi_{\mathrm{BD(grad)}}w\right\rangle_{L_2(\mathbb{R}^{d-1})} \\ &= \langle -\mathrm{i}\gamma u, v\rangle_{L_2(\mathbb{R}^{d-1})},\end{aligned}$$

which shows the remaining implication. □

12.6 Comments

The concept of abstract trace spaces has been introduced in [86] in order to study a multi-dimensional analogue for port-Hamiltonian systems. Also concerning differential equations at the boundary (so-called impedance type boundary conditions), the concept of abstract boundary valuc spaces has been employed, see [91].

A comparison between abstract and classical trace spaces has been provided in [37, 115] particularly concerning $H^{-1/2}(\mathbb{R}^{d-1})$. A good introduction for trace mappings for more complicated geometries can be found e.g. in [5]. The trace operator can also be suitably established for $H(\mathrm{curl}, \Omega)$-regular vector fields given that Ω is a so-called Lipschitz domain, see [18].

Exercises

Exercise 12.1 Let $\phi \in C_{\mathrm{c}}^{\infty}(\mathbb{R}^d)$, $f \in L_2(\mathbb{R}^d)$. Show that

$$\phi * f : x \mapsto \int_{\mathbb{R}^d} \phi(x-y) f(y)\, \mathrm{d}y$$

belongs to $H^1(\mathbb{R}^d)$ and that $\mathrm{grad}\,(\phi * f) = (\mathrm{grad}\,\phi) * f$. If, in addition, $f \in H^1(\mathbb{R}^d) = \mathrm{dom}(\mathrm{grad})$, then $\mathrm{grad}(\phi * f) = \phi * \mathrm{grad}\, f$, where the convolution is always taken component wise.

Exercise 12.2 Let $\Omega \subseteq \mathbb{R}^d$ be open. Let $f \in L_2(\Omega)$ and denote by $\widetilde{f} \in L_2(\mathbb{R}^d)$ the extension of f by zero. Let $v \in \mathbb{R}^d$, $\tau > 0$ and define $f_\tau := \widetilde{f}(\cdot + \tau v)|_{\Omega}$.

(a) Show that $f_\tau \to f$ in $L_2(\Omega)$ as $\tau \to 0$.
(b) Let now $f \in H^1(\Omega)$ and $\Omega + \tau v \subseteq \Omega$ for all $\tau > 0$. Show that $f_\tau \to f$ in $H^1(\Omega)$ as $\tau \to 0$.

Exercise 12.3 Prove Theorem 12.2.1.

Exercise 12.4 Let $\Omega \subseteq \mathbb{R}^d$ be open, $M\colon \mathrm{dom}(M) \subseteq \mathbb{C} \to L\big(L_2(\Omega) \times L_2(\Omega)^d\big)$ with $\mathrm{s_b}\,(M) < \nu_0$ for some $\nu_0 \in \mathbb{R}$, $c > 0$ such that for all $z \in \mathbb{C}_{\mathrm{Re} \geqslant \nu_0}$ we have $\mathrm{Re}\, zM(z) \geqslant c$, $\nu \geqslant \max\{\nu_0, 0\}$ and $\nu \neq 0$. Show that there exists a unique

$$\begin{pmatrix} v \\ q \end{pmatrix} \in H_\nu^1\left(\mathbb{R}; \mathrm{dom}\left(\begin{pmatrix} 0 & \mathrm{div} \\ \mathrm{grad} & 0 \end{pmatrix}\right)\right)$$

satisfying

$$\begin{cases} \left(\partial_{t,\nu} M(\partial_{t,\nu}) + \begin{pmatrix} 0 & \operatorname{div} \\ \operatorname{grad} & 0 \end{pmatrix}\right) \begin{pmatrix} v \\ q \end{pmatrix} = \begin{pmatrix} 0 \\ 0 \end{pmatrix} & \text{on } \Omega, \\ \pi_{\mathrm{BD(grad)}} v(t) = \phi(t) f & \text{for all } t \in \mathbb{R}, \end{cases}$$

for some bounded $\phi \in C^\infty(\mathbb{R})$ with $\inf \operatorname{spt} \phi > -\infty$ and $f \in \mathrm{BD(grad)}$.

Exercise 12.5 Let $\Omega = \mathbb{R}^{d-1} \times \mathbb{R}_{>0}$. Show that there exists a continuous linear operator $E : H^1(\Omega) \to H^1(\mathbb{R}^d)$ such that $E(\phi)|_\Omega = \phi$ for each $\phi \in H^1(\Omega)$.

Exercise 12.6 (Korn's Second Inequality) Let $\Omega = \mathbb{R}^{d-1} \times \mathbb{R}_{>0}$. Using Exercise 12.5 show that there exists $c > 0$ such that for all $\phi \in H^1(\Omega)^d$ we have

$$\|\phi\|_{H^1(\Omega)^d} \leqslant c \left(\|\phi\|_{L_2(\Omega)^d} + \|\operatorname{Grad} \phi\|_{L_2(\Omega)^{d\times d}}\right).$$

Thus, describe the space of boundary values of dom(Grad).

Hint: Prove a corresponding result for $\Omega = \mathbb{R}^d$ first after having shown that $C_c^\infty(\mathbb{R}^d)^d$ forms a dense subset of both $H^1(\Omega)^d$ and dom(Grad).

Exercise 12.7 Let $\Omega \subseteq \mathbb{R}^3$ be open. Compute $\mathrm{BD(curl)} := H_0(\mathrm{curl}, \Omega)^{\perp_{H(\mathrm{curl},\Omega)}}$ and show that $\mathrm{curl} : \mathrm{BD(curl)} \to \mathrm{BD(curl)}$ is well-defined, unitary and skew-selfadjoint.

References

5. W. Arendt et al., Form methods for evolution equations, and applications, in *18th Internet Seminar* (2015)
18. A. Buffa, M. Costabel, D. Sheen, On traces for **H(curl**, Ω) in Lipschitz domains. J. Math. Anal. Appl. **276**(2), 845–867 (2002)
37. A.F.M. ter Elst, G. Gorden, M. Waurick, The Dirichlet-to-Neumann operator for divergence form problems. Ann. Mat. Pura Appl. (4) **198**(1), 177–203 (2019)
60. J.-L. Lions, E. Magenes, *Non-Homogeneous Boundary Value Problems and Applications*, vol. I. Translated from the French by P. Kenneth, Die Grundlehren der mathematischen Wissenschaften, Band 181 (Springer, New York, 1972)
72. J. Nečas, *Direct Methods in the Theory of Elliptic Equations*. Springer Monographs in Mathematics. Translated from the 1967 French original by Gerard Tronel and Alois Kufner, Editorial coordination and preface by Šárka Nečasová and a contribution by Christian G. Simader (Springer, Heidelberg, 2012)
86. R. Picard, S. Trostorff, M. Waurick, On a comprehensive class of linear control problems. IMA J. Math. Control Inf. **33**(2), 257–291 (2016)
91. R. Picard et al., On Abstract grad-div Systems. J. Differ. Equ. **260**(6), 4888–4917 (2016)
115. S. Trostorff, A characterization of boundary conditions yielding maximal monotone operators. J. Funct. Anal. **267**(8), 2787–2822 (2014)

BY

Chapter 13
Continuous Dependence on the Coefficients I

The power of the functional analytic framework for evolutionary equations lies in its variety. In fact, as we have outlined in earlier chapters, it is possible to formulate many differential equations in the form

$$(\partial_t M(\partial_t) + A)\, U = F.$$

In this chapter we want to use this versatility and address continuity of the above expression (or more precisely of the solution operator) in $M(\partial_t)$. To see this more clearly, fix F and take a sequence of material laws $(M_n)_n$. We will address the following question: what are the conditions or notions of convergence of $(M_n)_n$ to some M in order that $(U_n)_n$ with U_n given as the solution of

$$(\partial_t M_n(\partial_t) + A)\, U_n = F$$

converges to U, which satisfies

$$(\partial_t M(\partial_t) + A)\, U = F?$$

In the first of two chapters on this subject, we shall specialise to $A = 0$; that is, we will discuss ordinary differential equations with infinite-dimensional state space. To begin with, we address the convergence of material laws pointwise in the Fourier–Laplace transformed domain and its relation to the convergence of material laws evaluated at the time derivative.

C. Seifert et al., *Evolutionary Equations*, Operator Theory: Advances and Applications 287, https://doi.org/10.1007/978-3-030-89397-2_13

13.1 Convergence of Material Laws

Throughout, let H be a Hilbert space. We briefly recall that a sequence $(T_n)_n$ in $L(H)$ converges in the *strong operator topology* to some $T \in L(H)$ if for all $x \in H$ we have

$$T_n x \to Tx \quad (n \to \infty).$$

$(T_n)_n$ is said to converge in the *weak operator topology* to $T \in L(H)$ if for all $x, y \in H$ we have

$$\langle y, T_n x\rangle \to \langle y, Tx\rangle \quad (n \to \infty).$$

We denote the set of material laws on H with abscissa of boundedness less than or equal to $\nu_0 \in \mathbb{R}$ by

$$\mathcal{M}(H, \nu_0) := \{M\colon \operatorname{dom}(M) \to L(H)\,;\ M \text{ material law}, \mathrm{s_b}(M) \leqslant \nu_0\}.$$

Remark 13.1.1 Let $\nu_0 \in \mathbb{R}$, $\nu > \nu_0$. Then $\mathcal{M}(H, \nu_0)$ is an algebra and $\mathcal{M}(H, \nu_0) \ni M \mapsto M(\partial_{t,\nu}) \in L\big(L_{2,\nu}(\mathbb{R}; H)\big)$ is an algebra homomorphism which is one-to-one by Theorem 8.2.1.

Definition Let $\nu_0 \in \mathbb{R}$. A sequence $(M_n)_{n\in\mathbb{N}}$ in $\mathcal{M}(H, \nu_0)$ is called *bounded* if

$$\sup_{n\in\mathbb{N}} \|M_n\|_{\infty, \mathbb{C}_{\mathrm{Re}>\nu_0}} < \infty.$$

Theorem 13.1.2 *Let $\nu_0 \in \mathbb{R}$, $(M_n)_n$ in $\mathcal{M}(H, \nu_0)$ be bounded. Assume that for all $z \in \mathbb{C}_{\mathrm{Re}>\nu_0}$ the sequence $(M_n(z))_n$ converges in the weak operator topology of $L(H)$ with limit $M(z)$ and let $\nu > \nu_0$. Then $M \in \mathcal{M}(H, \nu_0)$ and $M_n(\partial_{t,\nu}) \to M(\partial_{t,\nu})$ as $n \to \infty$ in the weak operator topology of $L\big(L_{2,\nu}(\mathbb{R}, H)\big)$.*

If, in addition, $(M_n(z))_n$ converges in the strong operator topology of $L(H)$ for all $z \in \mathbb{C}_{\mathrm{Re}>\nu_0}$, then, as $n \to \infty$, $M_n(\partial_{t,\nu}) \to M(\partial_{t,\nu})$ in the strong operator topology of $L\big(L_{2,\nu}(\mathbb{R}, H)\big)$.

Proof Let $z_0 \in \mathbb{C}_{\mathrm{Re}>\nu_0}$, $r \in (0, \mathrm{Re}\, z_0 - \nu_0)$. For $x, y \in H$, by Cauchy's integral formula, we deduce

$$\langle y, M_n(z_0)x\rangle = \frac{1}{2\pi \mathrm{i}} \int_{\partial B(z_0,r)} \frac{\langle y, M_n(z)x\rangle_H}{z - z_0}\, \mathrm{d}z \quad (n \in \mathbb{N}).$$

As $(M_n)_n$ is bounded, Lebesgue's dominated convergence theorem yields

$$\langle y, M(z_0)x\rangle = \frac{1}{2\pi \mathrm{i}} \int_{\partial B(z_0,r)} \frac{\langle y, M(z)x\rangle_H}{z - z_0}\, \mathrm{d}z.$$

Since

$$|\langle y, M(z)x\rangle|_H \leqslant \|x\|_H \|y\|_H \sup_{n\in\mathbb{N}} \|M_n\|_{\infty,\mathbb{C}_{\mathrm{Re}>\nu_0}} \quad (z \in \mathbb{C}_{\mathrm{Re}>\nu_0}), \tag{13.1}$$

$\langle y, M(\cdot)x\rangle_H$ is holomorphic in a neighbourhood of z_0. By Exercise 5.3 we obtain that $M\colon \mathbb{C}_{\mathrm{Re}>\nu_0} \to L(H)$ is holomorphic. In fact, the estimate (13.1) even implies that $M \in \mathcal{M}(H, \nu_0)$.

If $z \in \mathbb{C}_{\mathrm{Re}>\nu_0}$ and $(M_n(z))_n$ even converges in the strong operator topology, then the limit is clearly $M(z)$.

The convergence statements for $(M_n(\partial_{t,\nu}))_n$ (in the weak and strong operator topology) are then implied by Fourier–Laplace transformation. □

Remark 13.1.3 In Theorem 13.1.2, it suffices to assume that $(M_n(z))_n$ converges only for z belonging to a countable subset of $\mathbb{C}_{\mathrm{Re}>\nu_0}$ with an accumulation point in $\mathbb{C}_{\mathrm{Re}>\nu_0}$.

The next statement is essential for the convergence statement for "ordinary" evolutionary equations.

Proposition 13.1.4 *Let $(T_n)_n$ be a sequence in $L(H)$ converging in the strong operator topology to some $T \in L(H)$ with $0 \in \bigcap_{n\in\mathbb{N}} \rho(T_n)$, $\sup_{n\in\mathbb{N}} \|T_n^{-1}\| < \infty$ and $\operatorname{ran}(T) \subseteq H$ dense. Then T is continuously invertible and $(T_n^{-1})_n$ converges to T^{-1} in the strong operator topology.*

Proof We set $K := \sup_{n\in\mathbb{N}} \|T_n^{-1}\|$. We show that T is continuously invertible first. For this, let $x \in H$. Then

$$\|x\| = \left\|T_n^{-1}T_n x\right\| \leqslant K\,\|T_n x\| \to K\,\|Tx\| \quad (n \to \infty).$$

Hence, T is one-to-one and it follows that $\operatorname{ran}(T) \subseteq H$ is closed. Hence, $0 \in \rho(T)$. For $x \in H$ we conclude

$$\left\|T_n^{-1}x - T^{-1}x\right\| = \left\|T_n^{-1}(T - T_n)T^{-1}x\right\| \leqslant K\left\|(T - T_n)T^{-1}x\right\| \to 0$$

as $(n \to \infty)$. □

We are now in the position to obtain the first result on continuous dependence.

Theorem 13.1.5 *Let $\nu_0 \in \mathbb{R}$, $(M_n)_n$ a bounded sequence in $\mathcal{M}(H, \nu_0)$, $c > 0$ such that for all $n \in \mathbb{N}$ and $z \in \mathbb{C}_{\mathrm{Re}>\nu_0}$ we have*

$$\operatorname{Re} zM_n(z) \geqslant c.$$

If $(M_n(z))_n$ converges in the strong operator topology for all $z \in \mathbb{C}_{\mathrm{Re}>\nu_0}$ then for the limit $M(z)$ we have $M \in \mathcal{M}(H, \nu_0)$ with $\mathrm{Re}\, zM(z) \geqslant c$ for all $z \in \mathbb{C}_{\mathrm{Re}>\nu_0}$ and for $\nu > \nu_0$ we have

$$\left(\partial_{t,\nu} M_n(\partial_{t,\nu})\right)^{-1} \to \left(\partial_{t,\nu} M(\partial_{t,\nu})\right)^{-1}$$

in the strong operator topology.

Proof By Theorem 13.1.2, we observe $M \in \mathcal{M}(H, \nu_0)$. Let $z \in \mathbb{C}_{\mathrm{Re}>\nu_0}$. Then we have $\mathrm{Re}\, zM(z) = \lim_{n\to\infty} \mathrm{Re}\, zM_n(z) \geqslant c$ and hence $zM(z)$ is continuously invertible. Since $0 \in \bigcap_{n\in\mathbb{N}} \rho(zM_n(z))$ and $\left\|(zM_n(z))^{-1}\right\| \leqslant 1/c$ by Proposition 6.2.3(b), we deduce by Proposition 13.1.4 applied to $T_n = zM_n(z)$ that $(zM_n(z))^{-1} \to (zM(z))^{-1}$ in the strong operator topology. By Theorem 13.1.2, for $\nu > \nu_0$ we infer $\left(\partial_{t,\nu} M_n(\partial_{t,\nu})\right)^{-1} \to \left(\partial_{t,\nu} M(\partial_{t,\nu})\right)^{-1}$ in the strong operator topology. □

13.2 A Leading Example

We want to illustrate the findings of the previous section with the help of an ordinary differential equation. Also, we shall provide an argument on the limitations of the theory presented above. Let (Ω, Σ, μ) be a finite measure space.

Note that for $V \in L_\infty(\mu)$ with associated multiplication operator $V(\mathrm{m})$ as in Theorem 2.4.3 we have that

$$M : z \mapsto 1 + z^{-1}V(\mathrm{m}) \in L(L_2(\mu))$$

is a material law with $\mathrm{s_b}(M) = 0$ unless $V = 0$ (in case $V = 0$ we have $\mathrm{s_b}(M) = -\infty$). The corresponding evolutionary equation is given by

$$\partial_{t,\nu} u + V(\mathrm{m})u = f.$$

We want to study sequences of material laws of this form; that is, material laws induced by sequences $(V_n)_n$ in $L_\infty(\mu)$. First, we provide the following characterisation of the convergence of multiplication operators. We recall that for a Banach space X the weak* topology $\sigma(X', X)$ on X' is the coarsest topology such that all the mappings $X' \ni x' \mapsto x'(x)$ $(x \in X)$ are continuous.

Proposition 13.2.1 *Let $(V_n)_n$ in $L_\infty(\mu)$ and $V \in L_\infty(\mu)$. Then the following statements hold.*

(a) $V_n(\mathrm{m}) \to V(\mathrm{m})$ *in $L(L_2(\mu))$ if and only if $V_n \to V$ in $L_\infty(\mu)$.*
(b) $V_n(\mathrm{m}) \to V(\mathrm{m})$ *in the strong operator topology of $L(L_2(\mu))$ if and only if (V_n) is bounded in $L_\infty(\mu)$ and $V_n \to V$ in $L_1(\mu)$.*

(c) $V_n(\mathrm{m}) \to V(\mathrm{m})$ *in the weak operator topology of* $L(L_2(\mu))$ *if and only if* $V_n \to V$ *in the weak* topology* $\sigma\big(L_\infty(\mu), L_1(\mu)\big)$.

Proof

(a) This is a direct consequence of Proposition 2.4.6.
(b) Assume $V_n \to V$ in $L_1(\mu)$ and that $(V_n)_n$ is bounded in $L_\infty(\mu)$. Then $(V_n - V)_n$ is also bounded in $L_\infty(\mu)$. For $f \in L_\infty(\mu) \subseteq L_2(\mu)$ we obtain

$$\|V_n(\mathrm{m})f - V(\mathrm{m})f\|^2_{L_2(\mu)} = \int_\Omega |V_n - V|^2\,|f|^2\,\mathrm{d}\mu$$
$$\leqslant \sup_{n\in\mathbb{N}} \|V_n - V\|_{L_\infty(\mu)}\,\|f\|^2_{L_\infty(\mu)} \int_\Omega |V_n - V|\,d\mu \to 0.$$

Since $L_\infty(\mu)$ is dense in $L_2(\mu)$ and $(V_n(\mathrm{m}) - V(\mathrm{m}))_n$ is bounded by Proposition 2.4.6, we obtain $V_n(\mathrm{m}) \to V(\mathrm{m})$ in the strong operator topology of $L(L_2(\mu))$.

Now, let $V_n(\mathrm{m}) \to V(\mathrm{m})$ in the strong operator topology of $L(L_2(\mu))$. Then $(V_n(\mathrm{m}))_n$ is bounded in $L(L_2(\mu))$ by the uniform boundedness principle. Now Proposition 2.4.6 yields boundedness of $(V_n)_n$ in $L_\infty(\mu)$. Moreover, since $\mathbb{1}_\Omega \in L_2(\mu)$, we deduce $V_n = V_n(\mathrm{m})\mathbb{1}_\Omega \to V(\mathrm{m})\mathbb{1}_\Omega = V$ in $L_2(\mu)$. Since $L_2(\mu)$ embeds continuously into $L_1(\mu)$ we obtain $V_n \to V$ in $L_1(\mu)$.
(c) The assertion follows easily upon realising that $\phi \in L_1(\mu)$ if and only if there exists $\psi_1, \psi_2 \in L_2(\mu)$ such that $\phi = \psi_1\psi_2$. □

With the latter result at hand together with the results in the previous section, we easily deduce the next theorem on continuous dependence on the coefficients.

Theorem 13.2.2 *Let* $(V_n)_n$ *in* $L_\infty(\mu)$ *be bounded,* $V \in L_\infty(\mu)$*, and* $V_n \to V$ *in* $L_1(\mu)$*. Then there exists* $\nu > 0$ *such that*

$$\big(\partial_{t,\nu} + V_n(\mathrm{m})\big)^{-1} \to \big(\partial_{t,\nu} + V(\mathrm{m})\big)^{-1}$$

in the strong operator topology of $L\big(L_{2,\nu}(\mathbb{R}; L_2(\mu))\big)$.

Note that the convergence statement can be improved, see Exercise 13.3.

Proof By Proposition 13.2.1(b) we obtain $V_n(\mathrm{m}) \to V(\mathrm{m})$ in the strong operator topology of $L(L_2(\mu))$. Note that for $\nu \geqslant 1 + \sup_{n\in\mathbb{N}} \|V_n\|_{L_\infty(\mu)}$ we have

$$\mathrm{Re}(z + V_n(\mathrm{m})) \geqslant 1 \quad (z \in \mathbb{C}_{\mathrm{Re}>\nu}, n \in \mathbb{N}).$$

Now Theorem 13.1.5 applied to $M_n(z) = 1 + z^{-1}V_n(\mathrm{m})$ yields the assertion. □

Remark 13.2.3 Theorem 13.2.2 can be generalized in the following way. Let $(B_n)_n$ in $L(H)$, $B \in L(H)$, $B_n \to B$ in the strong operator topology. Then there exists $\nu > 0$ such that

$$\left(\partial_{t,\nu} + B_n\right)^{-1} \to \left(\partial_{t,\nu} + B\right)^{-1}$$

in the strong operator topology of $L\big(L_{2,\nu}(\mathbb{R}; L_2(\mu))\big)$.

In Theorem 13.2.2 we assumed strong convergence of the sequence of multiplication operators $(V_n(\mathrm{m}))_n$. A natural question to ask is whether the stated result can be improved to $(V_n)_n$ converging in the weak* topology $\sigma\big(L_\infty(\mu), L_1(\mu)\big)$ only. The answer is neither 'yes' nor 'no', but rather 'not quite', as we will show in the following. We start with a result on weak* limits of scaled periodic functions, which will serve as the prototypical example for a sequence converging in the weak* topology of L_∞.

Theorem 13.2.4 *Let* $f \in L_\infty(\mathbb{R}^d)$ *be* $[0, 1)^d$-periodic*; that is,*

$$f(\cdot + k) = f \quad (k \in \mathbb{Z}^d).$$

Then

$$f(n\cdot) \to \int_{[0,1)^d} f(x)\,\mathrm{d}x\,\mathbb{1}_{\mathbb{R}^d}$$

in the weak topology* $\sigma\big(L_\infty(\mathbb{R}^d), L_1(\mathbb{R}^d)\big)$ *as* $n \to \infty$.

Proof Without loss of generality, we may assume $\int_{[0,1)^d} f(x)\,\mathrm{d}x = 0$. By the density of simple functions in $L_1(\mathbb{R}^d)$ and the boundedness of $(f(n\cdot))_n$ in $L_\infty(\mathbb{R}^d)$, it suffices to show

$$\int_Q f(nx)\,\mathrm{d}x \to 0 \quad (n \to \infty)$$

for $Q = [a, b] := [a_1, b_1] \times \ldots \times [a_d, b_d]$ where $a = (a_1, \ldots, a_d), b = (b_1, \ldots, b_d) \in \mathbb{R}^d$. By translation and the periodicity of f we may assume $a = 0$. Thus, it suffices to show

$$\int_{[0,b]} f(nx)\,\mathrm{d}x \to 0 \quad (n \to \infty)$$

for all $b \in (0,\infty)^d$. So, let $b = (b_1, \ldots, b_d) \in (0,\infty)^d$. Let $n \in \mathbb{N}$. Then we find $z \in \mathbb{N}_0^d$ and $\zeta \in [0,1)^d$ such that $nb = z + \zeta$. We compute

$$\begin{aligned}
&\int_{[0,b]} f(nx)\,\mathrm{d}x \\
&= \frac{1}{n^d}\int_{[0,nb]} f(x)\,\mathrm{d}x \\
&= \frac{1}{n^d}\int_{[0,z_1]\times[0,nb_2]\times\ldots\times[0,nb_d]} f(x)\,\mathrm{d}x + \frac{1}{n^d}\int_{(z_1,z_1+\zeta_1]\times[0,nb_2]\times\ldots\times[0,nb_d]} f(x)\,\mathrm{d}x.
\end{aligned}$$

We now estimate

$$\begin{aligned}
\left|\frac{1}{n^d}\int_{(z_1,z_1+\zeta_1]\times[0,nb_2]\times\ldots\times[0,nb_d]} f(x)\,\mathrm{d}x\right| &\leqslant \frac{1}{n^d}\int_{(z_1,z_1+\zeta_1]\times[0,nb_2]\times\ldots\times[0,nb_d]} |f(x)|\,\mathrm{d}x \\
&\leqslant \frac{1}{n^d}\int_{(0,1]\times[0,nb_2]\times\ldots\times[0,nb_d]} \mathrm{d}x\,\|f\|_{L_\infty(\mu)} \\
&= \frac{1}{n} b_2\cdot\ldots\cdot b_d\,\|f\|_{L_\infty(\mu)}\,.
\end{aligned}$$

Continuing in this manner and using $z_j \leqslant nb_j$ for all $j \in \{1,\ldots,d\}$, we obtain

$$\left|\int_{[0,b]} f(nx)\,\mathrm{d}x\right| \leqslant \frac{1}{n^d}\left|\int_{[0,z]} f(x)\,\mathrm{d}x\right| + \frac{1}{n}\sum_{j=1}^{d}\frac{b_1\cdot\ldots\cdot b_d}{b_j}\,\|f\|_{L_\infty(\mu)}\,.$$

Since f is $[0,1)^d$-periodic and $z \in \mathbb{N}_0^d$ we observe

$$\int_{[0,z]} f(x)\,\mathrm{d}x = \prod_{j=1}^{d} z_j \int_{[0,1)^d} f(x)\,\mathrm{d}x = 0.$$

Thus,

$$\left|\int_{[0,b]} f(nx)\,\mathrm{d}x\right| \leqslant \frac{1}{n}\sum_{j=1}^{d}\frac{b_1\cdot\ldots\cdot b_d}{b_j}\,\|f\|_{L_\infty(\mu)}\,,$$

which tends to 0 as $n \to \infty$. □

Remark 13.2.5 Note that Theorem 13.2.4 also yields

$$f(n\cdot) \to \int_{[0,1)^d} f(x)\,\mathrm{d}x\,\mathbb{1}_\Omega$$

in the weak* topology $\sigma(L_\infty(\Omega), L_1(\Omega))$ for all measurable subsets $\Omega \subseteq \mathbb{R}^d$ with non-zero Lebesgue measure.

We now present an example which shows that weak* convergence of $(V_n)_n$ does not yield the result of Theorem 13.2.2.

Example 13.2.6 Let $(\Omega, \Sigma, \mu) = ((0,1), \mathcal{B}((0,1)), \lambda|_{(0,1)})$. For $n \in \mathbb{N}$ let V_n be given by $V_n(x) := \sin(2\pi n x)$ for $x \in (0,1)$. Then, by Theorem 13.2.4, we obtain $V_n \to 0$ in $\sigma\big(L_\infty((0,1)), L_1((0,1))\big)$ as $n \to \infty$. Let $\nu > 1$. Then $\big(\partial_{t,\nu} + V_n(\mathrm{m})\big)$ is continuously invertible as an operator in $L_{2,\nu}\big(\mathbb{R}; L_2((0,1))\big)$. Let $\widetilde{f} \in C([0,1])$ and denote $f\colon t \mapsto \mathbb{1}_{[0,\infty)}(t)\widetilde{f}$. Then $f \in L_{2,\nu}\big(\mathbb{R}; L_2((0,1))\big)$. The solution $u_n \in L_{2,\nu}\big(\mathbb{R}; L_2((0,1))\big)$ of

$$\big(\partial_{t,\nu} + V_n(\mathrm{m})\big)u_n = f$$

is given by the variations of constants formula; that is,

$$u_n(t,x) = \mathbb{1}_{[0,\infty)}(t)\int_0^t \exp\big(-(t-s)\sin(2\pi n x)\big)\,\mathrm{d}s\,\widetilde{f}(x) \quad (t \in \mathbb{R}, x \in (0,1)).$$

Thus, if a variant of Theorem 13.2.2 were true also in this case, $(u_n)_n$ needs to converge (in some sense) to the solution u of

$$\partial_{t,\nu} u = f,$$

which is given by

$$u(t,x) = \mathbb{1}_{[0,\infty)}(t) t \widetilde{f}(x) \quad (t \in \mathbb{R}, x \in (0,1)).$$

However, by Theorem 13.2.4, for $x \in (0,1)$ we deduce

$$\int_0^t \exp\big(-(t-s)\sin(2\pi n x)\big)\,\mathrm{d}s \to \int_0^t J(-(t-s))\,\mathrm{d}s \quad (n \to \infty)$$

in $\sigma\big(L_\infty((0,1)), L_1((0,1))\big)$ for each $t \geqslant 0$, where

$$J(s) := \int_0^1 \exp\big(s\sin(2\pi x)\big)\,\mathrm{d}x \quad (s \in \mathbb{R})$$

denotes the 0-th order modified Bessel function of the first kind, cf. [1, p. 9.6.19]. Moreover, for $\varphi \in C_c^\infty(\mathbb{R})$, $A \in \mathcal{B}((0,1))$ and using dominated convergence we obtain

$$\begin{aligned}
&\langle u_n, \varphi \mathbb{1}_A\rangle_{L_{2,\nu}(\mathbb{R};L_2((0,1)))} \\
&= \int_0^\infty \int_0^1 \int_0^t \exp\big(-(t-s)\sin(2\pi n x)\big)\,\mathrm{d}s\,\widetilde{f}(x)^*\mathbb{1}_A(x)\,\mathrm{d}x\,\varphi(t)\mathrm{e}^{-2\nu t}\,\mathrm{d}t \\
&\to \int_0^\infty \int_0^1 \int_0^t J(-(t-s))\,\mathrm{d}s\,\widetilde{f}(x)^*\mathbb{1}_A(x)\,\mathrm{d}x\,\varphi(t)\mathrm{e}^{-2\nu t}\,\mathrm{d}t \\
&= \langle \widetilde{u}, \varphi \mathbb{1}_A\rangle_{L_{2,\nu}(\mathbb{R};L_2((0,1)))}
\end{aligned}$$

with

$$\widetilde{u}(t,x) := \mathbb{1}_{[0,\infty)}(t)\int_0^t J(-(t-s))\,\mathrm{d}s\,\widetilde{f}(x) \quad (t \in \mathbb{R}, x \in (0,1)).$$

Since $(u_n)_n$ is bounded in $L_{2,\nu}\big(\mathbb{R}; L_2((0,1))\big)$ and, by Lemma 3.1.9, the set $\big\{\varphi\mathbb{1}_A\,;\, A \in \mathcal{B}((0,1)),\, \varphi \in C_c^\infty(\mathbb{R})\big\}$ is total in $L_{2,\nu}\big(\mathbb{R}; L_2((0,1))\big)$, we infer $u_n \to \widetilde{u}$ weakly in $L_{2,\nu}\big(\mathbb{R}; L_2((0,1))\big)$ as $n \to \infty$. In particular, $\widetilde{u} \neq u$. Furthermore, $\widetilde{u}$ is *not* of the form

$$\int_0^t \exp\big(-(t-s)\widetilde{V}(x)\big)\,\mathrm{d}s\,\widetilde{f}(x)$$

for some $\widetilde{V} \in L_\infty((0,1))$ and hence, we *cannot* hope for $\widetilde{u}$ to satisfy an equation of the type

$$\big(\partial_{t,\nu} + \widetilde{V}(\mathrm{m})\big)\widetilde{u} = f.$$

As we shall see next, in the framework of evolutionary equations it is possible to derive an equation involving suitable limits of $(V_n)_n$ and f as a right-hand side.

13.3 Convergence in the Weak Operator Topology

In this section, we consider a particular class of material laws and characterise convergence of the solution operators of the corresponding evolutionary equations in the weak operator topology. The main theorem that will serve to compute the limit equation satisfied by $\widetilde{u}$ in Example 13.2.6 reads as follows.

Theorem 13.3.1 *Let H be a Hilbert space, $(B_n)_n$ a bounded sequence in $L(H)$ and $\nu > \sup_{n\in\mathbb{N}} \|B_n\|$. Then $\left((\partial_{t,\nu} + B_n)^{-1}\right)_n$ converges in the weak operator topology of $L(L_{2,\nu}(\mathbb{R}; H))$ if and only if for all $k \in \mathbb{N}$ the sequence $(B_n^k)_n$ converges in the weak operator topology of $L(H)$. In either case, we have*

$$(\partial_{t,\nu} + B_n)^{-1} \to \sum_{k=0}^{\infty} \left(-\partial_{t,\nu}^{-1}\right)^k C_k \partial_{t,\nu}^{-1}$$

in the weak operator topology of $L(L_{2,\nu}(\mathbb{R}; H))$, where $C_k \in L(H)$ denotes the weak limit of $(B_n^k)_n$ for $k \in \mathbb{N}$ and $C_0 := 1_H$.

Remark 13.3.2 In the situation of Theorem 13.3.1, let $B_n^k \to C_k$ in the weak operator topology for all $k \in \mathbb{N}$. Let $L := \sup_{n\in\mathbb{N}} \|B_n\|$, $\nu > 2L$, and $f \in L_{2,\nu}(\mathbb{R}; H)$. By Theorem 13.3.1, if $(\partial_{t,\nu} + B_n)u_n = f$ for all $n \in \mathbb{N}$, then $(u_n)_n$ converges weakly in $L_{2,\nu}(\mathbb{R}; H)$ to some element $\widetilde{u} \in L_{2,\nu}(\mathbb{R}; H)$. In order to determine the differential equation satisfied by $\widetilde{u}$, we make the following observations: by weak convergence,

$$\|C_k\| \leqslant \liminf_{n\to\infty} \left\| B_n^k \right\| \leqslant L^k.$$

Hence, since $\left\|\partial_{t,\nu}^{-1}\right\|_{L_{2,\nu}} \leqslant \frac{1}{\nu}$ (see Sect. 3.2) we infer that

$$\sum_{k=1}^{\infty} \left(-\partial_{t,\nu}^{-1}\right)^k C_k$$

converges in $L(L_{2,\nu}(\mathbb{R}; H))$ and

$$\left\| \sum_{k=1}^{\infty} \left(-\partial_{t,\nu}^{-1}\right)^k C_k \right\| \leqslant \sum_{k=1}^{\infty} \left\|\partial_{t,\nu}^{-1}\right\|^k \|C_k\| < \sum_{k=1}^{\infty} \frac{1}{2^k} = 1.$$

Hence, since $C_0 = 1_H$ we deduce that $\sum_{k=0}^{\infty} \left(-\partial_{t,\nu}^{-1}\right)^k C_k$ is boundedly invertible by the Neumann series. Thus, we obtain

$$\begin{aligned} f &= \partial_{t,\nu} \left(\sum_{k=0}^{\infty} \left(-\partial_{t,\nu}^{-1}\right)^k C_k \right)^{-1} \widetilde{u} = \partial_{t,\nu} \left(1_H + \sum_{k=1}^{\infty} \left(-\partial_{t,\nu}^{-1}\right)^k C_k \right)^{-1} \widetilde{u} \\ &= \partial_{t,\nu} \sum_{\ell=0}^{\infty} \left(-\sum_{k=1}^{\infty} \left(-\partial_{t,\nu}^{-1}\right)^k C_k \right)^{\ell} \widetilde{u} = \partial_{t,\nu}\widetilde{u} + \partial_{t,\nu} \sum_{\ell=1}^{\infty} \left(-\sum_{k=1}^{\infty} \left(-\partial_{t,\nu}^{-1}\right)^k C_k \right)^{\ell} \widetilde{u}. \end{aligned}$$

Before we prove Theorem 13.3.1 we revisit Example 13.2.6.

Example 13.3.3 (Example 13.2.6 Continued) By Theorem 13.3.1, we need to compute the limit of $(\sin^k(2\pi n\cdot))_n$ in the weak* topology of $L_\infty((0,1))$ for all $k \in \mathbb{N}$. By Theorem 13.2.4, we obtain for all $k \in \mathbb{N}$

$$\begin{aligned}\lim_{n\to\infty} \sin^k(2\pi n\cdot) &= \int_0^1 \sin^k(2\pi\xi)\,\mathrm{d}\xi\,\mathbb{1}_{(0,1)} \\ &= \begin{cases} \frac{(2m)!}{(m!2^m)^2}\mathbb{1}_{(0,1)}, & k = 2m \text{ for some } m \in \mathbb{N}, \\ 0, & k \text{ odd}, \end{cases}\end{aligned}$$

in $\sigma\big(L_\infty((0,1)), L_1((0,1))\big)$. Hence, $u_n \to \widetilde{u}$ weakly, where $\widetilde{u}$ satisfies

$$\partial_{t,\nu}\widetilde{u} + \partial_{t,\nu}\sum_{\ell=1}^{\infty}\left(-\sum_{m=1}^{\infty}\partial_{t,\nu}^{-2m}\frac{(2m)!}{(m!2^m)^2}\right)^{\ell}\widetilde{u} = f$$

for $\nu > 2$ by Remark 13.3.2.

Proof of Theorem 13.3.1 Before we prove the equivalence, we make some observations. Since $\nu > \sup_{n\in\mathbb{N}}\|B_n\| =: L$, by a Neumann series argument we deduce that

$$\big(\partial_{t,\nu} + B_n\big)^{-1} = \sum_{k=0}^{\infty}\big(-\partial_{t,\nu}^{-1}B_n\big)^k\partial_{t,\nu}^{-1} = \sum_{k=0}^{\infty}\big(-\partial_{t,\nu}^{-1}\big)^k B_n^k\partial_{t,\nu}^{-1}.$$

The series $\sum_{k=0}^{\infty}\big(-\partial_{t,\nu}^{-1}\big)^k B_n^k\partial_{t,\nu}^{-1}$ is absolutely convergent in $L(L_{2,\nu}(\mathbb{R};H))$. Also note that for $M_n\colon \mathbb{C}_{\mathrm{Re}>L} \ni z \mapsto \sum_{k=0}^{\infty}(-\frac{1}{z})^k B_n^k\frac{1}{z}$ we have $M_n \in \mathcal{M}(H,\nu)$.

Assume now that $(B_n^k)_n$ converges in the weak operator topology to some C_k for all $k \in \mathbb{N}$. A little computation reveals that as $n \to \infty$,

$$M_n(z) \to \sum_{k=0}^{\infty}\left(-\frac{1}{z}\right)^k C_k\frac{1}{z} =: M(z) \quad (z \in \mathbb{C}_{\mathrm{Re}>L})$$

in the weak operator topology, where the series on the right-hand side converges in $L(H)$ since

$$\|C_k\| \leqslant \liminf_{n\to\infty}\left\|B_n^k\right\| \leqslant L^k \quad (k \in \mathbb{N}).$$

Moreover, since $\nu > L$, the sequence $(M_n)_n$ is bounded in $\mathcal{M}(H,\nu)$ and thus, $M \in \mathcal{M}(H,\nu)$ and

$$M_n(\partial_{t,\nu}) \to M(\partial_{t,\nu})$$

in the weak operator topology by Theorem 13.1.2.

Now, we assume that $\left((\partial_{t,\nu} + B_n)^{-1}\right)_n$ converges in the weak operator topology. Then $(M_n(\partial_{t,\nu}))_n$ converges in the weak operator topology. Let $k \in \mathbb{N}$. We need to show that for all $\phi, \psi \in H$ the sequence $(\langle \phi, B_n^k \psi \rangle_H)_n$ is convergent to some number $c_{k,\phi,\psi}$ as $n \to \infty$. The Riesz representation theorem then yields the existence of $C_k \in L(H)$ with $\langle \phi, C_k \psi \rangle = c_{k,\phi,\psi}$. So, let $\phi, \psi \in H$. Moreover, we consider the functions m_n and h_n given by

$$m_n(z) := \sum_{k=0}^{\infty} (-z)^k z \left\langle \phi, B_n^k \psi \right\rangle_H \quad (z \in B(0, 1/L), n \in \mathbb{N})$$

and

$$h_n(z) := \langle \phi, M_n(z)\psi \rangle_H = \sum_{k=0}^{\infty} \frac{1}{z} \left(-\frac{1}{z}\right)^k \left\langle \phi, B_n^k \psi \right\rangle_H \quad (z \in \mathbb{C}_{\mathrm{Re}>L}, n \in \mathbb{N}).$$

Clearly, m_n and h_n are holomorphic on their respective domains for each $n \in \mathbb{N}$ and the sequences $(m_n)_n$ and $(h_n)_n$ are uniformly bounded on compact subsets (in other words they form normal families). Moreover,

$$m_n(z) = h_n\left(\frac{1}{z}\right) \quad \left(z \in B\big(1/(2L), 1/(2L)\big), n \in \mathbb{N}\right).$$

We aim to show that the coefficients of the power series of m_n converge as n tends to infinity. The proof will be done in two steps. In step 1, we will prove that the sequence $(h_n)_n$ converges to a holomorphic function $h\colon \mathbb{C}_{\mathrm{Re}>L} \to \mathbb{C}$ uniformly on compact sets. Then, in the second step, we will use this to deduce that $(m_n)_n$ also converges uniformly on compact sets and prove the assertion with the help of Cauchy's integral formula.

Step 1: By Proposition 5.3.2, $(M_n(\mathrm{im} + \nu))_n$ converges in the weak operator topology of $L(L_2(\mathbb{R}; H))$. For $f, g \in L_2(\mathbb{R})$ we thus obtain that

$$\left(\langle f, h_n(\mathrm{im} + \nu) g \rangle_{L_2(\mathbb{R})}\right)_n = \left(\langle f\phi, M_n(\mathrm{im} + \nu) g\psi \rangle_{L_2(\mathbb{R}; H)}\right)_n$$

is convergent. Thus, using $L_2(\mathbb{R}) \cdot L_2(\mathbb{R}) = L_1(\mathbb{R})$, we obtain that

$$\Psi\colon L_1(\mathbb{R}) \ni u \mapsto \lim_{n\to\infty} \left(\int_{\mathbb{R}} h_n(\mathrm{i}t + \nu) u(t) \,\mathrm{d}t \right) \in \mathbb{C}$$

defines a linear functional, which is continuous, since

$$\sup_{n\in\mathbb{N}} \sup_{t\in\mathbb{R}} \|M_n(\mathrm{i}t + \nu)\|_{L(H)} = \sup_{n\in\mathbb{N}} \|M_n(\mathrm{im} + \nu)\|_{L(L_2(\mathbb{R}; H))} < \infty$$

by boundedness of $(B_n)_n$. Hence, since $L_1(\mathbb{R})' = L_\infty(\mathbb{R})$, we find a unique $\widetilde{h} \in L_\infty(\mathbb{R})$ with

$$\lim_{n\to\infty} \int_{\mathbb{R}} h_n(\mathrm{i}t + \nu)u(t)\,\mathrm{d}t = \int_{\mathbb{R}} \widetilde{h}(t)u(t)\,\mathrm{d}t \quad (u \in L_1(\mathbb{R})).$$

We now show that every subsequence $(h_{n_k})_k$ of $(h_n)_n$ has a subsequence $(h_{n_{k_l}})_l$ which converges locally uniformly to a holomorphic function $h\colon \mathbb{C}_{\mathrm{Re}>L} \to \mathbb{C}$ such that $h(\mathrm{i}\cdot +\nu) = \widetilde{h}$ a.e., and that this implies that the limit h does not depend on the subsequences. Then we conclude that $(h_n)_n$ itself converges locally uniformly to h.
So, let $(h_{n_k})_k$ be a subsequence of (h_n). By Montel's theorem (see [104, Theorem 6.2.2]), we find a subsequence $(h_{n_{k_l}})_l$ of $(h_{n_k})_k$ such that $h_{n_{k_l}} \to h$ as $l \to \infty$ uniformly on compact subsets of $\mathbb{C}_{\mathrm{Re}>L}$ for some holomorphic function $h\colon \mathbb{C}_{\mathrm{Re}>L} \to \mathbb{C}$. In particular, we obtain

$$\lim_{l\to\infty} \int_{\mathbb{R}} h_{n_{k_l}}(\mathrm{i}t + \nu)\varphi(t)\,\mathrm{d}t = \int_{\mathbb{R}} h(\mathrm{i}t + \nu)\varphi(t)\,\mathrm{d}t \quad (\varphi \in C_\mathrm{c}(\mathbb{R}))$$

by dominated convergence and hence, $h(\mathrm{i}t + \nu) = \widetilde{h}(t)$ for almost every $t \in \mathbb{R}$. This shows that the limit h is independent of choice of the subsequences $(h_{n_k})_k$ and $(h_{n_{k_l}})_l$. Indeed, if $\widehat{h}\colon \mathbb{C}_{\mathrm{Re}>L} \to \mathbb{C}$ is the limit of another subsubsequence of $(h_n)_n$ as above, then $\widehat{h}(\mathrm{i}\cdot+\nu) = \widetilde{h} = h(\mathrm{i}\cdot+\nu)$ a.e. Since $\widehat{h}$ and h are holomorphic, the identity theorem yields $\widehat{h} = h$.
Now, assume for a contradiction that $(h_n)_n$ does not converge locally uniformly to h. Then we find a subsequence $(h_{n_k})_k$ of $(h_n)_n$, a compact set $K \subseteq \mathbb{C}_{\mathrm{Re}>L}$ and $\varepsilon > 0$ such that

$$\left\|h_{n_k} - h\right\|_{\infty,K} \geqslant \varepsilon \quad (k \in \mathbb{N}). \tag{13.2}$$

However, the subsequence $(h_{n_k})_k$ has a subsequence $(h_{n_{k_l}})_l$ which converges locally uniformly to h, contradicting (13.2). Thus, $(h_n)_n$ itself converges locally uniformly to h, and, in particular, $h_n \to h$ pointwise on $\mathbb{C}_{\mathrm{Re}>L}$.

Step 2: By what we have shown in Step 1, the sequence $(m_n)_{n\in\mathbb{N}}$ converges pointwise on $B\big(1/(2L), 1/(2L)\big)$. Since $(m_n)_n$ is also uniformly bounded on compact subsets of $B(0, 1/L)$, we derive that $(m_n)_n$ converges uniformly on compact subsets of $B(0, 1/L)$ by Vitali's theorem (see [104, Theorem 6.2.8]). Choosing $0 < r < 1/L$, we thus obtain by Cauchy's integral formula

$$\left\langle \phi, B_n^k \psi \right\rangle_H = (-1)^k \frac{1}{2\pi\mathrm{i}} \int_{\partial B(0,r)} \frac{m_n(z)}{z^{k+2}}\,\mathrm{d}z.$$

Thus $(B_n^k)_n$ converges in the weak operator topology as $n \to \infty$. □

13.4 Comments

The problems discussed here are contained in [133, 138] for both the weak and the strong operator topology. The case of differential-algebraic equations has been invoked as well.

The appearance of memory effects; that is, the occurrence of higher order integral operators due to a weak convergence of the coefficients has been first observed by Tartar and can, for instance, be found in [113]. The limit equation, however, is described by a convolution term rather than a power series of integral operators. It is, however, possible to reformulate these resulting equations into one another, see [135].

The last characterisation of weak convergence in Theorem 13.3.1 was formulated for the first time in [89].

Exercises

Exercise 13.1 Let $(V_n)_n$ in $L_\infty(\mathbb{R}^d)$ and $V \in L_\infty(\mathbb{R}^d)$. Characterise convergence of $V_n(\mathrm{m}) \to V(\mathrm{m})$ in the strong operator topology of $L(L_2(\mathbb{R}^d))$ in terms of convergence of $(V_n)_n$ similar to as was done in Proposition 13.2.1.

Exercise 13.2 Show that there exists an unbounded sequence $(V_n)_n$ in $L_\infty((0,1))$ and $V \in L_\infty((0,1))$ with $V_n \to V$ in $L_1((0,1))$.

Exercise 13.3 Let (Ω, Σ, μ) be a finite measure space, $(V_n)_n$ a bounded sequence in $L_\infty(\mu)$ and assume that $V_n \to V$ in $L_1(\mu)$ for some $V \in L_\infty(\mu)$. Show that there exists $\nu > 0$ such that

$$\big(\partial_{t,\nu} + V_n(\mathrm{m})\big)^{-1} \to \big(\partial_{t,\nu} + V(\mathrm{m})\big)^{-1}$$

in the strong operator topology of $L\big(L_{2,\nu}(\mathbb{R}; L_2(\mu)), H_\nu^1(\mathbb{R}; L_2(\mu))\big)$.

Exercise 13.4 Let $D = \bigcup_{n\in\mathbb{Z}} [n + 1/2, n + 1]$, $V_n := \mathbb{1}_D(n\cdot)$. For suitable $\nu > 0$ compute the limit of

$$\big((\partial_{t,\nu} + V_n(\mathrm{m}))^{-1}\big)_n$$

in the weak operator topology of $L_{2,\nu}\big(\mathbb{R}; L_2((0,1))\big)$.

Exercise 13.5 Let H be a Hilbert space, $c > 0$ and $c \leqslant B_n = B_n^* \in L(H)$ for all $n \in \mathbb{N}$. Characterise, in terms of convergence of $(B_n)_n$ in a suitable sense, that

$$\big((\partial_{t,\nu} B_n)^{-1}\big)_n$$

converges in the weak operator topology. In the case of convergence, find its limit and a sufficient condition for which there exists a $B \in L(H)$ such that

$$(\partial_{t,\nu} B_n)^{-1} \to (\partial_{t,\nu} B)^{-1}$$

in the weak operator topology.

Exercise 13.6 Let H be a Hilbert space. Show that $B_{L(H)} := \{B \in L(H)\ ;\ \|B\| \leqslant 1\}$ is a compact subset under the weak operator topology. If, in addition, H is separable, show that $B_{L(H)}$ is also metrisable under the weak operator topology.

Exercise 13.7 Let H be a separable Hilbert space, $(B_n)_n$ in $L(H)$ bounded. Show that there exists a subsequence $(B_{n_k})_k$ of $(B_n)_n$, a material law $M\colon \operatorname{dom}(M) \to L(H)$ and $\nu > 0$ such that given $f \in L_{2,\nu}(\mathbb{R}; H)$ and $(u_k)_k$ in $L_{2,\nu}(\mathbb{R}; H)$ with

$$\partial_{t,\nu} u_k + B_{n_k} u_k = f \quad (k \in \mathbb{N}),$$

we deduce that $(u_k)_k$ converges weakly to some $u \in L_{2,\nu}(\mathbb{R}; H)$ with the property that

$$\partial_{t,\nu} M(\partial_{t,\nu}) u = f.$$

References

1. M. Abramowitz, I.A. Stegun, *Handbook of Mathematical Functions with Formulas, Graphs, and Mathematical Tables*, vol. 55. National Bureau of Standards Applied Mathematics Series. For sale by the Superintendent of Documents, U.S. Government Printing Office, Washington, D.C. (1964)
89. R. Picard, S. Trostorff, M. Waurick, Well-posedness via monotonicity. An overview, in *Operator Semigroups Meet Complex Analysis, Harmonic Analysis and Mathematical Physics. Operator Theory: Advances and Applications*, vol. 250 (2015), pp. 397–452
104. B. Simon, *Basic Complex Analysis. A Comprehensive Course in Analysis, Part 2A* (American Mathematical Society, Providence, 2015)
113. L. Tartar, *The General Theory of Homogenization*, vol. 7. Lecture Notes of the Unione Matematica Italiana. A personalized introduction (Springer, Berlin; UMI, Bologna, 2009)
133. M. Waurick, G-convergence of linear differential equations. J. Anal. Appl. **33**(4), 385–415 (2014)
135. M. Waurick, Limiting processes in evolutionary equations - a Hilbert space approach to homogenization. Dissertation. Technische Universität Dresden, 2011. http://nbn-resolving.de/urn:nbn:de:bsz:14-qucosa-67442
138. M. Waurick, On the continuous dependence on the coefficients of evolutionary equations. Habilitation. Technische Universität Dresden, 2016. http://arxiv.org/abs/1606.07731

Chapter 14
Continuous Dependence on the Coefficients II

This chapter is concerned with the study of problems of the form

$$\big(\partial_{t,\nu} M_n(\partial_{t,\nu}) + A\big)\, U_n = F$$

for a suitable sequence of material laws $(M_n)_n$ when $A \neq 0$. The aim of this chapter will be to provide the conditions required for convergence of the material law sequence to imply the existence of a limit material law M such that the limit $U = \lim_{n\to\infty} U_n$ exists and satisfies

$$\big(\partial_{t,\nu} M(\partial_{t,\nu}) + A\big)\, U = F.$$

Additionally, for material laws of the form $M_n(\partial_{t,\nu}) = M_{0,n} + \partial_{t,\nu}^{-1} M_{1,n}$ it will be desirable to have the respective limit material law satisfy $M(\partial_{t,\nu}) = M_0 + \partial_{t,\nu}^{-1} M_1$ for some $M_0, M_1 \in L(H)$. This cannot be expected (as we have seen in the guiding example in the previous chapter) if A is a bounded operator, the Hilbert space H is infinite-dimensional, and the material law sequence only converges pointwise in the weak operator topology. It will turn out, however, that if A is "strictly unbounded" then a suitable result can hold, even if we only assume weak convergence of the material law operators.

14.1 A Convergence Theorem

The main convergence theorem of this chapter will be presented next.

Theorem 14.1.1 *Let H be a Hilbert space, $\nu_0 \in \mathbb{R}$, $(M_n)_n$ in $\mathcal{M}(H, \nu_0)$ and $M \in \mathcal{M}(H, \nu_0)$. Assume there exists $c > 0$ such that for all $n \in \mathbb{N}$ we have*

$$\operatorname{Re} z M_n(z) \geqslant c \quad (z \in \mathbb{C}_{\operatorname{Re}>\nu_0}).$$

C. Seifert et al., *Evolutionary Equations*, Operator Theory: Advances and Applications 287, https://doi.org/10.1007/978-3-030-89397-2_14

Let $A\colon \operatorname{dom}(A) \subseteq H \to H$ *be skew-selfadjoint and assume* $\operatorname{dom}(A) \hookrightarrow H$ *compactly. If* $M_n(z) \to M(z)$ *as* $n \to \infty$ *in the weak operator topology for all* $z \in \mathbb{C}_{\operatorname{Re}>\nu_0}$*, then*

$$\left(\overline{\partial_{t,\nu} M_n(\partial_{t,\nu}) + A}\right)^{-1} \to \left(\overline{\partial_{t,\nu} M(\partial_{t,\nu}) + A}\right)^{-1}$$

in the strong operator topology of $L(L_{2,\nu}(\mathbb{R}; H))$ *for each* $\nu > \nu_0$.

For the proof of this theorem, we need a lemma first.

Lemma 14.1.2 *Let* H *be a Hilbert space,* $A\colon \operatorname{dom}(A) \subseteq H \to H$ *skew-selfadjoint,* $c > 0$*,* $(T_n)_n$ *in* $L(H)$ *with* $\operatorname{Re} T_n \geqslant c$ *for all* $n \in \mathbb{N}$*, and* $T \in L(H)$*. Assume* $\operatorname{dom}(A) \hookrightarrow H$ *compactly and* $T_n \to T$ *in the weak operator topology. Then* $0 \in \bigcap_{n\in\mathbb{N}} \rho(T_n + A) \cap \rho(T + A)$ *and*

$$(T_n + A)^{-1} \to (T + A)^{-1}$$

in the norm topology of $L(H)$.

Proof From $\operatorname{Re} T_n \geqslant c$ it follows that $0 \in \rho(T_n + A)$ $(n \in \mathbb{N})$ and $\left((T_n + A)^{-1}\right)_n$ is bounded in $L(H)$. Indeed, since $B := T_n + A$ satisfies $\operatorname{Re} B = \operatorname{Re} T_n \geqslant c$ and $\operatorname{dom}(B) = \operatorname{dom}(A) = \operatorname{dom}(B^*)$ due to the skew-selfadjointness of A, Proposition 6.3.1 yields the assertion. Moreover, since

$$A(T_n + A)^{-1} = 1 - T_n(T_n + A)^{-1}$$

for all $n \in \mathbb{N}$, it follows that $\left((T_n + A)^{-1}\right)_n$ is also bounded in $L(H, \operatorname{dom}(A))$ by the boundedness of $(T_n)_n$ in $L(H)$. Due to the convergence of $(T_n)_n$ to T, it follows that $\operatorname{Re} T \geqslant c$, and thus, $(T + A)^{-1} \in L(H, \operatorname{dom}(A))$. Before we come to a proof of the desired result, we will prove an auxiliary observation.

Claim: for all $(f_n)_n$ in H weakly converging to f, we have $(T_n + A)^{-1} f_n \to (T + A)^{-1} f$ in the norm topology of H.

For proving the claim, let $(f_n)_n$ in H be weakly convergent to some f. Consider $u_n := (T_n + A)^{-1} f_n$. Then $(u_n)_n$ is bounded in $\operatorname{dom}(A)$, since $\left((T_n + A)^{-1}\right)_n$ is bounded in $L(H, \operatorname{dom}(A))$ and $(f_n)_n$ is bounded in H. Hence, there exists a subsequence $(u_{n_k})_k$ which weakly converges to some u in $\operatorname{dom}(A)$. Since $\operatorname{dom}(A) \hookrightarrow H$ compactly, we infer $u_{n_k} \to u$ in the norm topology of H. Hence, in the equality

$$T_{n_k} u_{n_k} + A u_{n_k} = f_{n_k},$$

as $T_{n_k} \to T$ in the weak operator topology and $u_{n_k} \to u$ in H, we may let $k \to \infty$ and obtain for the weak limits

$$Tu + Au = f;$$

that is, $u = (T + A)^{-1} f$. Having identified the limit, a contradiction argument (here a so-called 'subsequence argument', see Exercise 14.3) concludes that $(u_n)_n$ itself converges weakly in $\mathrm{dom}(A)$ and strongly in H to u. Thus, the claim is proved.

Next, assume by contradiction that $\left((T_n + A)^{-1}\right)_n$ does not converge in operator norm to $(T + A)^{-1}$. Then we find an $\varepsilon > 0$ and a strictly increasing sequence of integers, $(n_k)_k$, and a sequence of unit vectors $(f_{n_k})_k$ in H such that

$$\left\| (T_{n_k} + A)^{-1} f_{n_k} - (T + A)^{-1} f_{n_k} \right\| \geqslant \varepsilon. \tag{14.1}$$

By possibly taking another subsequence, we may assume without loss of generality that $\left(f_{n_k}\right)_k$ converges weakly to some $f \in H$. By the claim proved above, we deduce $\left(T_{n_k} + A\right)^{-1} f_{n_k} \to (T + A)^{-1} f$ and $(T + A)^{-1} f_{n_k} \to (T + A)^{-1} f$, both in the norm topology of H as $k \to \infty$. Thus, we may let $k \to \infty$ in (14.1), and obtain the desired contradiction. □

Proof of Theorem 14.1.1 By Theorem 13.1.2 it suffices to show that for all $z \in \mathbb{C}_{\mathrm{Re}>\nu_0}$

$$(zM_n(z) + A)^{-1} \to (zM(z) + A)^{-1} \quad (n \to \infty)$$

in the strong operator topology. This, however, follows from Lemma 14.1.2 applied to $T_n = zM_n(z)$. □

Remark 14.1.3 Note that we only used convergence in the strong operator topology in the proof of Theorem 14.1.1. However, the assertion in Lemma 14.1.2 is about convergence in the norm topology. The reason that we cannot assert the convergence claimed in Theorem 14.1.1 in the norm topology is that the compact embedding of $\mathrm{dom}(A) \hookrightarrow H$ only works locally for fixed z, and not uniformly in z. This situation can, however, be rectified. We refer to Exercise 14.1 for this.

14.2 The Theorem of Rellich and Kondrachov

In order to apply Theorem 14.1.1, we need to provide a setting where the condition on the compactness of the embedding is satisfied. In fact, it is true that $H^1(\Omega)$ embeds compactly into $L_2(\Omega)$ given $\Omega \subseteq \mathbb{R}^d$ is bounded and has 'continuous boundary', see e.g. [5, Theorem 7.11]. In this chapter, we restrict ourselves to a proof of a less general statement.

A preparatory result needed to prove the compact embedding theorem is given next.

Proposition 14.2.1 *Let $I \subseteq \mathbb{R}$ be an open, bounded, non-empty interval. Then the mapping $H^1(\mathbb{R}) \ni f \mapsto f|_I \in H^1(I)$ is well-defined, continuous and onto. Moreover, there exists a continuous right inverse $H^1(I) \to H^1(\mathbb{R})$.*

For the proof of this proposition, we need an auxiliary result first.

Lemma 14.2.2 *Let $\Omega \subseteq \mathbb{R}^d$ be open and connected. Moreover, let $u \in H^1(\Omega)$ with* $\operatorname{grad} u = 0$. *Then u is constant.*

We leave the proof of this lemma as Exercise 14.2.

Proof of Proposition 14.2.1 The mapping $H^1(\mathbb{R}) \to H^1(I)$, $f \mapsto f|_I$ is readily confirmed to be continuous. It remains to prove that it is onto. Let $I = (a, b)$, $u \in H^1(I)$ and define the function v by

$$v(t) := \int_a^t \partial u(s)\,\mathrm{d}s \quad (t \in (a, b)).$$

Clearly, $v \in L_2((a, b))$ and we compute for each $\varphi \in C_c^\infty((a, b))$

$$\begin{aligned}\langle v, \varphi' \rangle_{L_2((a,b))} &= \int_a^b \left(\int_a^t \partial u(s)\,\mathrm{d}s \right)^* \varphi'(t)\,\mathrm{d}t = \int_a^b \int_s^b \varphi'(t)\,\mathrm{d}t\,\partial u(s)^*\,\mathrm{d}s \\ &= -\langle \partial u, \varphi \rangle_{L_2((a,b))}.\end{aligned}$$

This shows $v \in H^1((a, b))$ with $\partial v = \partial u$. Hence, by Lemma 14.2.2 there exists a constant $c \in \mathbb{C}$ with $u = c + v$. We now define f by

$$f(t) := \begin{cases} 0 & \text{if } t < a - 1 \text{ or } t > b + 1, \\ ct + c(1 - a) & \text{if } a - 1 \leqslant t \leqslant a, \\ u(t) & \text{if } a < t < b, \\ -(c + v(b))t + (c + v(b))(1 + b) & \text{if } b \leqslant t \leqslant b + 1. \end{cases}$$

We then easily see that $f \in H^1(\mathbb{R})$ and clearly $f|_{(a,b)} = u$. In order to see that $u \mapsto f$ is continuous, we need to establish that the value c depends continuously on u. This, however, follows from the estimate

$$\begin{aligned}|c| &= \frac{1}{\sqrt{b-a}} \left(\int_a^b |c|^2 \right)^{1/2} \leqslant \frac{1}{\sqrt{b-a}} (\|u\|_{L_2(a,b)} + \|v\|_{L_2(a,b)}) \\ &\leqslant \frac{1}{\sqrt{b-a}} (\|u\|_{L_2(a,b)} + (b-a)\,\|\partial u\|_{L_2(a,b)}) \\ &\leqslant \frac{\sqrt{2}\max\{1, (b-a)\}}{\sqrt{b-a}} \|u\|_{H^1(a,b)}.\end{aligned}$$

□

Theorem 14.2.3 *Let $I \subseteq \mathbb{R}$ be an open bounded interval. Then $H^1(I) \hookrightarrow L_2(I)$ compactly.*

Proof By Proposition 14.2.1, we find a continuous mapping $E\colon H^1(I) \to H^1(\mathbb{R})$ such that for all $u \in H^1(I)$ we have $E(u)|_I = u$. Moreover, by Exercise 4.3 the mapping $H^1(\mathbb{R}) \hookrightarrow C^{1/2}(\mathbb{R})$ is continuous. Thus,

$$H^1(I) \stackrel{E}{\to} H^1(\mathbb{R}) \hookrightarrow C^{1/2}(\mathbb{R}) \to C^{1/2}(I),$$

is a composition of continuous mappings, where the last mapping is the restriction to I. Since $C^{1/2}(I) \hookrightarrow C(I)$ compactly by the Arzelà–Ascoli theorem, and $C(I) \hookrightarrow L_2(I)$ continuously, we infer $H^1(I) \hookrightarrow L_2(I)$ compactly. □

We now have the opportunity to study the limit behaviour of a periodic mixed type problem.

Example 14.2.4 (Highly Oscillatory Problems) Let $s_1, s_2\colon \mathbb{R} \to [0, 1]$ be 1-periodic, measurable functions. Then for $\nu > 0$, we set

$$S^{(n)} := \overline{\left(\partial_{t,\nu}\begin{pmatrix} s_1(n\mathrm{m}) & 0 \\ 0 & s_2(n\mathrm{m})\end{pmatrix} + \begin{pmatrix} 1 - s_1(n\mathrm{m}) & 0 \\ 0 & 1 - s_2(n\mathrm{m})\end{pmatrix} + \begin{pmatrix} 0 & \partial \\ \partial_0 & 0\end{pmatrix}\right)}^{-1},$$

where $\partial = \operatorname{div}$ and $\partial_0 = \operatorname{grad}_0$ are regarded as operators in $L_2((0,1))$ with respective domains $H^1((0,1))$ and $H_0^1((0,1))$. Then, by Theorem 14.2.3, the operator $A := \begin{pmatrix} 0 & \partial \\ \partial_0 & 0\end{pmatrix}$ satisfies the assumptions of Theorem 14.1.1. Moreover, Theorem 13.2.4 implies that the remaining assumptions of Theorem 14.1.1 are satisfied. Hence, we deduce that $\left(S^{(n)}\right)_n$ converges in the strong operator topology on $L\big(L_{2,\nu}(\mathbb{R}; L_2((0,1)))\big)$ to the limit

$$\overline{\left(\partial_{t,\nu}\begin{pmatrix} \int_0^1 s_1 & 0 \\ 0 & \int_0^1 s_2\end{pmatrix} + \begin{pmatrix} 1 - \int_0^1 s_1 & 0 \\ 0 & 1 - \int_0^1 s_2\end{pmatrix} + \begin{pmatrix} 0 & \partial \\ \partial_0 & 0\end{pmatrix}\right)}^{-1}.$$

Next, we aim to provide an application to more than one spatial dimension. For this, we will also need a corresponding compactness statement. This is the subject of the rest of this section.

Theorem 14.2.5 (Rellich–Kondrachov) *Let $\Omega \subseteq \mathbb{R}^d$ be open and bounded. Then $H_0^1(\Omega) \hookrightarrow L_2(\Omega)$ compactly.*

Proof Without loss of generality (by shifting and shrinking of Ω and extending by 0), we may assume that $\Omega = (0,1)^d$. We carry out the proof by induction on the spatial dimension d. The case $d = 1$ has been dealt with in Theorem 14.2.3. Assume the statement is true for some $d-1$. Using that $C_c^\infty((0,1)^d)$ is dense in $H_0^1((0,1)^d)$, we infer the continuity of the injection

$$R\colon H_0^1((0,1)^d) \to H^1\big(\mathbb{R}; L_2((0,1)^{d-1})\big) \cap L_2\big(\mathbb{R}; H_0^1((0,1)^{d-1})\big)$$
$$\phi \mapsto \big(t \mapsto \big(\omega \mapsto \phi(t,\omega)\big)\big),$$

where we identify ϕ with its extension to $\mathbb{R}^d$ by 0. The range space is endowed with the usual sum scalar product.

Let $(\phi_n)_n$ be a weakly convergent nullsequence in $H_0^1((0,1)^d)$. In particular, $(R\phi_n)_n$ is bounded in $H^1\big(\mathbb{R}; L_2((0,1)^{d-1})\big)$ and hence, it is also bounded in $C_{\mathrm{b}}\big(\mathbb{R}; L_2((0,1)^{d-1})\big)$ by Theorem 4.1.2 (and Corollary 4.1.3); that is,

$$\sup_{t\in[0,1],n\in\mathbb{N}} \|\phi_n(t,\cdot)\|_{L_2((0,1)^{d-1})} < \infty. \tag{14.2}$$

Let $f \in L_2((0,1)^{d-1})$. Then $(\phi_{n,f})_n$ given by

$$\phi_{n,f}\colon t \mapsto \langle \phi_n(t,\cdot), f\rangle_{L_2((0,1)^{d-1})}$$

is a weakly convergent nullsequence in $H^1((0,1))$. We obtain by Theorem 14.2.3 that $\phi_{n,f} \to 0$ in $L_2((0,1))$ as $n \to \infty$. By separability of $L_2((0,1)^{d-1})$ we find $D \subseteq L_2((0,1)^{d-1})$ countable and dense, a subsequence (again labeled by n) and a nullset $N \subseteq \mathbb{R}$ such that $\phi_{n,f}(t) \to 0$ for all $t \in \mathbb{R}\setminus N$ and $f \in D$ as $n \to \infty$. By (14.2), we deduce $\phi_{n,f}(t) \to 0$ for all $t \in \mathbb{R}\setminus N$ and $f \in L_2((0,1)^{d-1})$ as $n \to \infty$, or, in other words, $\phi_n(t,\cdot) \to 0$ weakly in $L_2((0,1)^{d-1})$ for each $t \in \mathbb{R}\setminus N$ as $n \to \infty$.

Next, we show that there exists a nullset $N \subseteq N_1 \subseteq \mathbb{R}$ such that $\phi_n(t,\cdot) \to 0$ in $L_2((0,1)^{d-1})$ for all $t \in \mathbb{R}\setminus N_1$. For this, since $(R\phi_n)_n$ in $L_2\big(\mathbb{R}; H_0^1((0,1)^{d-1})\big)$ is bounded, we find a nullset $N \subseteq N_1 \subseteq \mathbb{R}$ such that $(\phi_n(t,\cdot))_n$ is bounded in $H_0^1((0,1)^{d-1})$ for all $t \in \mathbb{R}\setminus N_1$. Let $t \in \mathbb{R}\setminus N_1$. Then there exists a further subsequence $(\phi_{n_k}(t,\cdot))_k$ which converges weakly in $H_0^1((0,1)^{d-1})$. By the induction hypothesis, $(\phi_{n_k}(t,\cdot))_{n_k}$ converges strongly in $L_2((0,1)^{d-1})$, and since we have already seen that it is a weak nullsequence in $L_2((0,1)^{d-1})$, we derive $\phi_{n_k}(t,\cdot) \to 0$ in $L_2((0,1)^{d-1})$. By a subsequence argument we derive that

$$\phi_n(t,\cdot) \to 0$$

in $L_2((0,1)^{d-1})$ for all $t \in \mathbb{R}\setminus N_1$.

Now, for $n \in \mathbb{N}$ we deduce

$$\|\phi_n\|^2_{L_2((0,1)^d)} = \int_0^1 \|\phi_n(t,\cdot)\|^2_{L_2((0,1)^{d-1})}\,\mathrm{d}t \to 0,$$

where we have used dominated convergence, which is possible due to (14.2). □

14.3 The Periodic Gradient

In this section we investigate the gradient on periodic functions on $\mathbb{R}^d$. Throughout, we set $Y := [0, 1)^d$.

Definition (Periodic Gradient) We define

$$C_\sharp^\infty(Y) := \left\{\phi|_Y\,;\ \phi \in C^\infty(\mathbb{R}^d),\ \phi(\cdot + k) = \phi\ (k \in \mathbb{Z}^d)\right\}$$

and

$$\begin{aligned} \operatorname{grad}_{\sharp,\infty} \colon C_\sharp^\infty(Y) \subseteq L_2(Y) &\to L_2(Y)^d \\ \phi &\mapsto \operatorname{grad} \phi. \end{aligned}$$

Moreover, we set $\operatorname{div}_\sharp := -\operatorname{grad}_{\sharp,\infty}^*$ and $\operatorname{grad}_\sharp := -\operatorname{div}_\sharp^* = \overline{\operatorname{grad}_{\sharp,\infty}}$.

Remark 14.3.1 The operators just introduced can easily be shown to lie between the operator realisations we have introduced in earlier chapters. Indeed, it is easy to see that

$$\operatorname{div}_0 \subseteq \operatorname{div}_\sharp \text{ and } \operatorname{grad}_0 \subseteq \operatorname{grad}_\sharp$$

and, consequently, we also have

$$\operatorname{grad}_\sharp \subseteq \operatorname{grad} \text{ and } \operatorname{div}_\sharp \subseteq \operatorname{div}.$$

The corresponding domains for the operators $\operatorname{grad}_\sharp$ and $\operatorname{div}_\sharp$ will be denoted by $H_\sharp^1(Y)$ and $H_\sharp(\operatorname{div}, Y)$, respectively.

For the next results, we define the periodic extension operator. For $\phi \in L_2(Y)^m$ we put

$$\phi_{\mathrm{pe}}(x + k) := \phi(x)$$

for almost every $x \in Y$ and all $k \in \mathbb{Z}^d$.

We start with the following two observations.

Lemma 14.3.2 *Let $f \in L_2(Y)$ and $(\rho_k)_k$ be a δ-sequence in $C_c^\infty(\mathbb{R}^d)$ (cf. Exercise 3.1). Define*

$$f_k := (\rho_k * f_{\mathrm{pe}})|_Y \quad (k \in \mathbb{N}).$$

Then $f_k \in C_\sharp^\infty(Y)$ for each $k \in \mathbb{N}$ and $f_k \to f$ in $L_2(Y)$ as $k \to \infty$.

Proof It follows as in Exercise 3.2 that $\rho_k * f_{\mathrm{pe}}$ is in C^∞. Moreover, one easily sees that $\rho_k * f_{\mathrm{pe}}$ is $[0,1)^d$-periodic, and hence, $f_k \in C_\sharp^\infty(Y)$ for each $k \in \mathbb{N}$. For the convergence we observe

$$\big(\rho_k * (\mathbb{1}_{Y+B(0,1)} f_{\mathrm{pe}})\big)(x) = f_k(x) \quad (x \in Y, k \in \mathbb{N}).$$

Moreover, by Exercise 3.2 we have $\rho_k * (\mathbb{1}_{Y+B(0,1)} f_{\mathrm{pe}}) \to \mathbb{1}_{Y+B(0,1)} f_{\mathrm{pe}}$ in $L_2(\mathbb{R}^d)$ as $k \to \infty$, and thus,

$$f_k = \big(\rho_k * (\mathbb{1}_{Y+B(0,1)} f_{\mathrm{pe}})\big)|_Y \to (\mathbb{1}_{Y+B(0,1)} f_{\mathrm{pe}})|_Y = f \quad (k \to \infty) \quad \text{in } L_2(Y). \qquad \square$$

Lemma 14.3.3 *$C_\sharp^\infty(Y)^d$ is a core for* $\operatorname{div}_\sharp$.

Proof First we note that $C_\sharp^\infty(Y)^d \subseteq \operatorname{dom}(\operatorname{div}_\sharp)$. To see this, for $\phi \in C_\sharp^\infty(Y)$, $\Psi \in C_\sharp^\infty(Y)^d$ we compute

$$\begin{aligned} \langle \operatorname{grad}\phi, \Psi\rangle_{L_2(Y)^d} &= \int_Y \langle \operatorname{grad}\phi(x), \Psi(x)\rangle_{\mathbb{K}^d}\,\mathrm{d}x = -\int_Y \phi(x)^* \operatorname{div}\Psi(x)\,\mathrm{d}x \\ &= \langle \phi, -\operatorname{div}\Psi\rangle_{L_2(Y)} \end{aligned}$$

by integration by parts (note that the boundary values cancel out due to the periodicity of ϕ and Ψ). Now, let $q \in \operatorname{dom}(\operatorname{div}_\sharp)$ and $(\rho_k)_k$ be a δ-sequence in $C_c^\infty(\mathbb{R}^d)$. For $k \in \mathbb{N}$ we define

$$q_k := (\rho_k * q_{\mathrm{pe}})|_Y,$$

and obtain $q_k \in C_\sharp^\infty(Y)^d$ and $q_k \to q$ in $L_2(Y)^d$ as $k \to \infty$ by Lemma 14.3.2. It is left to show that $\operatorname{div} q_k \to \operatorname{div}_\sharp q$ in $L_2(Y)$ as $k \to \infty$. For doing so, we show that $\operatorname{div} q_k = \big(\rho_k * (\operatorname{div}_\sharp q)_{\mathrm{pe}}\big)|_Y$, which would then yield the assertion again by Lemma 14.3.2. So, let $k \in \mathbb{N}$ and $\phi \in C_\sharp^\infty(Y)$. We compute

$$\begin{aligned} \langle q_k, \operatorname{grad}\phi\rangle_{L_2(Y)^d} &= \int_Y \Big\langle \int_{\mathbb{R}^d} \rho_k(y) q_{\mathrm{pe}}(x-y)\,\mathrm{d}y, \operatorname{grad}\phi(x)\Big\rangle_{\mathbb{K}^d} \mathrm{d}x \\ &= \int_{\mathbb{R}^d} \rho_k(y) \int_Y \big\langle q_{\mathrm{pe}}(x-y), \operatorname{grad}\phi(x)\big\rangle_{\mathbb{K}^d}\,\mathrm{d}x\,\mathrm{d}y \\ &= \int_{\mathbb{R}^d} \rho_k(y) \int_{Y-y} \big\langle q_{\mathrm{pe}}(x), (\operatorname{grad}\phi)_{\mathrm{pe}}(x+y)\big\rangle_{\mathbb{K}^d}\,\mathrm{d}x\,\mathrm{d}y \\ &= \int_{\mathbb{R}^d} \rho_k(y) \int_Y \big\langle q(x), (\operatorname{grad}\phi)_{\mathrm{pe}}(x+y)\big\rangle_{\mathbb{K}^d}\,\mathrm{d}x\,\mathrm{d}y \end{aligned}$$

$$
\begin{aligned}
&= \int_{\mathbb{R}^d} \rho_k(y) \int_Y \langle q(x), (\operatorname{grad} \phi_{\mathrm{pe}}(\cdot + y))(x)\rangle_{\mathbb{K}^d} \,\mathrm{d}x \,\mathrm{d}y \\
&= -\int_{\mathbb{R}^d} \rho_k(y) \int_Y \langle \operatorname{div}_\sharp q(x), \phi_{\mathrm{pe}}(x+y)\rangle_{\mathbb{K}^d} \,\mathrm{d}x \,\mathrm{d}y \\
&= -\int_{\mathbb{R}^d} \rho_k(y) \int_{Y+y} \langle (\operatorname{div}_\sharp q)_{\mathrm{pe}}(x-y), \phi_{\mathrm{pe}}(x)\rangle_{\mathbb{K}^d} \,\mathrm{d}x \,\mathrm{d}y \\
&= -\langle (\rho_k * (\operatorname{div}_\sharp q)_{\mathrm{pe}})|_Y, \phi\rangle_{L_2(Y)},
\end{aligned}
$$

where we have used periodicity as well as $\phi_{\mathrm{pe}}(\cdot + y) \in C_\sharp^\infty(Y)$. □

Remark 14.3.4 The proof of Lemma 14.3.3 reveals that every $q \in \ker(\operatorname{div}_\sharp)$ can be approximated by elements in $C_\sharp^\infty(Y)^d \cap \ker(\operatorname{div}_\sharp)$.

Proposition 14.3.5 *Let $\Omega \subseteq \mathbb{R}^d$ be open, bounded, $u \in H_\sharp^1(Y)$ and $q \in H_\sharp(\operatorname{div}, Y)$. Then $u_{\mathrm{pe}}|_\Omega \in H^1(\Omega)$, $q_{\mathrm{pe}}|_\Omega \in H(\operatorname{div}, \Omega)$ and*

$$
\operatorname{grad}\left(u_{\mathrm{pe}}|_\Omega\right) = \left(\operatorname{grad}_\sharp u\right)_{\mathrm{pe}}|_\Omega \text{ and } \operatorname{div}\left(q_{\mathrm{pe}}|_\Omega\right) = \left(\operatorname{div}_\sharp q\right)_{\mathrm{pe}}|_\Omega.
$$

Proof Let first $\phi \in C_\sharp^\infty(Y)$. Then by definition $\phi_{\mathrm{pe}} \in C^\infty(\mathbb{R}^d)$ and we easily see

$$
\operatorname{grad} \phi_{\mathrm{pe}} = (\operatorname{grad} \phi)_{\mathrm{pe}} = (\operatorname{grad}_\sharp \phi)_{\mathrm{pe}}.
$$

Moreover, since Ω is bounded, we infer $\phi_{\mathrm{pe}} \in H^1(\Omega)$. By definition of $H_\sharp^1(Y)$ we find a sequence $(\phi_k)_{k\in\mathbb{N}}$ in $C_\sharp^\infty(Y)$ such that $\phi_k \to u$ in $L_2(Y)$ and $\operatorname{grad}_\sharp \phi_k \to \operatorname{grad}_\sharp u$ in $L_2(Y)^d$ as $k \to \infty$. Since

$$
L_2(Y) \to L_2(\Omega), \quad f \mapsto f_{\mathrm{pe}}
$$

is bounded due to the boundedness of Ω, we also derive $\phi_{k,\mathrm{pe}} \to u_{\mathrm{pe}}$ in $L_2(\Omega)$ and $(\operatorname{grad}_\sharp \phi_k)_{\mathrm{pe}} \to (\operatorname{grad}_\sharp u)_{\mathrm{pe}}$ in $L_2(\Omega)^d$ as $k \to \infty$. By what we have shown above, we infer

$$
\operatorname{grad} \phi_{k,\mathrm{pe}} = (\operatorname{grad}_\sharp \phi_k)_{\mathrm{pe}} \to (\operatorname{grad}_\sharp u)_{\mathrm{pe}} \quad (k \to \infty)
$$

in $L_2(\Omega)^d$, and thus, $u_{\mathrm{pe}} \in H^1(\Omega)$ with $\operatorname{grad} u_{\mathrm{pe}} = (\operatorname{grad}_\sharp u)_{\mathrm{pe}}$ by the closedness of grad. The proof for q follows by the same argument with Lemma 14.3.3 as an additional resource. □

The extension result just established yields the following compactness statement.

Theorem 14.3.6 (Rellich–Kondrachov II) *The embedding $H_\sharp^1(Y) \hookrightarrow L_2(Y)$ is compact.*

Proof Let $(\phi_n)_n$ be a bounded sequence in $H^1_\sharp(Y)$. Let $\Omega \subseteq \mathbb{R}^d$ be open and bounded such that $\overline{Y} \subseteq \Omega$. By Proposition 14.3.5, we deduce that $\left(\phi_{n,\mathrm{pe}}|_\Omega\right)_n$ is bounded in $H^1(\Omega)$. Let $\psi \in C_c^\infty(\Omega)$ with $\psi = 1$ on $\overline{Y}$. Then $\left(\psi\phi_{n,\mathrm{pe}}\right)_n$ is bounded in $H_0^1(\Omega)$. By Theorem 14.2.5, we find an $L_2(\Omega)$-convergent subsequence. This sequence also converges in $L_2(Y)$. Since $\psi = 1$ on Y, we obtain the assertion. □

Next, we provide a Poincaré-type inequality for the periodic gradient.

Proposition 14.3.7 *There exists $c \geqslant 0$ such that for all $u \in H^1_\sharp(Y)$*

$$\left\| u - \int_Y u \right\|_{L_2(Y)} \leqslant c \left\| \operatorname{grad}_\sharp u \right\|_{L_2(Y)^d}.$$

In particular, $\operatorname{ran}(\operatorname{grad}_\sharp) \subseteq L_2(Y)^d$ *is closed,* $\ker(\operatorname{grad}_\sharp) = \operatorname{lin}\{\mathbb{1}_Y\}$ *and the operator*

$$\operatorname{grad}_\sharp \colon H^1_\sharp(Y) \cap \{\mathbb{1}_Y\}^\perp \to \operatorname{ran}(\operatorname{grad}_\sharp)$$

is an isomorphism.

Proof The proof is left as Exercise 14.4. □

We are now in a position to formulate the particular example we have in mind. Problems of this type with highly oscillatory coefficients are also referred to as *homogenisation problems*. We refer to the comments section for more details on this.

Example 14.3.8 (Homogenisation Problem for the Wave Equation) Let $c > 0$, $a \colon \mathbb{R}^d \to \mathbb{K}^{d\times d}$ be bounded, measurable, $a(x) = a(x)^* \geqslant c$ for all $x \in \mathbb{R}^d$. Furthermore, assume that a is $[0,1)^d$-periodic. Let $\nu > 0$, $f \in L_{2,\nu}(\mathbb{R}; L_2(Y))$ and for $n \in \mathbb{N}$ consider the problem of finding $u_n \in L_{2,\nu}(\mathbb{R}; L_2(Y))$ such that

$$\partial_{t,\nu}^2 u_n - \operatorname{div}_\sharp a(n\mathrm{m}) \operatorname{grad}_\sharp u_n = f. \tag{14.3}$$

We have already established that there exists a uniquely determined solution, u_n. Employing the same trick as in Sect. 11.3, we shall rewrite (14.3) using $v_n := \partial_{t,\nu} u_n$, the canonical embedding $\iota_\sharp \colon \operatorname{ran}(\operatorname{grad}_\sharp) \hookrightarrow L_2(Y)^d$ as well as $q_n := -\iota_\sharp^* a(n\mathrm{m})\iota_\sharp\iota_\sharp^* \operatorname{grad}_\sharp u_n$ to obtain

$$\left(\partial_{t,\nu}\begin{pmatrix}1 & 0\\ 0 & \left(\iota_\sharp^* a(n\mathrm{m})\iota_\sharp\right)^{-1}\end{pmatrix} + \begin{pmatrix}0 & \operatorname{div}_\sharp \iota_\sharp\\ \iota_\sharp^* \operatorname{grad}_\sharp & 0\end{pmatrix}\right)\begin{pmatrix}v_n\\ q_n\end{pmatrix} = \begin{pmatrix}f\\ 0\end{pmatrix}.$$

Note that we have used that $\left(\iota_\sharp^* a(n\mathrm{m})\iota_\sharp\right) \colon \operatorname{ran}(\operatorname{grad}_\sharp) \to \operatorname{ran}(\operatorname{grad}_\sharp)$ is continuously invertible and strictly positive definite (uniformly in n); see Proposition 11.3.5. Also note that $\iota_\sharp^* a(n\mathrm{m})\iota_\sharp$ is selfadjoint. As in Exercise 11.3 we see that $\left(\iota_\sharp^* \operatorname{grad}_\sharp\right)^* = -\operatorname{div}_\sharp \iota_\sharp$. Thus, the operator

$$S^{(n)} := \left(\overline{\partial_{t,\nu} \begin{pmatrix} 1 & 0 \\ 0 & \left(\iota_\sharp^* a(n\mathrm{m})\iota_\sharp\right)^{-1} \end{pmatrix} + \begin{pmatrix} 0 & \operatorname{div}_\sharp \iota_\sharp \\ \iota_\sharp^* \operatorname{grad}_\sharp & 0 \end{pmatrix}}\right)^{-1}$$

is well-defined and bounded in $L_{2,\nu}(\mathbb{R}; L_2(Y) \times \operatorname{ran}(\operatorname{grad}_\sharp))$. We aim to find the limit of $(S^{(n)})_n$ as $n \to \infty$. For this, we want to apply Theorem 14.1.1. We readily see using Theorem 14.3.6 and Exercise 14.5 that

$$A := \begin{pmatrix} 0 & \operatorname{div}_\sharp \iota_\sharp \\ \iota_\sharp^* \operatorname{grad}_\sharp & 0 \end{pmatrix}$$

satisfies the assumptions in Theorem 14.1.1. Thus, it is left to analyse $\left(\left(\iota_\sharp^* a(n\mathrm{m})\iota_\sharp\right)^{-1}\right)_n$. This is the subject of the next section. For this reason, we define

$$\mathfrak{a}_n := \left(\iota_\sharp^* a(n\mathrm{m})\iota_\sharp\right)^{-1} \quad (n \in \mathbb{N}).$$

14.4 The Limit of $(\mathfrak{a}_n)_n$

In this section, we shall apply our earlier findings to higher-dimensional problems. Again, we fix $Y := [0,1)^d$ as well as $\iota_\sharp\colon \operatorname{ran}(\operatorname{grad}_\sharp) \hookrightarrow L_2(Y)^d$, the canonical embedding. Before we are able to present the central result of this section, we need a preliminary result.

Throughout, let $a\colon \mathbb{R}^d \to \mathbb{K}^{d\times d}$ be measurable, bounded and $[0,1)^d$-periodic such that $\operatorname{Re} a(x) \geqslant c$ for each $x \in \mathbb{R}^d$ for some $c > 0$.

Lemma 14.4.1 *Let $\xi \in \mathbb{K}^d$. Then there exists a unique $v_\xi \in L_2(Y)^d$ with $v_\xi - \xi \in \operatorname{ran}(\operatorname{grad}_\sharp)$ and $a(\mathrm{m})v_\xi \in \ker(\operatorname{div}_\sharp)$.*

Proof Take $w \in H_\sharp^1(Y)$ such that

$$\operatorname{grad}_\sharp w = -\iota_\sharp \left(\iota_\sharp^* a(\mathrm{m})\iota_\sharp\right)^{-1} \iota_\sharp^* a(\mathrm{m})\xi = -\iota_\sharp \mathfrak{a}_n \iota_\sharp^* a(\mathrm{m})\xi.$$

This is possible, since the right-hand side belongs to $\operatorname{ran}(\operatorname{grad}_\sharp)$ by definition. We put $v_\xi := \operatorname{grad}_\sharp w + \xi$. Then $v_\xi - \xi \in \operatorname{ran}(\operatorname{grad}_\sharp)$ and we have

$$\begin{aligned} \iota_\sharp^* a(\mathrm{m})v_\xi &= \iota_\sharp^* a(\mathrm{m}) \left(\operatorname{grad}_\sharp w + \xi\right) = \iota_\sharp^* a(\mathrm{m}) \left(-\iota_\sharp \mathfrak{a}_n \iota_\sharp^* a(\mathrm{m})\xi + \xi\right) \\ &= -\iota_\sharp^* a(\mathrm{m})\iota_\sharp \mathfrak{a}_n \iota_\sharp^* a(\mathrm{m})\xi + \iota_\sharp^* a(\mathrm{m})\xi = 0. \end{aligned}$$

The latter gives $a(\mathrm{m})v_\xi \in \operatorname{ran}(\operatorname{grad}_\sharp)^\perp = \ker(\operatorname{div}_\sharp)$. For the uniqueness, we assume $v \in \operatorname{ran}(\operatorname{grad}_\sharp)$ with $a(\mathrm{m})v \in \ker(\operatorname{div}_\sharp)$. Then

$$(\iota_\sharp^* a(\mathrm{m})\iota_\sharp)\iota_\sharp^* v = \iota_\sharp^* a(\mathrm{m})v = 0,$$

which implies $\iota_\sharp^* v = 0$ since $\iota_\sharp^* a(\mathrm{m})\iota_\sharp$ is invertible. Thus $v = 0$. □

The previous result induces the linear mapping

$$a_{\mathrm{hom}} \colon \mathbb{K}^d \ni \xi \mapsto \int_Y a v_\xi \in \mathbb{K}^d,$$

where $v_\xi \in L_2(Y)^d$ is the unique vector field from Lemma 14.4.1.

Remark 14.4.2 We gather some elementary facts on a_{hom}.

(a) We have $(a^*)_{\mathrm{hom}} = a_{\mathrm{hom}}^*$. In particular, if a is pointwise selfadjoint then so is a_{hom}. Indeed, let $\xi, \zeta \in \mathbb{K}^d$ and v_ξ and $v_\zeta \in L_2(Y)^d$ be the corresponding functions for a^* and a, respectively, according to Lemma 14.4.1. Then there exist $w_\xi, w_\zeta \in \operatorname{dom}(\operatorname{grad}_\sharp)$ with $v_\xi - \xi = \operatorname{grad}_\sharp w_\xi$ and $v_\zeta - \zeta = \operatorname{grad}_\sharp w_\zeta$. We compute

$$\begin{aligned}
\left\langle (a^*)_{\mathrm{hom}}\xi, \zeta\right\rangle_{\mathbb{K}^d} &= \int_Y \left\langle \left(a^* v_\xi\right)(y), v_\zeta(y) - \operatorname{grad}_\sharp w_\zeta(y)\right\rangle_{\mathbb{K}^d} \,\mathrm{d}y \\
&= \int_Y \left\langle \left(a^* v_\xi\right)(y), v_\zeta(y)\right\rangle_{\mathbb{K}^d} \,\mathrm{d}y \\
&\quad - \int_Y \left\langle \left(a^* v_\xi\right)(y), \operatorname{grad}_\sharp w_\zeta(y)\right\rangle_{\mathbb{K}^d} \,\mathrm{d}y \\
&= \int_Y \left\langle v_\xi(y), \left(a v_\zeta\right)(y)\right\rangle_{\mathbb{K}^d} \,\mathrm{d}y - \left\langle a^* v_\xi, \operatorname{grad}_\sharp w_\zeta\right\rangle_{L_2(Y)^d} \\
&= \int_Y \left\langle v_\xi(y), \left(a v_\zeta\right)(y)\right\rangle_{\mathbb{K}^d} \,\mathrm{d}y \\
&= \int_Y \left\langle \operatorname{grad}_\sharp w_\xi(y) + \xi, \left(a v_\zeta\right)(y)\right\rangle_{\mathbb{K}^d} \,\mathrm{d}y \\
&= \int_Y \left\langle \xi, \left(a v_\zeta\right)(y)\right\rangle_{\mathbb{K}^d} \,\mathrm{d}y = \langle \xi, a_{\mathrm{hom}}\zeta\rangle_{\mathbb{K}^d} .
\end{aligned}$$

(b) $\operatorname{Re} a_{\mathrm{hom}}$ is strictly positive definite. As above, one shows

$$\operatorname{Re}\langle \xi, a_{\mathrm{hom}}\xi\rangle_{\mathbb{K}^d} = \operatorname{Re}\int_Y \left\langle v_\xi(y), (a v_\xi)(y)\right\rangle_{\mathbb{K}^d} \,\mathrm{d}y \geqslant c \left\| v_\xi \right\|_{L_2(Y)^d}^2 \quad (\xi \in \mathbb{K}^d)$$

and since the right-hand side is strictly positive if $\xi \neq 0$ by Lemma 14.4.1, we derive the assertion.

The construction of a_{hom} now allows us to formulate the main result of this section.

Theorem 14.4.3 *We have*

$$\mathfrak{a}_n = \left(\iota_\sharp^* a(n\mathrm{m})\iota_\sharp\right)^{-1} \to \left(\iota_\sharp^* a_{\text{hom}}\iota_\sharp\right)^{-1} =: \mathfrak{a}_{\text{hom}} \quad (n \to \infty)$$

in the weak operator topology of $L(\operatorname{ran}(\operatorname{grad}_\sharp))$.

The proof of Theorem 14.4.3 requires some more preparation. One of the results needed is a variant of Theorem 13.2.4 for $L_2(Y)$. However, it will be beneficial to finish Example 14.3.8 first.

Example 14.4.4 (Example 14.3.8 Continued) The operator sequence $(S^{(n)})_n$ converges in the strong operator topology of $L\big(L_{2,\nu}\big(\mathbb{R}; L_2(Y) \times \operatorname{ran}(\operatorname{grad}_\sharp)\big)\big)$ to the following limit

$$\left(\overline{\partial_{t,\nu}\begin{pmatrix} 1 & 0 \\ 0 & \mathfrak{a}_{\text{hom}} \end{pmatrix} + \begin{pmatrix} 0 & \operatorname{div}_\sharp \iota_\sharp \\ \iota_\sharp^* \operatorname{grad}_\sharp & 0 \end{pmatrix}}\right)^{-1}.$$

Lemma 14.4.5 *Let* $f \colon \mathbb{R}^d \to \mathbb{K}$ *be measurable and* $[0,1)^d$*-periodic. Let* $\Omega \subseteq \mathbb{R}^d$ *be open, bounded and assume* $f|_Y \in L_2(Y)$*. Then*

$$f(n\cdot) \to \left(\int_Y f\right)\mathbb{1}_\Omega$$

weakly in $L_2(\Omega)$ *as* $n \to \infty$.

Proof Due to the boundedness of Ω we find a finite set $F \subseteq \mathbb{Z}^d$ such that $\Omega \subseteq \bigcup_{k\in F} k + Y$. Thus, by periodicity, it suffices to restrict our attention to the case when $\Omega = Y$. We define

$$X := \left\{f \colon \mathbb{R}^d \to \mathbb{K}\,;\ f \text{ is } [0,1)^d\text{-periodic, } f|_Y \in L_2(Y)\right\}$$

endowed with the norm $\|f\|_X := \|f|_Y\|_{L_2(Y)}$. It is not difficult to see that X is a Hilbert space. For $n \in \mathbb{N}$, we define $T_n \colon X \to L_2(Y)$ by $T_n f := f(n\cdot)$. Then, for all $n \in \mathbb{N}$, T_n is an isometry. Indeed, for $f \in X$, we compute

$$\int_Y |f(nx)|^2\,\mathrm{d}x = \frac{1}{n^d}\int_{nY} |f(y)|^2\,\mathrm{d}y = \frac{1}{n^d} n^d \int_Y |f(y)|^2\,\mathrm{d}y = \|f\|_{L_2(Y)}^2,$$

where we used periodicity again. Recall that $S(Y)$ denotes the simple functions on Y and consider

$$D := \{f \in X\,;\ f|_Y \in S(Y)\}.$$

Then D is dense in X. Also, if $h \in D$, then $h \in L_\infty(\mathbb{R}^d)$. By Theorem 13.2.4, we note

$$\langle T_n h, g\rangle_{L_2(Y)} = \langle h(n\cdot), g\rangle_{L_2(Y)} \to \left\langle \left(\int_Y h\right)\mathbb{1}_Y, g\right\rangle_{L_2(Y)} \quad (n \to \infty)$$

for all $g \in L_2(Y) \subseteq L_1(Y)$. Hence, $T_n h \to Th$ weakly in $L_2(Y)$ as $n \to \infty$, where for $f \in X$, we define $Tf := \left(\int_Y f\right)\mathbb{1}_Y \in L_2(Y)$.

Next, if $f \in X$, $h \in D$ and $g \in L_2(Y)$, then

$$\begin{aligned}|\langle T_n f - Tf, g\rangle| &\leqslant |\langle T_n f - T_n h, g\rangle| + |\langle T_n h - Th, g\rangle| + |\langle Th - Tf, g\rangle| \\ &\leqslant \|f-h\|_X \|g\|_{L_2(Y)} + |\langle T_n h - Th, g\rangle| \\ &\quad + \|T\|\|g\|_{L_2(Y)}\|f-h\|_X.\end{aligned}$$

Hence, for $\varepsilon > 0$, by density of D in X, we find $h \in D$ such that

$$\|f-h\|_X\|g\|_{L_2(Y)} + \|T\|\|g\|_{L_2(Y)}\|f-h\| \leqslant \frac{\varepsilon}{2}.$$

Then, we find $n_0 \in \mathbb{N}$ so that for all $n \geqslant n_0$, $|\langle T_n h - Th, g\rangle| \leqslant \varepsilon/2$ resulting in $|\langle T_n f - Tf, g\rangle| \leqslant \varepsilon$. □

Lemma 14.4.6 *Let $(q_n)_n$ and $(r_n)_n$ be weakly convergent sequences in a Hilbert space H with weak limits $q, r \in H$, respectively. Moreover, let $X \subseteq H$ be a closed subspace and $\iota\colon X \to H$ the canonical embedding. Assume that*

$$q_n \in X \text{ for each } n \in \mathbb{N} \text{ and } \left(\iota^* r_n\right)_n \text{ is strongly convergent in } X.$$

Then

$$\lim_{n\to\infty} \langle r_n, q_n\rangle_H = \langle r, q\rangle_H .$$

Proof Since $\iota^*\colon H \to X$ is continuous it is also weakly continuous, and thus,

$$\iota^* r_n \to \iota^* r \quad (n \to \infty)$$

strongly in X. For $n \in \mathbb{N}$ we compute

$$\langle r_n, q_n\rangle_H = \left\langle r_n, \iota\iota^* q_n\right\rangle_H = \left\langle \iota^* r_n, \iota^* q_n\right\rangle_X \to \left\langle \iota^* r, \iota^* q\right\rangle_X .$$

Since X is a closed subspace, it is also weakly closed and thus $q \in X$ which yields

$$\left\langle \iota^* r, \iota^* q\right\rangle_X = \langle r, q\rangle_H . \qquad \square$$

The next theorem is a version of the so-called ‘div-curl lemma’.

Theorem 14.4.7 *Let $(q_n)_n$ and $(r_n)_n$ be weakly convergent sequences in $L_2(Y)^d$ to some $q, r \in L_2(Y)^d$, respectively. Assume that*

$$q_n \in \operatorname{ran}(\operatorname{grad}_\sharp) \text{ for each } n \in \mathbb{N} \text{ and } \left(\iota_\sharp^* r_n\right)_n \text{ is strongly convergent in } \operatorname{ran}(\operatorname{grad}_\sharp).$$

Then

$$\int_Y \langle r_n(x), q_n(x)\rangle_{\mathbb{K}^d}\,\phi(x)\,\mathrm{d}x \to \int_Y \langle r(x), q(x)\rangle_{\mathbb{K}^d}\,\phi(x)\,\mathrm{d}x$$

for all $\phi \in C_c^\infty(Y)$ as $n \to \infty$.

Proof Let $\phi \in C_c^\infty(Y)$, $n \in \mathbb{N}$. Since $q_n \in \operatorname{ran}(\operatorname{grad}_\sharp)$, we find a unique $w_n \in H_\sharp^1(Y)$ with $w_n \in \{\mathbb{1}_Y\}^\perp = \ker(\operatorname{grad}_\sharp)^\perp$ such that

$$\operatorname{grad}_\sharp w_n = q_n.$$

Moreover, since $\operatorname{grad}_\sharp\colon H_\sharp^1(Y) \cap \{\mathbb{1}_Y\}^\perp \to \operatorname{ran}(\operatorname{grad}_\sharp)$ is an isomorphism by Proposition 14.3.7, we infer that $(w_n)_n$ is a weakly convergent sequence in $H_\sharp^1(Y)$ and denote its weak limit by $w \in H_\sharp^1(Y)$. By Theorem 14.3.6, we deduce $w_n \to w$ strongly in $L_2(Y)^d$. Moreover, note that $(\phi w_n)_n$ weakly converges to ϕw in $H_\sharp^1(Y)$. In particular, $\operatorname{grad}_\sharp(\phi w_n) \to \operatorname{grad}_\sharp(\phi w)$ weakly in $L_2(Y)^d$. For $n \in \mathbb{N}$, we compute

$$\begin{aligned}\int_Y \langle r_n(x), q_n(x)\rangle_{\mathbb{K}^d}\,\phi(x)\,\mathrm{d}x &= \langle r_n, q_n\phi\rangle_{L(Y)^d} = \left\langle r_n, (\operatorname{grad}_\sharp w_n)\phi\right\rangle_{L(Y)^d} \\ &= \left\langle r_n, \operatorname{grad}_\sharp(\phi w_n)\right\rangle_{L(Y)^d} - \left\langle r_n, w_n \operatorname{grad}_\sharp \phi\right\rangle_{L_2(Y)^d}.\end{aligned}$$

Now, the first term on the right-hand side of this equality tends to $\left\langle r, \operatorname{grad}_\sharp(\phi w)\right\rangle_{L_2(Y)^d}$ by Lemma 14.4.6 applied to $X = \operatorname{ran}(\operatorname{grad}_\sharp)$, which is closed by Proposition 14.3.7. The second term tends to $\left\langle r, w \operatorname{grad}_\sharp \phi\right\rangle_{L_2(Y)^d}$ by strong convergence of $(w_n)_n$ and weak convergence of $(r_n)_n$ in $L_2(Y)^d$. Thus, we obtain

$$\begin{aligned}\int_Y \langle r_n(x), q_n(x)\rangle_{\mathbb{K}^d}\,\phi(x)\,\mathrm{d}x &\to \left\langle r, \operatorname{grad}_\sharp(\phi w)\right\rangle_{L_2(Y)^d} - \left\langle r, w \operatorname{grad}_\sharp \phi\right\rangle_{L_2(Y)^d} \\ &= \int_Y \langle r(x), q(x)\rangle_{\mathbb{K}^d}\,\phi(x)\,\mathrm{d}x \quad (n \to \infty).\end{aligned}$$ □

We will apply the latter theorem to the concrete case when $r_n = a(n\mathrm{m})q_n$ in order to determine the weak limit of $(a(n\mathrm{m})q_n)_n$.

Lemma 14.4.8 *Let* $(q_n)_n$ *and* $(a(n\mathrm{m})q_n)_n$ *be weakly convergent in* $L_2(Y)^d$ *to some* q *and* r*, respectively. Assume that*

$$q_n \in \mathrm{ran}(\mathrm{grad}_\sharp) \text{ for each } n \in \mathbb{N} \text{ and } \left(\iota_\sharp^* a(n\mathrm{m})q_n\right)_n \text{ is strongly convergent in } \mathrm{ran}(\mathrm{grad}_\sharp).$$

Then $r = a_{\mathrm{hom}}q$.

Proof Let $\xi \in \mathbb{K}^d$ and choose $v := v_\xi \in L_2(Y)^d$ according to Lemma 14.4.1 for a^* instead of a; that is, $v - \xi \in \mathrm{ran}(\mathrm{grad}_\sharp)$ and $a^*(\mathrm{m})v \in \ker(\mathrm{div}_\sharp)$. For $n \in \mathbb{N}$, we define $v_n := v_{\mathrm{pe}}(n\cdot) \in L_2(Y)^d$. Next, let $g \in C_\sharp^\infty(Y)$. Then we compute

$$\begin{aligned}\left\langle a^*(n\mathrm{m})v_n, \mathrm{grad}_\sharp\, g\right\rangle_{L_2(Y)^d} &= \int_Y \left\langle a^*(nx)v_{\mathrm{pe}}(nx), \mathrm{grad}_\sharp\, g(x)\right\rangle_{\mathbb{K}^d} \mathrm{d}x \\ &= \frac{1}{n^d}\int_{nY} \left\langle a^*(y)v_{\mathrm{pe}}(y), \left(\mathrm{grad}_\sharp\, g\right)(y/n)\right\rangle_{\mathbb{K}^d} \mathrm{d}y \\ &= \frac{1}{n^{d-1}}\int_{nY} \left\langle a^*(y)v_{\mathrm{pe}}(y), (\mathrm{grad}\, g(\cdot/n))(y)\right\rangle_{\mathbb{K}^d} \mathrm{d}y.\end{aligned}$$

In order to compute the last integral, we employ Lemma 14.3.3 and Remark 14.3.4 to find a sequence $(\phi_k)_{k\in\mathbb{N}}$ in $C_\sharp^\infty(Y)^d \cap \ker(\mathrm{div}_\sharp)$ such that $\phi_k \to a^*(\mathrm{m})v$ as $k \to \infty$ in $L_2(Y)^d$. The latter implies $(\phi_k)_{\mathrm{pe}} \to a^*(\mathrm{m})v_{\mathrm{pe}}$ as $k \to \infty$ in $L_2(nY)^d$ for each $n \in \mathbb{N}$ and $\mathrm{div}(\phi_k)_{\mathrm{pe}} = 0$ for all $k \in \mathbb{N}$ by Proposition 14.3.5. Thus, we obtain with integration by parts (note that the boundary terms vanish due to the periodicity of ϕ_k and g)

$$\begin{aligned}\left\langle a^*(n\mathrm{m})v_n, \mathrm{grad}_\sharp\, g\right\rangle_{L_2(Y)^d} &= \frac{1}{n^{d-1}}\left\langle a^*(\mathrm{m})v_{\mathrm{pe}}, (\mathrm{grad}\, g(\cdot/n))\right\rangle_{L_2(nY)^d} \\ &= \frac{1}{n^{d-1}}\lim_{k\to\infty}\left\langle (\phi_k)_{\mathrm{pe}}, (\mathrm{grad}\, g(\cdot/n))\right\rangle_{L_2(nY)^d} = 0.\end{aligned}$$

Since $C_\sharp^\infty(Y)$ is a core for $\mathrm{grad}_\sharp$, we infer that $a^*(n\mathrm{m})v_n \in \mathrm{ran}(\mathrm{grad}_\sharp)^\perp$ and hence,

$$\iota_\sharp^* a^*(n\mathrm{m})v_n = 0 \quad (n \in \mathbb{N}).$$

Moreover, we have $a^*(n\mathrm{m})v_n \to \int_Y a^*v = (a^*)_{\mathrm{hom}}\xi$ weakly in $L_2(Y)^d$ as $n \to \infty$ by Lemma 14.4.5. Thus, by Theorem 14.4.7 applied to q_n and $r_n := a^*(n\mathrm{m})v_n$, we deduce that for all $\phi \in C_c^\infty(Y)$

$$\lim_{n\to\infty}\int_Y \left\langle a^*(nx)v_n(x), q_n(x)\right\rangle_{\mathbb{K}^d} \phi(x)\,\mathrm{d}x = \int_Y \left\langle (a^*)_{\mathrm{hom}}\xi, q(x)\right\rangle_{\mathbb{K}^d} \phi(x)\,\mathrm{d}x.$$

On the other hand, $v_n \to \left(\int_Y v\right)\mathbb{1}_Y = \xi\mathbb{1}_Y$ weakly in $L_2(Y)^d$ as $n \to \infty$ by Lemma 14.4.5, where $\int_Y v = \xi$ follows from $v - \xi \in \mathrm{ran}(\mathrm{grad}_\sharp)$. Thus, we can apply Theorem 14.4.7 to $q_n := v_n$ and $r_n := a(nm)q_n$ and obtain for all $\phi \in C_c^\infty(Y)$

$$\int_Y \langle a^*(nx)v_n(x), q_n(x)\rangle_{\mathbb{K}^d} \phi(x)\,\mathrm{d}x = \int_Y \langle v_n(x), a(nx)q_n(x)\rangle_{\mathbb{K}^d} \phi(x)\,\mathrm{d}x$$
$$\to \int_Y \langle \xi, r(x)\rangle_{\mathbb{K}^d} \phi(x)\,\mathrm{d}x$$

as $n \to \infty$. Thus, we have

$$\int_Y \langle (a^*)_{\mathrm{hom}}\xi, q(x)\rangle_{\mathbb{K}^d} \phi(x)\,\mathrm{d}x = \int_Y \langle \xi, r(x)\rangle_{\mathbb{K}^d} \phi(x)\,\mathrm{d}x$$

for each $\phi \in C_c^\infty(Y)$. Hence, we infer

$$\langle \xi, r(x)\rangle_{\mathbb{K}^d} = \langle (a^*)_{\mathrm{hom}}\xi, q(x)\rangle_{\mathbb{K}^d} = \langle \xi, a_{\mathrm{hom}}q(x)\rangle_{\mathbb{K}^d}$$

for almost every $x \in Y$, where we have used Remark 14.4.2(a). Since the latter holds for each $\xi \in \mathbb{K}^d$, we deduce $r = a_{\mathrm{hom}}q$. □

Proof of Theorem 14.4.3 Let $n \in \mathbb{N}$ and for $u \in \operatorname{ran}(\operatorname{grad}_\sharp)$ we put $q_n := \mathfrak{a}_n u$. We need to show that $(q_n)_n$ weakly converges to $\mathfrak{a}_{\mathrm{hom}}u$. For this, we choose subsequences (without relabeling) such that both $(q_n)_n$ and $(a(nm)q_n)_n$ weakly converge to some q and r, respectively. By definition, we have $q_n \in \operatorname{ran}(\operatorname{grad}_\sharp)$ and $\iota_\sharp^* a(nm)q_n = u$ for each $n \in \mathbb{N}$. Hence, by Lemma 14.4.8, we deduce $a_{\mathrm{hom}}q = r$. As $\operatorname{ran}(\operatorname{grad}_\sharp)$ is closed, it is also weakly closed, and hence, $q \in \operatorname{ran}(\operatorname{grad}_\sharp)$. Thus, we have

$$\iota_\sharp^* a_{\mathrm{hom}} \iota_\sharp q = \iota_\sharp^* r,$$

or equivalently

$$q = \mathfrak{a}_{\mathrm{hom}} \iota_\sharp^* r.$$

Now, since $u = \iota_\sharp^* a(nm)q_n \to \iota_\sharp^* r$ weakly, we infer

$$q = \mathfrak{a}_{\mathrm{hom}} u.$$

A subsequence argument now yields the claim. □

14.5 Comments

The theory of finding partial differential equations as appropriate limit problems of partial differential equations with highly oscillatory coefficients is commonly referred to as 'homogenisation'. The mathematical theory of homogenisation goes back to the late 1960s and early 70s. We refer to [11] as an early monograph wrapping up the available theory to that date.

The usual way of addressing homogenisation problems is to look at static (i.e., time-independent) problems first. The corresponding elliptic equation is then intensively studied. Even though it might be hidden in the derivations above, the ‘study of the elliptic problem’ essentially boils down to addressing the limit behaviour of $\mathfrak{a}_n$ as $n \to \infty$; see [37, 132]. Consequently, generalisations of the periodic case have been introduced. The periodic case (and beyond) is covered in [11, 21]; non-periodic cases and corresponding notions have been introduced in [108, 109] and, independently, in [70, 71].

An important technical tool to obtain results in this direction is the div-curl lemma or the notion of ‘compensated compactness’. In the above presented material, this is Theorem 14.4.7; the main difficulty to overcome is that of finding a limit of a product $(\langle q_n, r_n\rangle)_n$ of weakly convergent sequences $(q_n)_n$, $(r_n)_n$ in $L_2(\Omega)^3$ for some open $\Omega \subseteq \mathbb{R}^3$. It turns out that if $(\operatorname{curl} q_n)_n$ and $(\operatorname{div} r_n)_n$ converge strongly in an appropriate sense, then $\int_\Omega \langle q_n, r_n\rangle \phi$ converges to the desired limit for all $\phi \in C_c^\infty(\Omega)$. In Theorem 14.4.7 the curl-condition is strengthened in as much as we ask q_n to be a gradient, which results in $\operatorname{curl} q_n = 0$. The div-condition is replaced by the condition involving $\iota_\sharp^*$, which can in fact be shown to be equivalent, see [130]. The restriction to periodic boundary value problems is a mere convenience. It can be shown that the arguments work similarly for non-periodic boundary conditions, and even with the same limit, see [113, Lemma 10.3].

There are many generalisations of the div-curl lemma. For this, we refer to [17] (and the references given there) and to the rather recently found operator-theoretic perspective, with plenty of applications not solely restricted to the operators div and curl, see [80, 130].

We shortly comment on the term ‘compensated compactness’. In general, one cannot expect for two weakly convergent sequences $(q_n)_n$ and $(r_n)_n$ in $L_2(\Omega)^3$ that the sequence of their scalar product $\langle q_n, r_n\rangle$ to converge to the scalar product of the limits. If, however, either $(q_n)_n$ or $(r_n)_n$ are bounded in a space compactly embedded into $L_2(\Omega)^3$, then either of those sequence converge in norm in $L_2(\Omega)^3$ and $\lim_{n\to\infty} \langle q_n, r_n\rangle = \langle \lim_{n\to\infty} q_n, \lim_{n\to\infty} r_n\rangle$ follows. However, even though neither $H_0(\operatorname{curl}, \Omega)$ nor $H(\operatorname{div}, \Omega)$ are compactly embedded into $L_2(\Omega)^3$, one can still conclude that for bounded sequences $(q_n)_n$ in $H_0(\operatorname{curl}, \Omega)$ and $(r_n)_n$ in $H(\operatorname{div}, \Omega)$ we have

$$\lim_{n\to\infty} \langle q_n, r_n\rangle = \Big\langle \lim_{n\to\infty} q_n, \lim_{n\to\infty} r_n \Big\rangle .$$

Thus, one might argue that the respectively missing compactness of the embeddings of $H_0(\operatorname{curl}, \Omega)$ and $H(\operatorname{div}, \Omega)$ into $L_2(\Omega)^3$ is somehow ‘compensated’. Following the core arguments in [130], one might also argue that the deeper reason for the convergence of the scalar products is more closely related to (general) Helmholtz decompositions.

The way of deriving the homogenised equation (i.e., the limit of $\mathfrak{a}_n$) is akin to some derivations in [21, 128]. Further reading on homogenisation problems can also be found in these references. The first step of combining homogenisation processes and evolutionary equations has been made in [135] and has had some profound developments for both quantitative and qualitative results; see [23, 42, 136, 138].

Exercises

Exercise 14.1 Under the same assumptions of Theorem 14.1.1 show

$$\left\| \left(\left(\overline{\partial_{t,\nu} M_n(\partial_{t,\nu}) + A}\right)^{-1} - \left(\overline{\partial_{t,\nu} M(\partial_{t,\nu}) + A}\right)^{-1} \right) \partial_{t,\nu}^{-1} \right\|_{L(L_{2,\nu}(\mathbb{R};H))} \to 0.$$

Exercise 14.2 Let $\Omega \subseteq \mathbb{R}^d$ be open and $\varepsilon > 0$. We define the set

$$\Omega_\varepsilon := \{x \in \Omega\,;\ \operatorname{dist}(x, \partial\Omega) > \varepsilon\}.$$

(a) Let $(\phi_k)_{k\in\mathbb{N}}$ in $C_c^\infty(\mathbb{R}^d)$ be a δ-sequence (cf. Exercise 3.1) and $u \in H^1(\Omega)$. We identify each function on Ω by its extension to $\mathbb{R}^d$ by 0. Prove that for $k \in \mathbb{N}$ large enough, $\phi_k * u \in H^1(\Omega_\varepsilon)$ with

$$\operatorname{grad}(\phi_k * u) = \phi_k * \operatorname{grad} u \text{ on } \Omega_\varepsilon.$$

(b) Use (a) to prove Lemma 14.2.2.

Exercise 14.3 Prove the 'subsequence argument': Let X be a topological space and $(x_n)_n$ a sequence in X. Assume that there exists $x \in X$ such that each subsequence of $(x_n)_n$ has a subsequence converging to x. Show that $x_n \to x$ as $n \to \infty$.

Exercise 14.4 Let H_0, H_1 be Hilbert spaces and $C\colon \operatorname{dom}(C) \subseteq H_0 \to H_1$ be a closed linear operator such that $\operatorname{dom}(C) \hookrightarrow H_0$ compactly. Let $P_{\ker(C)^\perp}\colon H_0 \to H_0$ denote the orthogonal projection onto the closed subspace $\ker(C)^\perp$. Prove that there exists $c > 0$ such that

$$\forall u \in \operatorname{dom}(C) : \left\| P_{\ker(C)^\perp} u \right\|_{H_0} \leqslant c \left\| Cu \right\|_{H_1}.$$

Apply this result to prove Proposition 14.3.7.

Exercise 14.5 Let H_0, H_1 be Hilbert spaces. Let $C\colon \operatorname{dom}(C) \subseteq H_0 \to H_1$ be closed and densely defined. Assume that $\operatorname{dom}(C) \cap \ker(C)^\perp \hookrightarrow H_0$ compactly. Show that, then, $\operatorname{dom}(C^*) \cap \ker(C^*)^\perp \hookrightarrow H_1$ compactly.

Exercise 14.6 Let $\nu > 0$, $\Omega = [0,1)^d$, $s \in L_\infty(\mathbb{R})$ be 1-periodic, $0 \leqslant s \leqslant 1$, and a as in Example 14.3.8. Show that $(u_n)_n$ in $L_{2,\nu}(\mathbb{R}; L_2(Y))$ satisfying

$$\partial_{t,\nu}^2 s(n\mathrm{m})u_n + \partial_{t,\nu}(1 - s(n\mathrm{m}))u_n - \operatorname{div}_\sharp a(n\mathrm{m}) \operatorname{grad}_\sharp u_n = f$$

for some $f \in L_{2,\nu}(\mathbb{R}; L_2(Y))$ is convergent to some $u \in L_{2,\nu}(\mathbb{R}; L_2(Y))$. Which limit equation is satisfied by u?

Exercise 14.7 Let $(\alpha_n)_n$ be a nullsequence in $[0, 1]$ and let a be as in Example 14.3.8. Show

$$\left(\overline{\begin{pmatrix} \partial_{t,\nu} & 0 \\ 0 & \partial_{t,\nu}^{\alpha_n} \mathfrak{a}_n \end{pmatrix} + \begin{pmatrix} 0 & \operatorname{div}_\sharp \iota_\sharp \\ \iota_\sharp^* \operatorname{grad}_\sharp & 0 \end{pmatrix}}\right)^{-1} \to \left(\overline{\begin{pmatrix} \partial_{t,\nu} & 0 \\ 0 & \mathfrak{a}_{\mathrm{hom}} \end{pmatrix} + \begin{pmatrix} 0 & \operatorname{div}_\sharp \iota_\sharp \\ \iota_\sharp^* \operatorname{grad}_\sharp & 0 \end{pmatrix}}\right)^{-1}$$

in the strong operator topology. Show that if $f \in L_{2,-\mu}(\mathbb{R}; L_2(Y)_\perp)$, where $L_2(Y)_\perp := \left\{\phi \in L_2(Y)\,;\, \int_Y \phi = 0\right\}$ for some small enough $\mu > 0$, we have

$$\left(\overline{\begin{pmatrix} \partial_{t,\nu} & 0 \\ 0 & \mathfrak{a}_{\mathrm{hom}} \end{pmatrix} + \begin{pmatrix} 0 & \operatorname{div}_\sharp \iota_\sharp \\ \iota_\sharp^* \operatorname{grad}_\sharp & 0 \end{pmatrix}}\right)^{-1} \begin{pmatrix} f \\ 0 \end{pmatrix} \in L_{2,-\mu}\big(\mathbb{R}; L_2(Y) \times \operatorname{ran}(\operatorname{grad}_\sharp)\big).$$

References

5. W. Arendt et al., Form methods for evolution equations, and applications, in *18th Internet Seminar* (2015)
11. A. Bensoussan, J.-L. Lions, G. Papanicolaou, *Asymptotic Analysis for Periodic Structures*, vol. 5. Studies in Mathematics and Its Applications (North-Holland, Amsterdam, 1978)
17. M. Briane, J. Casado-Díaz, F. Murat, The div-curl lemma "trente ans après": an extension and an application to the G-convergence of unbounded monotone operators. J. Math. Pures Appl. (9) **91**(5), 476–494 (2009)
21. D. Cioranescu, P. Donato, *An Introduction to Homogenization*, vol. 17. Oxford Lecture Series in Mathematics and Its Applications (The Clarendon Press, Oxford University Press, New York, 1999)
23. S. Cooper, M. Waurick, Fibre homogenisation. J. Funct. Anal. **276**(11), 3363–3405 (2019)
37. A.F.M. ter Elst, G. Gorden, M. Waurick, The Dirichlet-to-Neumann operator for divergence form problems. Ann. Mat. Pura Appl. (4) **198**(1), 177–203 (2019)
42. S. Franz, M. Waurick, Resolvent estimates and numerical implementation for the homogenisation of one-dimensional periodic mixed type problems. Z. Angew. Math. Mech. **98**(7), 1284–1294 (2018)
70. F. Murat, Compacité par compensation. Ann. Scuola Norm. Sup. Pisa Cl. Sci. (4) **5**(3), 489–507 (1978)
71. F. Murat, L. Tartar, H-convergence, in *Topics in the Mathematical Modelling of Composite Materials*, vol. 31. Progr. Nonlinear Differential Equations Appl. (Birkhäuser, Boston, 1997), 21–43

80. D. Pauly, A global div-curl-lemma for mixed boundary conditions in weak Lipschitz domains and a corresponding generalized $A_0^* - A_1$-lemma in Hilbert spaces'. Analysis (Berlin) **39**(2), 33–58 (2019)
108. S. Spagnolo, Sulla convergenza di soluzioni di equazioni paraboliche ed ellittiche. Ann. Scuola Norm. Sup. Pisa (3) **22**, 571–597 (1968); errata, ibid. (3) **22**, 673 (1968)
109. S. Spagnolo, Sul limite delle soluzioni di problemi di Cauchy relativi all'equazione del calore. Ann. Scuola Norm. Sup. Pisa (3) **21**, 657–699 (1967)
113. L. Tartar, *The General Theory of Homogenization*, vol. 7. Lecture Notes of the Unione Matematica Italiana. A Personalized Introduction (Springer, Berlin; UMI, Bologna, 2009)
128. V.V. Jikov, S.M. Kozlov, O.A. Oleinik, *Homogenization of Differential Operators and Integral Functionals* (Springer, Berlin, 1994)
130. M. Waurick, A Functional Analytic Perspective to the div-curl Lemma. J. Oper. Theory **80**(1), 95–111 (2018)
132. M. Waurick, G-convergence and the weak operator topology. Proc. Appl. Math. Mech. **16**, 521–522 (2016)
135. M. Waurick, Limiting processes in evolutionary equations - a Hilbert space approach to homogenization. Dissertation. Technische Universität Dresden, 2011. http://nbn-resolving.de/urn:nbn:de:bsz:14-qucosa-67442
136. M. Waurick, Nonlocal H-convergence. Calc. Var. Partial Differ. Equ. **57**(6), 46 (2018)
138. M. Waurick, On the continuous dependence on the coefficients of evolutionary equations. Habilitation. Technische Universität Dresden, 2016. http://arxiv.org/abs/1606.07731

Chapter 15
Maximal Regularity

In this chapter, we address the issue of maximal regularity. More precisely, we provide a criterion on the 'structure' of the evolutionary equation

$$\left(\overline{\partial_{t,\nu} M(\partial_{t,\nu}) + A}\right) U = F$$

in question and the right-hand side F in order to obtain $U \in \operatorname{dom}(\partial_{t,\nu} M(\partial_{t,\nu})) \cap \operatorname{dom}(A)$. If $F \in L_{2,\nu}(\mathbb{R}; H)$, $U \in \operatorname{dom}(\partial_{t,\nu} M(\partial_{t,\nu})) \cap \operatorname{dom}(A)$ is the optimal regularity one could hope for. However, one cannot expect U to be as regular since $\left(\partial_{t,\nu} M(\partial_{t,\nu}) + A\right)$ is simply not closed in general. Hence, in all the cases where $\left(\partial_{t,\nu} M(\partial_{t,\nu}) + A\right)$ is *not* closed, the desired regularity property does not hold for $F \in L_{2,\nu}(\mathbb{R}; H)$. However, note that by Picard's theorem, $F \in \operatorname{dom}(\partial_{t,\nu})$ implies the desired regularity property for U given the positive definiteness condition for the material law is satisfied and A is skew-selfadjoint. In this case, one even has $U \in \operatorname{dom}(\partial_{t,\nu}) \cap \operatorname{dom}(A)$, which is more regular than expected. Thus, in the general case of an unbounded, skew-selfadjoint operator A neither the condition $F \in \operatorname{dom}(\partial_{t,\nu})$ nor $F \in L_{2,\nu}(\mathbb{R}; H)$ yields precisely the regularity $U \in \operatorname{dom}(\partial_{t,\nu} M(\partial_{t,\nu})) \cap \operatorname{dom}(A)$ since

$$\operatorname{dom}(\partial_{t,\nu}) \cap \operatorname{dom}(A) \subseteq \operatorname{dom}(\partial_{t,\nu} M(\partial_{t,\nu})) \cap \operatorname{dom}(A) \subseteq \operatorname{dom}(\overline{\partial_{t,\nu} M(\partial_{t,\nu}) + A}),$$

where the inclusions are proper in general. It is the aim of this chapter to provide an example case, where less regularity of F actually yields *more* regularity for U. If one focusses on time-regularity only, this improvement of regularity is in stark contrast to the general theory developed in the previous chapters. Indeed, in this regard, one can coin the (time) regularity asserted in Picard's theorem as "U is as regular as F". For a more detailed account on the usual perspective of maximal regularity (predominantly) for parabolic equations, we refer to the Comments section of this chapter.

C. Seifert et al., *Evolutionary Equations*, Operator Theory: Advances and Applications 287, https://doi.org/10.1007/978-3-030-89397-2_15

15.1 Guiding Examples and Non-Examples

Before we present the abstract theory, we motivate the general setting looking at a particular example. Traditionally, in the discussion of partial differential equations and their classification, people focus on regularity theory. Thus, one finds the non-exhaustive categories 'elliptic', 'parabolic', and 'hyperbolic'. Since we do not want to dive into the intricacies of this classification much less their regularity, we only name some examples of the said subclasses. Laplace's equation from Chap. 1 falls into the class of elliptic PDEs, the heat equation is a paradigm example of a parabolic equation and Maxwell's equations or the transport equation are hyperbolic.

Since we predominantly treat time-dependent equations and elliptic PDEs usually are time-independent, we only look at examples for hyperbolic and parabolic equations more closely. As for the hyperbolic case, we consider the transport equation next and highlight that any 'gain' in regularity as hinted at in the introduction of this chapter is not possible.

Example 15.1.1 We define $\partial\colon H^1(\mathbb{R}) \subseteq L_2(\mathbb{R}) \to L_2(\mathbb{R}), \phi \mapsto \phi'$. Then, by Corollary 3.2.6, $\partial^* = -\partial$; that is, ∂ is skew-selfadjoint. We consider for $\nu > 0$ the operator

$$\partial_{t,\nu} + \partial$$

in $L_{2,\nu}(\mathbb{R}; L_2(\mathbb{R}))$. Then, by Picard's theorem, $0 \in \rho\big(\overline{\partial_{t,\nu} + \partial}\big)$; that is, $\big(\overline{\partial_{t,\nu} + \partial}\big)^{-1} \in L(L_{2,\nu}(\mathbb{R}; L_2(\mathbb{R})))$. Next, consider the functions

$$u\colon (t,x) \mapsto \mathbb{1}_{\mathbb{R}_{\geq 0}}(t) t \mathrm{e}^{-t} h(x-t)$$
$$f\colon (t,x) \mapsto \mathbb{1}_{\mathbb{R}_{\geq 0}}(t)(1-t) \mathrm{e}^{-t} h(x-t)$$

for some $h \in L_2(\mathbb{R})$. Then it is not difficult to see that $u, f \in L_{2,\nu}(\mathbb{R}; L_2(\mathbb{R}))$. If $h \in C_{\mathrm{c}}^\infty(\mathbb{R})$, then

$$u \in H_\nu^1(\mathbb{R}; H^1(\mathbb{R})) \subseteq \operatorname{dom}(\partial_{t,\nu} + \partial)$$

and

$$(\partial_{t,\nu} + \partial)u = f.$$

If $h \in L_2(\mathbb{R}) \setminus H^1(\mathbb{R})$, then one can show that $u \in \operatorname{dom}\big(\overline{\partial_{t,\nu} + \partial}\big)$, $\big(\overline{\partial_{t,\nu} + \partial}\big)u = f$ and

$$u \notin \operatorname{dom}(\partial_{t,\nu}) \cap \operatorname{dom}(\partial).$$

For this observation, we refer to Exercise 15.1. Thus, being in the domain of $\overline{\partial_{t,\nu} + \partial}$ does not necessarily imply being in the domain of either $\operatorname{dom}(\partial_{t,\nu})$ or $\operatorname{dom}(\partial)$.

The last example has shown that we cannot expect an improvement of regularity for the considered transport equation. In fact, it is possible to provide an example of a similar type for the wave equation (and similar hyperbolic type equations including Maxwell's equations). Thus, in order to have an improvement of regularity one needs to further restrict the class of evolutionary equations. We now provide a guiding example, where we discuss an abstract variant of the heat equation.

Example 15.1.2 Let ℓ_2 be the space of square summable sequences indexed by $n \in \mathbb{N}$. We note that ℓ_2 is isomorphic to $L_2(\#_{\mathbb{N}})$, where $\#_{\mathbb{N}}$ is the counting measure on $\mathbb{N}$. We introduce $\mathrm{m}\colon \operatorname{dom}(\mathrm{m}) \subseteq \ell_2 \to \ell_2$ the operator of multiplying by the argument. Then, m is an unbounded, selfadjoint operator. Next, we consider the operator

$$\partial_{t,\nu}\begin{pmatrix} 1 & 0 \\ 0 & 0 \end{pmatrix} + \begin{pmatrix} 0 & 0 \\ 0 & 1 \end{pmatrix} + \begin{pmatrix} 0 & -\mathrm{m} \\ \mathrm{m} & 0 \end{pmatrix}$$

on $L_{2,\nu}(\mathbb{R}; \ell_2)$. Then, Picard's theorem applies and we obtain

$$0 \in \rho\left(\overline{\partial_{t,\nu}\begin{pmatrix} 1 & 0 \\ 0 & 0 \end{pmatrix} + \begin{pmatrix} 0 & 0 \\ 0 & 1 \end{pmatrix} + \begin{pmatrix} 0 & -\mathrm{m} \\ \mathrm{m} & 0 \end{pmatrix}}\right).$$

For $f \in L_{2,\nu}(\mathbb{R}; \ell_2)$ define

$$\begin{pmatrix} u \\ q \end{pmatrix} := \left(\overline{\partial_{t,\nu}\begin{pmatrix} 1 & 0 \\ 0 & 0 \end{pmatrix} + \begin{pmatrix} 0 & 0 \\ 0 & 1 \end{pmatrix} + \begin{pmatrix} 0 & -\mathrm{m} \\ \mathrm{m} & 0 \end{pmatrix}}\right)^{-1} \begin{pmatrix} f \\ 0 \end{pmatrix}.$$

Then $u \in \operatorname{dom}(\partial_{t,\nu}) \cap \operatorname{dom}(\mathrm{m})$ and $q \in \operatorname{dom}(\mathrm{m})$. We ask the reader to fill in the details in Exercise 15.2.

Remark 15.1.3 The last example is in fact an abstract version of the heat equation on bounded domains. We refer to [90, Section 2.2.2] for a corresponding reasoning for the Schrödinger equation.

Let us compare the two different examples, the transport equation and the abstract parabolic equation. From the perspective of evolutionary equations; that is, looking at equations of the form

$$(\partial_{t,\nu} M_0 + M_1 + A)U = F,$$

for the transport equation we have $M_0 = 1$ and $M_1 = 0$. In the case of the abstract parabolic equation, M_0 has a nontrivial kernel, which is compensated in M_1. Moreover, the decomposition of kernel and range of M_0 is comparable to the block structure of A. Thus, we may hope for an improvement of regularity as in

Example 15.1.2 if these abstract conditions are met. This observation is the starting point of parabolic evolutionary pairs to be defined in the next section.

15.2 The Maximal Regularity Theorem and Fractional Sobolev Spaces

In order to be able to formulate the main theorem of this chapter, we need the notion of fractional Sobolev spaces. For this, we recall from Example 5.3.4 and Sect. 7.2 that we already dealt with fractional powers of the time-derivative. For α, $\nu \geq 0$, we thus consistently define

$$\partial_{t,\nu}^{\alpha} := \mathcal{L}_{\nu}^{*}(\mathrm{im}+\nu)^{\alpha}\mathcal{L}_{\nu},$$

with maximal domain in $L_{2,\nu}(\mathbb{R}; H)$, where we agree with setting $\mathcal{L}_0 := \mathcal{F}$. Note that in this case, using Proposition 7.2.1, $0 \in \rho(\partial_{t,\nu}^{\alpha})$ given $\nu > 0$. Hence, the following construction yields Hilbert spaces; for this also recall that $\langle\cdot,\cdot\rangle_A$ denotes the graph inner product of a linear operator A defined in a Hilbert space.

Definition Let α, $\nu \geq 0$. Then we define

$$H_{\nu}^{\alpha}(\mathbb{R}; H) := \left(\mathrm{dom}(\partial_{t,\nu}^{\alpha}), (f, g) \mapsto \langle \partial_{t,\nu}^{\alpha} f, \partial_{t,\nu}^{\alpha} g\rangle_{L_{2,\nu}(\mathbb{R};H)}\right)$$

for $\nu > 0$ and

$$H_{0}^{\alpha}(\mathbb{R}; H) := \left(\{f \in L_2(\mathbb{R}; H);\ \mathcal{F}f \in \mathrm{dom}((\mathrm{im})^{\alpha})\}, (f, g) \mapsto \langle \mathcal{F}f, \mathcal{F}g\rangle_{(\mathrm{im})^{\alpha}}\right).$$

Lemma 15.2.1 *For all α, $\nu \geq 0$ the space $H_{\nu}^{\alpha}(\mathbb{R}; H)$ is a Hilbert space. Moreover, $H_{\nu}^{\alpha}(\mathbb{R}; H) \hookrightarrow L_{2,\nu}(\mathbb{R}; H)$ continuously and densely.*

Proof We only show the claim for $\nu > 0$. By Fourier–Laplace transformation, the claim follows if we show that

$$(\mathrm{im}+\nu)^{\alpha}\colon \mathrm{dom}((\mathrm{im}+\nu)^{\alpha}) \subseteq L_2(\mathbb{R}; H) \to L_2(\mathbb{R}; H)$$

is densely defined and continuously invertible. For this, we find $n \in \mathbb{N}$ and $\beta \in [0, 1)$ such that $\alpha = n + \beta$. It is easy to see that $(\mathrm{im}+\nu)^{\alpha} = (\mathrm{im}+\nu)^{n}(\mathrm{im}+\nu)^{\beta}$. Thus, continuous invertibility readily follows from the continuous invertibility of $(\mathrm{im}+\nu)$ and $(\mathrm{im}+\nu)^{\beta}$ (for the latter, see also Proposition 7.2.1). For the case when $H = \mathbb{K}$, it follows from Theorem 2.4.3 that $(\mathrm{im}+\nu)^{\alpha}$ is densely defined. Thus, it follows from Lemma 3.1.8 that $(\mathrm{im}+\nu)^{\alpha}$ is densely defined also for general H. □

In order to state our main theorem, we introduce the notion of parabolic pairs.

Definition Let $M\colon \operatorname{dom}(M) \subseteq \mathbb{C} \to L(H)$ be a material law, $A\colon \operatorname{dom}(A) \subseteq H \to H$ and $\alpha \in (0,1]$. We call (M, A) an *(α-)fractional parabolic pair* if the following conditions are met: there exist $\nu > \max\{0, \mathrm{s_b}(M)\}$ and $c > 0$ such that

$$\operatorname{Re} zM(z) \geqslant c \quad (z \in \mathbb{C}_{\mathrm{Re}>\nu}),$$

and moreover, we find a closed subspace $H_0 \subseteq H$, $H_1 := H_0^{\perp}$, $C\colon \operatorname{dom}(C) \subseteq H_0 \to H_1$ closed and densely defined, and $M_{00} \in \mathcal{M}(H_0;\nu)$, $N \in \mathcal{M}(H;\nu)$ such that

$$M(z) = \begin{pmatrix} M_{00}(z) & 0 \\ 0 & 0 \end{pmatrix} + z^{-1}N(z), \quad A = \begin{pmatrix} 0 & -C^* \\ C & 0 \end{pmatrix},$$

and

$$\operatorname{Re} z^{1-\alpha} M_{00}(z) \geqslant c' \quad (z \in \mathbb{C}_{\mathrm{Re}>\nu})$$

for some $c' > 0$, and $\mathbb{C}_{\mathrm{Re}>\nu} \ni z \mapsto z^{1-\alpha}M_{00}(z) \in L(H_0)$ is bounded. A 1-fractional parabolic pair is called *parabolic*.

Remark 15.2.2

(a) If (M, A) is α-fractional parabolic and β-fractional parabolic with the same decomposition $H = H_0 \oplus H_1$, then $\alpha = \beta$. Indeed, assume that $\alpha < \beta$. Then

$$z^{1-\beta}M_{00}(z) = z^{\alpha-\beta}z^{1-\alpha}M_{00}(z) \to 0 \quad (|z| \to \infty, z \in \mathbb{C}_{\mathrm{Re}>\nu})$$

contradicting the real-part condition.

(b) If (M, A) is α-fractional parabolic, then there exists $\mu > \nu$ such that for all $z \in \mathbb{C}_{\mathrm{Re}>\mu}$

$$\operatorname{Re} z^{1-\alpha}\Big(M_{00}(z) + z^{-1}N_{00}(z)\Big) \geqslant c'/2 \tag{15.1}$$

for some $c' > 0$, where $N_{00}(z) := \iota_{H_0}^* N(z)\iota_{H_0} \in L(H_0)$. Indeed, this follows from the fact that $z^{-\alpha}N_{00}(z) \to 0$ as $\operatorname{Re} z \to \infty$.

The main theorem of this chapter is the following:

Theorem 15.2.3 *Let $\alpha \in (0,1]$ and (M, A) be α-fractional parabolic (with $H = H_0 \oplus H_1$ and C from H_0 to H_1) and assume that* (15.1) *holds for all $z \in \mathbb{C}_{Re>\nu}$ for some $\nu > \max\{0, \mathrm{s_b}(M)\}$. Let $f \in L_{2,\nu}(\mathbb{R}; H_0)$ and $g \in H_\nu^{\alpha/2}(\mathbb{R}; H_1)$. Then the solution $(u, v) := \left(\overline{\partial_{t,\nu}M(\partial_{t,\nu}) + A}\right)^{-1}(f, g) \in L_{2,\nu}(\mathbb{R}; H)$ satisfies*

$$u \in H_\nu^\alpha(\mathbb{R}; H_0) \cap H_\nu^{\alpha/2}\big(\mathbb{R}; \operatorname{dom}(C)\big)$$
$$v \in H_\nu^{\alpha/2}(\mathbb{R}; H_1) \cap L_{2,\nu}\big(\mathbb{R}; \operatorname{dom}(C^*)\big).$$

More precisely,

$$\left(\overline{\partial_{t,\nu} M(\partial_{t,\nu}) + A}\right)^{-1} : L_{2,\nu}(\mathbb{R}; H_0) \oplus H_\nu^{\alpha/2}(\mathbb{R}; H_1)$$
$$\to \left(H_\nu^\alpha(\mathbb{R}; H_0) \cap H_\nu^{\alpha/2}(\mathbb{R}; \mathrm{dom}(C))\right) \oplus \left(H_\nu^{\alpha/2}(\mathbb{R}; H_1) \cap L_{2,\nu}(\mathbb{R}; \mathrm{dom}(C^*))\right)$$

is continuous.

Example 15.2.4 (Heat Equation) Let us recall the heat equation from Theorem 6.2.4. For $\Omega \subseteq \mathbb{R}^d$ open, we let $a \in L(L_2(\Omega)^d)$ such that

$$\operatorname{Re} a \geqslant c$$

in the sense of positive definiteness. It is not difficult to see that

$$\left(z \mapsto \begin{pmatrix} 1 & 0 \\ 0 & az^{-1} \end{pmatrix}, \begin{pmatrix} 0 & \operatorname{div}_0 \\ \operatorname{grad} & 0 \end{pmatrix}\right),$$

is parabolic; with the obvious orthogonal decomposition of the underlying Hilbert space. Let $f \in L_{2,\nu}(\mathbb{R}; L_2(\Omega))$. Then

$$\begin{pmatrix} \theta \\ q \end{pmatrix} := \left(\partial_{t,\nu} \begin{pmatrix} 1 & 0 \\ 0 & 0 \end{pmatrix} + \begin{pmatrix} 0 & 0 \\ 0 & a^{-1} \end{pmatrix} + \begin{pmatrix} 0 & \operatorname{div}_0 \\ \operatorname{grad} & 0 \end{pmatrix}\right)^{-1} \begin{pmatrix} f \\ 0 \end{pmatrix}$$

particularly satisfies the regularity statement

$$\theta \in H_\nu^1(\mathbb{R}; L_2(\Omega)) \cap L_{2,\nu}\left(\mathbb{R}; H^1(\Omega)\right) \text{ and } q \in L_{2,\nu}\left(\mathbb{R}; H_0(\operatorname{div}, \Omega)\right).$$

The next example deals with a parabolic variant of the equations introduced in (7.3) and (7.4) describing fractional elasticity. We modify the equations at hand by considering $\alpha \in [1, 2]$.

Example 15.2.5 (Parabolic Fractional Viscoelasticity) Let $\Omega \subseteq \mathbb{R}^d$ open and recall the differential operators Div and Grad_0 from Sect. 7.1 defined in the spaces $L_2(\Omega)^{d\times d}_{\mathrm{sym}}$ and $L_2(\Omega)^d$, respectively. Let $c > 0$ and $D \in L\left(L_2(\Omega)^{d\times d}_{\mathrm{sym}}\right)$, $\rho = \rho^* \in L(L_2(\Omega)^d)$. For $\nu > 0$ and $f \in L_{2,\nu}(\mathbb{R}; L_2(\Omega)^d)$ consider the problem of finding $u\colon \mathbb{R} \times \Omega \to \mathbb{R}^d$ such that

$$\partial_{t,\nu} \rho \partial_{t,\nu} u - \operatorname{Div} T = f \tag{15.2}$$

$$T = D\partial_{t,\nu}^\alpha \operatorname{Grad}_0 u, \tag{15.3}$$

for some $\alpha \in [1, 2)$, where $\rho \geqslant c$ and $\operatorname{Re} D \geqslant c$ in the sense of positive definiteness. We rewrite the system just introduced by using $v := \partial_{t,\nu}^{\alpha} u$ to (formally) obtain

$$\partial_{t,\nu}\rho\partial_{t,\nu}^{1-\alpha}v - \operatorname{Div} T = f$$
$$T = D\operatorname{Grad}_0 v.$$

Note that $\gamma := 1 + (1-\alpha) \in (0, 1]$. Thus, using the selfadjointness and positive definiteness of ρ as well as Proposition 7.2.1, we infer

$$\operatorname{Re}(z^{\gamma}\rho) \geqslant \nu^{\gamma} c \quad (z \in \mathbb{C}_{\operatorname{Re}\geqslant\nu}).$$

Consequently, applying Proposition 6.2.3(b) to $a = D$, we get that

$$\left(z \mapsto \begin{pmatrix} z^{\gamma-1}\rho & 0 \\ 0 & z^{-1}D^{-1} \end{pmatrix}, \begin{pmatrix} 0 & -\operatorname{Div} \\ -\operatorname{Grad}_0 & 0 \end{pmatrix}\right)$$

is γ-fractional parabolic. In consequence, the solution (v, T) of

$$\overline{\left(\partial_{t,\nu}\begin{pmatrix} \partial_{t,\nu}^{\gamma-1}\rho & 0 \\ 0 & \partial_{t,\nu}^{-1}D^{-1} \end{pmatrix} + \begin{pmatrix} 0 & -\operatorname{Div} \\ -\operatorname{Grad}_0 & 0 \end{pmatrix}\right)}\begin{pmatrix} v \\ T \end{pmatrix} = \begin{pmatrix} f \\ 0 \end{pmatrix}$$

additionally satisfies the following regularity properties

$$v \in H_{\nu}^{\gamma}\big(\mathbb{R}; L_2(\Omega)^d\big) \cap H_{\nu}^{\gamma/2}\big(\mathbb{R}; \operatorname{dom}(\operatorname{Grad}_0)\big),$$
$$T \in H_{\nu}^{\gamma/2}\big(\mathbb{R}; L_2(\Omega)_{\mathrm{sym}}^{d\times d}\big) \cap L_{2,\nu}\big(\mathbb{R}; \operatorname{dom}(\operatorname{Div})\big).$$

Rephrasing this for $u = \partial_{t,\nu}^{-\alpha} v$, we even have

$$u \in H_{\nu}^{2}\big(\mathbb{R}; L_2(\Omega)^d\big) \cap H_{\nu}^{1+\alpha/2}\big(\mathbb{R}; \operatorname{dom}(\operatorname{Grad}_0)\big),$$

which, since $\alpha/2 \leqslant 1$, particularly implies that the equations (15.2) and (15.3) are equalities valid in $L_{2,\nu}\big(\mathbb{R}; L_2(\Omega)^d\big)$ and $L_{2,\nu}\big(\mathbb{R}; L_2(\Omega)_{\mathrm{sym}}^{d\times d}\big)$, respectively.

15.3 The Proof of Theorem 15.2.3

The decisive estimate in connection to the proof of Theorem 15.2.3 is contained in the following statement. For the entire rest of the section, we shall denote the norm and scalar product in $H_{\nu}^{\alpha}(\mathbb{R}; K)$, K some Hilbert space, by $\|\cdot\|_{\alpha}$ and $\langle\cdot,\cdot\rangle_{\alpha}$, respectively.

Lemma 15.3.1 *Let H_0, H_1 be Hilbert spaces, $C\colon \operatorname{dom}(C) \subseteq H_0 \to H_1$ densely defined and closed. Let $\alpha \in [0,1]$, $M_j\colon \operatorname{dom}(M_j) \subseteq \mathbb{C} \to L(H_j)$ material laws for $j \in \{0,1\}$, $\nu > \max\{s_b(M_0), s_b(M_1), 0\}$ with*

$$\mathbb{C}_{\mathrm{Re}\geqslant\nu} \ni z \mapsto z^{1-\alpha}M_0(z) \in L(H_0)$$

bounded. Assume there exists $c > 0$ such that for all $z \in \mathbb{C}_{\mathrm{Re}\geqslant\nu}$

$$\operatorname{Re} zM_0(z) \geqslant c, \quad \operatorname{Re} M_1(z) \geqslant c, \quad \operatorname{Re} z^{1-\alpha}M_0(z) \geqslant c.$$

Let $f \in L_{2,\nu}(\mathbb{R}; H_0)$, $g \in H_\nu^{\alpha/2}(\mathbb{R}; H_1)$ as well as $u \in H_\nu^1\big(\mathbb{R}; \operatorname{dom}(C)\big)$ and $v \in H_\nu^1\big(\mathbb{R}; \operatorname{dom}(C^*)\big)$. Assume the equalities

$$\begin{aligned}\partial_{t,\nu}M_0(\partial_{t,\nu})u - C^*v &= f,\\ v + M_1(\partial_{t,\nu})Cu &= g.\end{aligned}$$

Then

$$\begin{aligned}&\|u\|_\alpha^2 + \|Cu\|_{\alpha/2}^2 + \|v\|_{\alpha/2}^2 + \left\|C^*v\right\|_0^2\\ &\leqslant 2\left(1 + \left(m_1^2 + m_0^2 + \frac{1}{2}\right)\left(\frac{2}{c} + \frac{m_1}{c^2}\right)^2\right)\left(\|f\|_0^2 + \|g\|_{\alpha/2}^2\right)\end{aligned}$$

with $m_1 := \|M_1\|_{\infty,\mathbb{C}_{\mathrm{Re}>\nu}}$ and $m_0 := \left\|z \mapsto z^{1-\alpha}M_0(z)\right\|_{\infty,\mathbb{C}_{\mathrm{Re}>\nu}}$.

Proof We compute

$$\begin{aligned}c\,\|Cu\|_{\alpha/2}^2 &\leqslant c\,\|Cu\|_{\alpha/2}^2 + c\,\|u\|_{\alpha/2}^2\\ &\leqslant \operatorname{Re}\left\langle M_1(\partial_{t,\nu})Cu, Cu\right\rangle_{\alpha/2} + \operatorname{Re}\left\langle \partial_{t,\nu}M_0(\partial_{t,\nu})u, u\right\rangle_{\alpha/2}\\ &= \operatorname{Re}\left\langle g - v, Cu\right\rangle_{\alpha/2} + \operatorname{Re}\left\langle \partial_{t,\nu}M_0(\partial_{t,\nu})u, u\right\rangle_{\alpha/2}\\ &\leqslant \|g\|_{\alpha/2}\,\|Cu\|_{\alpha/2} + \operatorname{Re}\left\langle \partial_{t,\nu}M_0(\partial_{t,\nu})u - C^*v, u\right\rangle_{\alpha/2}\\ &= \|g\|_{\alpha/2}\,\|Cu\|_{\alpha/2} + \operatorname{Re}\left\langle f, \left(\partial_{t,\nu}^*\right)^{\alpha/2}\left(\partial_{t,\nu}\right)^{\alpha/2}u\right\rangle_0\\ &\leqslant \|g\|_{\alpha/2}\,\|Cu\|_{\alpha/2} + \|f\|_0\,\|u\|_\alpha,\end{aligned}$$

where we used that

$$\begin{aligned}\left\| \left(\partial_{t,\nu}^*\right)^{\alpha/2} \left(\partial_{t,\nu}\right)^{\alpha/2} u \right\|_0 &= \left\| (-\mathrm{i}m+\nu)^{\alpha/2} (\mathrm{i}m+\nu)^{\alpha/2} u \right\|_{L_2(\mathbb{R};H_0)} \\ &= \left\| \frac{(-\mathrm{i}m+\nu)^{\alpha/2}}{(\mathrm{i}m+\nu)^{\alpha/2}} (\mathrm{i}m+\nu)^{\alpha} u \right\|_{L_2(\mathbb{R};H_0)} \\ &\leqslant \| (\mathrm{i}m+\nu)^{\alpha} u \|_{L_2(\mathbb{R};H_0)} = \|u\|_\alpha .\end{aligned}$$

Moreover,

$$\begin{aligned} c\,\|u\|_\alpha^2 &\leqslant \operatorname{Re}\left\langle \partial_{t,\nu}^{1-\alpha} M_0(\partial_{t,\nu})\partial_{t,\nu}^{\alpha} u, \partial_{t,\nu}^{\alpha} u\right\rangle_0 \\ &= \operatorname{Re}\left\langle \partial_{t,\nu} M_0(\partial_{t,\nu}) u, \partial_{t,\nu}^{\alpha} u\right\rangle_0 \\ &= \operatorname{Re}\left\langle f + C^* v, \partial_{t,\nu}^{\alpha} u\right\rangle_0 \\ &\leqslant \|f\|_0 \|u\|_\alpha + \operatorname{Re}\left\langle \left(\partial_{t,\nu}^*\right)^{\alpha/2} v, \partial_{t,\nu}^{\alpha/2} C u\right\rangle_0 \\ &\leqslant \|f\|_0 \|u\|_\alpha + \|v\|_{\alpha/2} \|Cu\|_{\alpha/2} \\ &= \|f\|_0 \|u\|_\alpha + \left\|g - M_1(\partial_{t,\nu})Cu\right\|_{\alpha/2} \|Cu\|_{\alpha/2} \\ &\leqslant \|f\|_0 \|u\|_\alpha + \|g\|_{\alpha/2} \|Cu\|_{\alpha/2} + m_1 \|Cu\|_{\alpha/2}^2 \\ &\leqslant \left(1 + \frac{m_1}{c}\right)\left(\|f\|_0 \|u\|_\alpha + \|g\|_{\alpha/2} \|Cu\|_{\alpha/2} \right). \end{aligned}$$

Thus, we obtain for $\varepsilon > 0$

$$\begin{aligned} & c\left(\|u\|_\alpha^2 + \|Cu\|_{\alpha/2}^2\right) \\ & \leqslant \left(2 + \frac{m_1}{c}\right)\left(\|f\|_0 \|u\|_\alpha + \|g\|_{\alpha/2} \|Cu\|_{\alpha/2}\right) \\ & \leqslant \frac{1}{2}\left(2 + \frac{m_1}{c}\right)\left(\frac{1}{\varepsilon}\left(\|f\|_0^2 + \|g\|_{\alpha/2}^2\right) + \varepsilon\left(\|u\|_\alpha^2 + \|Cu\|_{\alpha/2}^2\right)\right). \end{aligned}$$

Choosing $\varepsilon = c^2/(2c + m_1)$ and subtracting the term involving u and Cu on both sides of the inequality, we deduce

$$\begin{aligned} \frac{c}{2}\left(\|u\|_\alpha^2 + \|Cu\|_{\alpha/2}^2\right) &\leqslant \frac{1}{2}\left(2 + \frac{m_1}{c}\right)\frac{1}{\varepsilon}\left(\|f\|_0^2 + \|g\|_{\alpha/2}^2\right) \\ &= \frac{1}{2c}\left(2 + \frac{m_1}{c}\right)^2\left(\|f\|_0^2 + \|g\|_{\alpha/2}^2\right) \end{aligned}$$

and therefore

$$\left(\|u\|_\alpha^2 + \|Cu\|_{\alpha/2}^2\right) \leqslant \left(\frac{2}{c} + \frac{m_1}{c^2}\right)^2 \left(\|f\|_0^2 + \|g\|_{\alpha/2}^2\right).$$

Finally, we compute

$$\begin{aligned}\frac{1}{2}\|v\|_{\alpha/2}^2 &\leqslant \|g\|_{\alpha/2}^2 + \left\|M_1(\partial_{t,\nu})Cu\right\|_{\alpha/2}^2 \\ &\leqslant \|g\|_{\alpha/2}^2 + m_1^2\left(\frac{2}{c} + \frac{m_1}{c^2}\right)^2 \left(\|f\|_0^2 + \|g\|_{\alpha/2}^2\right)\end{aligned}$$

and

$$\begin{aligned}\frac{1}{2}\left\|C^*v\right\|_0^2 &\leqslant \left\|\partial_{t,\nu}M_0(\partial_{t,\nu})u\right\|_0^2 + \|f\|_0^2 \\ &\leqslant \left\|\partial_{t,\nu}^{1-\alpha}M_0(\partial_{t,\nu})\partial_{t,\nu}^{\alpha}u\right\|_0^2 + \|f\|_0^2 \\ &\leqslant m_0^2\|u\|_\alpha^2 + \|f\|_0^2 \\ &\leqslant m_0^2\left(\frac{2}{c} + \frac{m_1}{c^2}\right)^2 \left(\|f\|_0^2 + \|g\|_{\alpha/2}^2\right) + \|f\|_0^2.\end{aligned}$$

□

The next preliminary finding is a refinement of the surjectivity statement in Picard's theorem.

Proposition 15.3.2 *Let H be a Hilbert space, $M\colon \mathrm{dom}(M) \subseteq \mathbb{C} \to L(H)$ a material law, $\nu > \mathrm{s_b}(M)$, with $\nu > 0$, and $A\colon \mathrm{dom}(A) \subseteq H \to H$ skew-selfadjoint. Assume there exists $c > 0$ such that for all $z \in \mathbb{C}_{\mathrm{Re}>\nu}$ we have*

$$\mathrm{Re}\, zM(z) \geqslant c.$$

Let $\beta \in [0, 1]$.

(a) *The inclusion*

$$\left(\partial_{t,\nu}M(\partial_{t,\nu}) + A\right)\left[H_\nu^2\left(\mathbb{R}; \mathrm{dom}(A)\right)\right] \subseteq H_\nu^\beta(\mathbb{R}; H)$$

is dense.

(b) *Let $H_0 \subseteq H$ be a closed subspace and $H_1 := H_0^\perp$. Then*

$$\left(\partial_{t,\nu}M(\partial_{t,\nu}) + A\right)\left[H_\nu^2\left(\mathbb{R}; \mathrm{dom}(A)\right)\right] \subseteq L_{2,\nu}(\mathbb{R}; H_0) \oplus H_\nu^\beta(\mathbb{R}; H_1)$$

is dense.

Proof

(a) Since $H_\nu^1(\mathbb{R}; H)$ is dense in $H_\nu^\beta(\mathbb{R}; H)$ (this is a consequence of Lemma 15.2.1), it suffices to show the claim for $\beta = 1$. Next, by Picard's theorem, for $f \in \operatorname{dom}(\partial_{t,\nu})$, we obtain $u = \left(\partial_{t,\nu} M(\partial_{t,\nu}) + A\right)^{-1} f \in \operatorname{dom}(\partial_{t,\nu}) \cap L_{2,\nu}\big(\mathbb{R}; \operatorname{dom}(A)\big)$. In particular, it follows that

$$\left(\partial_{t,\nu} M(\partial_{t,\nu}) + A\right)\left[H_\nu^1(\mathbb{R}; H) \cap L_{2,\nu}\big(\mathbb{R}; \operatorname{dom}(A)\big)\right] \subseteq L_{2,\nu}(\mathbb{R}; H)$$

is dense. Multiplying this inclusion by $\partial_{t,\nu}^{-1}$, we infer that

$$\left(\partial_{t,\nu} M(\partial_{t,\nu}) + A\right)\left[H_\nu^2(\mathbb{R}; H) \cap H_\nu^1\big(\mathbb{R}; \operatorname{dom}(A)\big)\right] \subseteq H_\nu^1(\mathbb{R}; H)$$

is dense. Hence, for $f \in H_\nu^1(\mathbb{R}; H)$, we find $(u_n)_n$ in $H_\nu^2(\mathbb{R}; H) \cap H_\nu^1(\mathbb{R}; \operatorname{dom}(A))$ such that $f_n := \left(\partial_{t,\nu} M(\partial_{t,\nu}) + A\right) u_n \to f$ in $H_\nu^1(\mathbb{R}; H)$ as $n \to \infty$. Next, for $\varepsilon > 0$, $(1 + \varepsilon\partial_{t,\nu})^{-1} u \in H_\nu^2(\mathbb{R}; \operatorname{dom}(A))$ given $u \in H_\nu^1(\mathbb{R}; \operatorname{dom}(A))$. Moreover, $(1 + \varepsilon\partial_{t,\nu})^{-1} f \to f$ in $H_\nu^1(\mathbb{R}; H)$ as $\varepsilon \to 0$, by Lemma 9.3.3(b) and the fact that $\partial_{t,\nu}^{-1}$ commutes with $(1 + \varepsilon\partial_{t,\nu})^{-1}$. Thus, we compute for $\varepsilon > 0$ and $n \in \mathbb{N}$

$$\begin{aligned}
&\left\|\left(\partial_{t,\nu} M(\partial_{t,\nu}) + A\right)(1 + \varepsilon\partial_{t,\nu})^{-1} u_n - f\right\|_1 \\
&\leqslant \left\|(1 + \varepsilon\partial_{t,\nu})^{-1} f_n - (1 + \varepsilon\partial_{t,\nu})^{-1} f\right\|_1 + \left\|(1 + \varepsilon\partial_{t,\nu})^{-1} f - f\right\|_1 \\
&\leqslant \|f_n - f\|_1 + \left\|(1 + \varepsilon\partial_{t,\nu})^{-1} f - f\right\|_1 \to 0
\end{aligned}$$

as $n \to \infty$ and $\varepsilon \to 0$, which concludes the proof of (a).

(b) By (a), it suffices to show that

$$H_\nu^\beta(\mathbb{R}; H) = H_\nu^\beta(\mathbb{R}; H_0) \oplus H_\nu^\beta(\mathbb{R}; H_1) \subseteq L_{2,\nu}(\mathbb{R}; H_0) \oplus H_\nu^\beta(\mathbb{R}; H_1)$$

is dense (note that the first equality follows from the fact that $H \ni u \mapsto (u_0, u_1) \in H_0 \oplus H_1$ is unitary). The desired density result thus follows from Lemma 15.2.1. □

Next, we shall proceed with a proof of our main theorem in this chapter.

Proof of Theorem 15.2.3 For $i, j \in \{0, 1\}$ we set $N_{ij}(z) := \iota_{H_i}^* N(z) \iota_{H_j}$. Let $(f, g) \in \left(\overline{\partial_{t,\nu} M(\partial_{t,\nu}) + A}\right)\left[H_\nu^2\big(\mathbb{R}; \operatorname{dom}(C) \oplus \operatorname{dom}(C^*)\big)\right]$. Defining

$$(u, v) := \left(\overline{\partial_{t,\nu} M(\partial_{t,\nu}) + A}\right)^{-1} (f, g) \in H_\nu^2\big(\mathbb{R}; \operatorname{dom}(C) \oplus \operatorname{dom}(C^*)\big),$$

we have

$$\begin{aligned} \partial_{t,\nu} M_{00}(\partial_{t,\nu})u + N_{00}(\partial_{t,\nu})u - C^* v &= f - N_{01}(\partial_{t,\nu})v, \\ N_{11}(\partial_{t,\nu})v + Cu &= g - N_{10}(\partial_{t,\nu})u. \end{aligned}$$

Since $\operatorname{Re} zM(z) \geqslant c$, we infer

$$\operatorname{Re} N_{11}(\partial_{t,\nu}) \geqslant c.$$

Thus, by Proposition 6.2.3(b), we deduce that $M_1(\partial_{t,\nu}) := N_{11}(\partial_{t,\nu})^{-1}$ satisfies the real-part condition imposed on M_1 in Lemma 15.3.1. Moreover, since (M, A) is α-fractional parabolic,

$$M_0(z) := M_{00}(z) + z^{-1} N_{00}(z)$$

fulfills the real part and boundedness assumptions in Lemma 15.3.1. Introducing

$$\begin{aligned} \widetilde{f} &:= f - N_{01}(\partial_{t,\nu})v \in H_\nu^1(\mathbb{R}; H_0) \subseteq L_{2,\nu}(\mathbb{R}; H_0) \\ \widetilde{g} &:= M_1(\partial_{t,\nu})g - M_1(\partial_{t,\nu})N_{10}(\partial_{t,\nu})u \in H_\nu^1(\mathbb{R}; H_1) \subseteq H_\nu^{\alpha/2}(\mathbb{R}; H_1), \end{aligned}$$

we get

$$\begin{aligned} \partial_{t,\nu} M_0(\partial_{t,\nu})u - C^* v &= \widetilde{f}, \\ v + M_1(\partial_{t,\nu})Cv &= \widetilde{g}. \end{aligned}$$

Thus, using Lemma 15.3.1, we find $\kappa \geqslant 0$ in terms of M_0, M_1 and the positivity constants such that (recall that $m_1 := \|M_1\|_{\infty,\mathbb{C}_{\mathrm{Re}>\nu}}$)

$$\begin{aligned} &\|u\|_\alpha^2 + \|Cu\|_{\alpha/2}^2 + \|v\|_{\alpha/2}^2 + \left\|C^* v\right\|_0^2 \\ &\leqslant \kappa\left(\|\widetilde{f}\|_0^2 + \|\widetilde{g}\|_{\alpha/2}^2 \right) \\ &\leqslant 2\kappa\left(\|f\|_0^2 + \|N\|_{\infty,\mathbb{C}_{\mathrm{Re}>\nu}}^2 \|v\|_0^2 + m_1^2 \|g\|_{\alpha/2}^2 + m_1^2 \|N\|_{\infty,\mathbb{C}_{\mathrm{Re}>\nu}}^2 \|u\|_{\alpha/2}^2 \right) \\ &\leqslant 2\kappa\left(\|f\|_0^2 + \|N\|_{\infty,\mathbb{C}_{\mathrm{Re}>\nu}}^2 \|v\|_0^2 + m_1^2 \|g\|_{\alpha/2}^2 + 2m_1^2 \|N\|_{\infty,\mathbb{C}_{\mathrm{Re}>\nu}}^2 \left(\varepsilon \|u\|_\alpha^2 + \frac{1}{\varepsilon} \|u\|_0^2\right)\right) \end{aligned}$$

for all $\varepsilon > 0$, where in the last estimate, we used

$$\|u\|_{\alpha/2}^2 = \left\langle \partial_{t,\nu}^{\alpha/2} u, \partial_{t,\nu}^{\alpha/2} u \right\rangle_0 = \left\langle u, (\partial_{t,\nu}^{\alpha/2})^* \partial_{t,\nu}^{\alpha/2} u \right\rangle_0 \leqslant \|u\|_0 \|u\|_\alpha .$$

Hence, choosing $\varepsilon > 0$ small enough and using that $\overline{\left(\partial_{t,\nu}M(\partial_{t,\nu}) + A\right)}^{-1}$ is continuous from $L_{2,\nu}(\mathbb{R}; H)$ into itself, we find $\kappa' \geqslant 0$ such that

$$\|u\|_{\alpha}^{2} + \|Cu\|_{\alpha/2}^{2} + \|v\|_{\alpha/2}^{2} + \left\|C^{*}v\right\|_{0}^{2} \leqslant \kappa'\left(\|f\|_{0}^{2} + \|g\|_{\alpha/2}^{2}\right),$$

which establishes the assertion (using the density result in Proposition 15.3.2(b)).

□

15.4 Comments

The issue of maximal regularity (in Hilbert spaces for simplicity) is a priori formulated for equations of the type

$$u' + Au = f,$$

where f lies in some $L_2\big((0,T); H\big)$ and A is an unbounded operator in H. The question of maximal regularity then addresses, whether a solution u to this equation exists and satisfies $u \in L_2\big((0,T); \operatorname{dom}(A)\big) \cap H^1\big((0,T); H\big)$. In Hilbert spaces, whether or not this question can be answered in the affirmative solely relies on the properties of A. Hence, one shortens this question to whether A 'has maximal regularity'. The present situation is conveniently understood: A has maximal regularity if and only if $-A$ is the generator of a holomorphic semigroup, see [33, Theorem 2.2] and [105, Lemma 3,1]. One major example class is the class of operators that are defined with the help of forms, see [5] for an introductory text. People then studied the situation of time-dependent A. It has then been shown in various contexts and under suitable conditions on the (smoothness of the) time-dependence of A, whether A has maximal regularity or not. For this, we refer to [2, 8, 30] for an account of possible conditions. The evolutionary equations case, which is addressed for the first time in [88] in the time-independent and in [123] for the non-autonomous case, is different in as much as the focus of the underlying rationale is shifted away from the spatial derivative operator towards the material law. The proof of Theorem 15.2.3 outlined above is the autonomous version of [123].

Exercises

Exercise 15.1 Consider the situation of Example 15.1.1.

(a) Show that $0 \in \rho(\overline{\partial_{t,\nu} + \partial})$ for all $\nu > 0$. Next, let u be as in Example 15.1.1 and show that $u \notin \operatorname{dom}(\partial_{t,\nu})$.
(b) Let $\nu > 0$ and show using Picard's theorem that

$$0 \in \rho\left(\overline{\partial_{t,\nu}\begin{pmatrix}1 & 0\\ 0 & 1\end{pmatrix} + \begin{pmatrix}0 & \partial\\ \partial & 0\end{pmatrix}}\right).$$

Show that there exist $f, g \in L_{2,\nu}(\mathbb{R}; L_2(\mathbb{R}))$ such that for

$$\begin{pmatrix}u_f\\ v_f\end{pmatrix} := \left(\overline{\partial_{t,\nu}\begin{pmatrix}1 & 0\\ 0 & 1\end{pmatrix} + \begin{pmatrix}0 & \partial\\ \partial & 0\end{pmatrix}}\right)^{-1}\begin{pmatrix}f\\ 0\end{pmatrix}$$

and

$$\begin{pmatrix}u_g\\ v_g\end{pmatrix} := \left(\overline{\partial_{t,\nu}\begin{pmatrix}1 & 0\\ 0 & 1\end{pmatrix} + \begin{pmatrix}0 & \partial\\ \partial & 0\end{pmatrix}}\right)^{-1}\begin{pmatrix}0\\ g\end{pmatrix}$$

we have $u_f, u_g \notin \operatorname{dom}(\partial_{t,\nu})$.

Exercise 15.2 Let u and q be defined as in Example 15.1.2. Show that $u \in \operatorname{dom}(\partial_{t,\nu})$ and $q \in \operatorname{dom}(\mathrm{m})$ by explicit computation (not using Theorem 15.2.3). *Hint:* Find an ordinary differential equation satisfied by u. Use the explicit solution of this ordinary differential equation to show the claim.

Exercise 15.3 Let $\alpha \geqslant 0$ and $\nu > 0$. Show that

$$\begin{aligned}\partial_{t,\nu} \colon \operatorname{dom}(\partial_{t,\nu}^{\lceil\alpha\rceil+1}) \subseteq H_\nu^\alpha(\mathbb{R}) &\to H_\nu^\alpha(\mathbb{R})\\ u &\mapsto \partial_{t,\nu} u\end{aligned}$$

is densely defined closable with continuous invertible closure.

Exercise 15.4 (Local Maximal Regularity) Let H_0, H_1 be Hilbert spaces, $a \in L(H_1)$ be such that $\operatorname{Re} a \geqslant c$ for some $c > 0$. Furthermore, let $C \colon \operatorname{dom}(C) \subseteq H_0 \to H_1$ be densely defined and closed. Let $T > 0$. Show that for every $f \in L_2\big((0,T); H_0\big)$ there exists a unique $u \in H^1\big((0,T); H_0\big) \cap L_2\big((0,T); \operatorname{dom}(C^*aC)\big)$ with $u(0) = 0$ such that

$$u'(t) + C^*aCu(t) = f(t) \quad (\text{a.e. } t \in (0,T)).$$

Hint: Reformulate the equation satisfied by u into an evolutionary equation, apply Theorem 15.2.3.

Exercise 15.5 Let H_0, H_1 be Hilbert spaces, $a \in L(H_1)$ be such that $\operatorname{Re} a \geqslant c$ for some $c > 0$. Furthermore, let $C\colon \operatorname{dom}(C) \subseteq H_0 \to H_1$ be densely defined and closed. Let $T > 0$. Define $\partial_0\colon \operatorname{dom}(\partial_0) \subseteq L_2\big((0,T)\,;H_0\big) \to L_2\big((0,T)\,;H_0\big)$ with $\partial_0 u = u'$ and

$$\operatorname{dom}(\partial_0) = \Big\{u \in H^1\big((0,T)\,;H_0\big)\,;\ u(0) = 0\Big\}.$$

Show that for $u \in H^1\big((0,T)\,;H_0\big)$ the point-evaluation $u(0) = 0$ is well-defined. Then show that $\partial_0 + C^*aC$ is continuously invertible and closed as an operator in $L_2\big((0,T)\,;H_0\big)$.
Hint: For the first part use Theorem 12.1.3. For the second part, apply the result of Exercise 15.4. Show that in the situation of the previous exercise, there exists $\kappa > 0$ independently of f and u with

$$\|u\|_{H^1\left((0,T);H_0\right)\cap L_2\left((0,T);\operatorname{dom}(C^*aC)\right)} \leqslant \kappa\|f\|_{L_2\left((0,T);H_0\right)}.$$

Exercise 15.6 Recall Maxwell's equations from Theorem 6.2.8:

$$\partial_{t,\nu}\begin{pmatrix}\varepsilon & 0\\ 0 & \mu\end{pmatrix} + \begin{pmatrix}\sigma & 0\\ 0 & 0\end{pmatrix} + \begin{pmatrix}0 & -\operatorname{curl}\\ \operatorname{curl}_0 & 0\end{pmatrix}$$

in $L_{2,\nu}\left(\mathbb{R}; L_2(\Omega)^3 \times L_2(\Omega)^3\right)$ with $\varepsilon, \mu, \sigma\colon \Omega \to \mathbb{R}^{3\times 3}$ satisfying the following property: there exist $c > 0$ and $\nu_0 > 0$ such that for all $\nu \geqslant \nu_0$ we have

$$\nu\varepsilon(x) + \operatorname{Re}\sigma(x) \geqslant c, \quad \mu(x) \geqslant c \quad (x \in \Omega).$$

By Theorem 6.2.8, for $\nu \geqslant \nu_0$ and $j_0 \in L_{2,\nu}(\mathbb{R}; L_2(\Omega)^3)$, there exists a unique pair $(E, H) \in L_{2,\nu}(\mathbb{R}; L_2(\Omega)^6)$ such that

$$\begin{pmatrix}E\\ H\end{pmatrix} := \left(\overline{\partial_{t,\nu}\begin{pmatrix}\varepsilon & 0\\ 0 & \mu\end{pmatrix} + \begin{pmatrix}\sigma & 0\\ 0 & 0\end{pmatrix} + \begin{pmatrix}0 & -\operatorname{curl}\\ \operatorname{curl}_0 & 0\end{pmatrix}}\right)^{-1}\begin{pmatrix}j_0\\ 0\end{pmatrix}.$$

Assume there exist open sets $\Omega_0, \Omega_1 \subseteq \Omega$ such that $\overline{\Omega_0} \subseteq \Omega_1 \subseteq \overline{\Omega_1} \subseteq \Omega$ with $\operatorname{spt} j_0(t) \subseteq \Omega_0$ for a.e. $t \in \mathbb{R}$. Moreover, $j_0 \in H_\nu^{1/2}\left(\mathbb{R}; L_2(\Omega_1)^3\right)$. Furthermore, assume $\varepsilon = 0$ on $\overline{\Omega_1}$. Show that $t \mapsto H(t)|_{\Omega_0} \in H_\nu^1\left(\mathbb{R}; L_2(\Omega_0)^3\right)$.

Exercise 15.7 Let H_0, H_1 be Hilbert spaces, $a, b \in L(H_1)$ be such that $\operatorname{Re} b \geqslant c$ for some $c > 0$. Furthermore, let $C\colon \operatorname{dom}(C) \subseteq H_0 \to H_1$ be densely defined and closed. Let $f \in L_2(\mathbb{R}; H_0)$ with $\inf\operatorname{spt} f > -\infty$. Show that for $\nu > 0$ large enough, there exists for a unique $u \in H^2_\nu(\mathbb{R}; H_0) \cap \mathrm{dom}\left(C^*(a+b\partial_{t,\nu})C\right)$ satisfying

$$\partial_{t,\nu}^2 u + C^*(a + b\partial_{t,\nu})Cu = f.$$

Hint: Use the substitution $w := \partial_{t,\nu} u$ and $q := -(a + b\partial_{t,\nu})Cu$ to reformulate the equation in question as an evolutionary equation. Then apply Theorem 15.2.3.

References

2. M. Achache, E.M. Ouhabaz, 'Lions' maximal regularity problem with $H^{\frac{1}{2}}$-regularity in time. J. Differ. Equ. **266**(6), 3654–3678 (2019)
5. W. Arendt et al., *Form Methods for Evolution Equations, and Applications*. 18th Internet Seminar, 2015
8. P. Auscher, M. Egert, On non-autonomous maximal regularity for elliptic operators in divergence form. Arch. Math. (Basel) **107**(3), 271–284 (2016)
30. D. Dier, R. Zacher, Non-autonomous maximal regularity in Hilbert spaces. J. Evol. Equ. **17**(3), 883–907 (2017)
33. G. Dore, L_p regularity for abstract differential equations. Funct. Anal. Relat. Top. 1991 **1540**, 25–38 (1993)
88. R. Picard, S. Trostorff, M. Waurick, On maximal regularity for a class of evolutionary equations. J. Math. Anal. Appl. **449**(2), 1368–1381 (2017)
90. R. Picard et al., *A Primer for a Secret Shortcut to PDEs of Mathematical Physics*, vol. 140. Frontiers in Mathematics (Birkhäuser, Basel, 2020)
105. L. de Simon, Un'applicazione della teoria degli integrali singolari allo studio delle equazioni differenziali lineari astratte del primo ordine. Rend. Sem. Mat. Univ. Padova **34**, 205–223 (1964)
123. S. Trostorff, M. Waurick, Maximal regularity for non-autonomous evolutionary equations. Integr. Equ. Oper. Theory **93**(3), Id/No 30, 37 (2021)

Chapter 16
Non-Autonomous Evolutionary Equations

Previously, we focussed on evolutionary equations of the form

$$\left(\overline{\partial_{t,\nu}M(\partial_{t,\nu}) + A}\right) U = F.$$

In this chapter, where we turn back to well-posedness issues, we replace the material law operator $M(\partial_{t,\nu})$, which is invariant under translations in time, by an operator of the form

$$\mathcal{M} + \partial_{t,\nu}^{-1}\mathcal{N},$$

where both $\mathcal{M}$ and $\mathcal{N}$ are bounded linear operators in $L_{2,\nu}(\mathbb{R}; H)$. Thus, it is the aim in the following to provide criteria on $\mathcal{M}$ and $\mathcal{N}$ under which the operator

$$\partial_{t,\nu}\mathcal{M} + \mathcal{N} + A \tag{16.1}$$

is closable with continuously invertible closure in $L_{2,\nu}(\mathbb{R}; H)$. In passing, we shall also replace the skew-selfadjointness of A by a suitable real part condition. Under additional conditions on $\mathcal{M}$ and $\mathcal{N}$, we will also see that the solution operator is causal. Finally, we will put the autonomous version of Picard's theorem into perspective of the non-autonomous variant developed here.

In order to get grip on the domain of the anticipated operator sum, we need to assume a commutator condition of the coefficient operators and the time-derivative. Thus, the replacement for the assumption of the coefficient to be a "material law operator" (i.e., a bounded analytic function of the time-derivative) is to be evolutionary and to have a bounded commutator with the time-derivative (in a suitable sense). Since we proved in Theorem 8.2.1 that bounded analytic functions of the time-derivative are exactly the ones that are causal and autonomous (and evolutionary), one may view the following theorem as a direct generalisation of Picard's theorem in the way that "autonomous" is dropped.

C. Seifert et al., *Evolutionary Equations*, Operator Theory: Advances and Applications 287, https://doi.org/10.1007/978-3-030-89397-2_16

16.1 Examples

In principle finding examples for the non-autonomous theory is relatively simple. The prototype case focusses on time-dependent multiplication operators. In order to illustrate our findings below, we shall revisit the heat equation and Maxwell's equations.

Non-Autonomous Heat Equation
Let $\Omega \subseteq \mathbb{R}^d$ be open and $a\colon \mathbb{R} \times \Omega \to \mathbb{R}^{d\times d}$ bounded and measurable. Assume there exists $c > 0$ such that

$$\operatorname{Re} a(t,x) \geqslant c \quad \text{(a.e. } (t,x) \in \mathbb{R} \times \Omega).$$

Then the non-autonomous variant of the equations describing heat conduction are

$$\begin{aligned} \partial_{t,\nu}\theta + \operatorname{div}_0 q &= Q \\ q(t,x) &= a(t,x) \operatorname{grad}\theta(t,x) \quad ((t,x) \in \mathbb{R} \times \Omega). \end{aligned}$$

The resulting block operator matrix

$$\partial_{t,\nu} \begin{pmatrix} 1 & 0 \\ 0 & 0 \end{pmatrix} + \begin{pmatrix} 0 & 0 \\ 0 & a^{-1} \end{pmatrix} + \begin{pmatrix} 0 & \operatorname{div}_0 \\ \operatorname{grad} & 0 \end{pmatrix}$$

is then closable and continuously invertible in $L_{2,\nu}\big(\mathbb{R}; L_2(\Omega) \times L_2(\Omega)^d\big)$ for all $\nu > 0$ by Theorem 16.3.1.

Non-Autonomous Maxwell's Equations
Let $\Omega \subseteq \mathbb{R}^3$ be open and $\varepsilon, \mu, \sigma\colon \mathbb{R}\times\Omega \to \mathbb{R}^{3\times 3}$ bounded and measurable. Assume that ε and μ are Lipschitz continuous w.r.t. the temporal variables uniformly in space; that is, there exists $L \geqslant 0$ such that

$$\|\varepsilon(s,x) - \varepsilon(t,x)\|_{\mathbb{R}^{3\times 3}} + \|\mu(s,x) - \mu(t,x)\|_{\mathbb{R}^{3\times 3}} \leqslant L\,|t-s| \quad (s,t \in \mathbb{R}, x \in \Omega).$$

Assume $\varepsilon(t,x)^\top = \varepsilon(t,x)$ and $\mu(t,x)^\top = \mu(t,x)$ for all $t \in \mathbb{R}$, $x \in \Omega$. Furthermore, assume there exist $c, \nu_0 > 0$ such that for all $\nu \geqslant \nu_0$ we have

$$\mu(t,x) \geqslant c, \text{ and } \nu\varepsilon(t,x) + \frac{1}{2}\varepsilon'(t)(x) + \operatorname{Re}\sigma(t,x) \geqslant c \quad ((t,x) \in \mathbb{R} \times \Omega).$$

Then it will not be difficult to see that the operator

$$\partial_{t,\nu} \begin{pmatrix} \varepsilon(\mathrm{m}_t, \mathrm{m}_x) & 0 \\ 0 & \mu(\mathrm{m}_t, \mathrm{m}_x) \end{pmatrix} + \begin{pmatrix} \sigma(\mathrm{m}_t, \mathrm{m}_x) & 0 \\ 0 & 0 \end{pmatrix} + \begin{pmatrix} 0 & -\operatorname{curl} \\ \operatorname{curl}_0 & 0 \end{pmatrix}$$

is closable and continuously invertible in $L_{2,\nu}\big(\mathbb{R}; L_2(\Omega)^3 \times L_2(\Omega)^3\big)$ for all $\nu \geqslant \nu_0$ by Theorem 16.3.1; see also Exercise 16.1.

16.2 Non-Autonomous Picard's Theorem—The ODE Case

Let H be a Hilbert space and $\nu > 0$. In this section we will focus on the ODE-case first, which is modelled by $A = 0$ in (16.1).

Theorem 16.2.1 *Let $\mathcal{M}, \mathcal{M}', \mathcal{N} \in L(L_{2,\nu}(\mathbb{R}; H))$ with $\mathcal{M}$, $\mathcal{N}$ causal and $\operatorname{Re}\mathcal{M} \geqslant 0$. Assume*

$$\mathcal{M}\partial_{t,\nu} \subseteq \partial_{t,\nu}\mathcal{M} - \mathcal{M}'$$

and

$$\operatorname{Re}\big\langle\phi, \big(\partial_{t,\nu}\mathcal{M} + \mathcal{N}\big)\,\phi\big\rangle \geqslant c\,\langle\phi, \phi\rangle$$

for some $c > 0$ and all $\phi \in \operatorname{dom}\big(\partial_{t,\nu}\mathcal{M}\big)$. Then

$$0 \in \rho\left(\partial_{t,\nu}\mathcal{M} + \mathcal{N}\right),$$

$\left\|(\partial_{t,\nu}\mathcal{M} + \mathcal{N})^{-1}\right\| \leqslant 1/c$, and $\big(\partial_{t,\nu}\mathcal{M} + \mathcal{N}\big)^{-1}$ is causal. Moreover,

$$\operatorname{Re}\big\langle\phi, \big(\partial_{t,\nu}\mathcal{M} + \mathcal{N}\big)^*\,\phi\big\rangle \geqslant c\,\langle\phi, \phi\rangle \qquad \big(\phi \in \operatorname{dom}\big((\partial_{t,\nu}\mathcal{M} + \mathcal{N})^*\big)\big).$$

Remark 16.2.2 The only non-trivial condition in Theorem 16.2.1 is the commutator condition

$$\mathcal{M}\partial_{t,\nu} \subseteq \partial_{t,\nu}\mathcal{M} - \mathcal{M}'.$$

This condition is satisfied for multiplication operators induced by a Lipschitz continuous function, see also Exercise 16.1.

We leave the proof of $0 \in \rho\left(\partial_{t,\nu}\mathcal{M} + \mathcal{N}\right)$ and the norm estimate as Exercise 16.4. For the proof of causality, we need some preparations. The first result will also be of some value in the next chapter. It deals with a reformulation of causality for resolvents.

Proposition 16.2.3 *Let $\mathcal{B}\colon \operatorname{dom}(\mathcal{B}) \subseteq L_{2,\nu}(\mathbb{R}; H) \to L_{2,\nu}(\mathbb{R}; H)$ be linear, $0 \in \rho(\mathcal{B})$, and assume that there exists $c > 0$ such that for all $\phi \in \operatorname{dom}(\mathcal{B})$ we have*

$$\operatorname{Re}\,\langle\phi, \mathcal{B}\phi\rangle_{L_{2,\nu}(\mathbb{R};H)} \geqslant c\,\langle\phi, \phi\rangle_{L_{2,\nu}(\mathbb{R};H)}\,.$$

Then the following two statements are equivalent:

(i) $\mathcal{B}^{-1}$ *is causal.*
(ii) *For all* $\phi \in \operatorname{dom}(\mathcal{B})$ *and all* $a \in \mathbb{R}$ *we have*

$$\operatorname{Re}\left\langle \mathbb{1}_{(-\infty,a]}\phi, \mathcal{B}\phi \right\rangle_{L_{2,\nu}(\mathbb{R};H)} \geqslant c \left\langle \mathbb{1}_{(-\infty,a]}\phi, \phi \right\rangle_{L_{2,\nu}(\mathbb{R};H)}.$$

Proof (ii)⇒(i): Let $f \in L_{2,\nu}(\mathbb{R}; H)$ and $a \in \mathbb{R}$ with $\operatorname{spt} f \subseteq [a, \infty)$. Then, using (ii), for $\phi := \mathcal{B}^{-1} f \in \operatorname{dom}(\mathcal{B})$ we have

$$\begin{aligned} 0 = \operatorname{Re}\left\langle \mathbb{1}_{(-\infty,a]}\phi, f \right\rangle_{L_{2,\nu}(\mathbb{R};H)} &= \operatorname{Re}\left\langle \mathbb{1}_{(-\infty,a]}\phi, \mathcal{B}\phi \right\rangle_{L_{2,\nu}(\mathbb{R};H)} \\ \geqslant c \left\langle \mathbb{1}_{(-\infty,a]}\phi, \phi \right\rangle_{L_{2,\nu}(\mathbb{R};H)} &= c \left\| \mathbb{1}_{(-\infty,a]}\phi \right\|^2_{L_{2,\nu}(\mathbb{R};H)}, \end{aligned}$$

which yields $\operatorname{spt}\phi \subseteq [a, \infty)$. Thus, $\mathcal{B}^{-1}$ is causal.
(i)⇒(ii): Let $a \in \mathbb{R}$, $\phi \in \operatorname{dom}(\mathcal{B})$, and $f := \mathcal{B}\phi$. Then $\phi_1 := \mathcal{B}^{-1}\mathbb{1}_{(-\infty,a]} f \in \operatorname{dom}(\mathcal{B})$ and, using causality of $\mathcal{B}^{-1}$, we obtain

$$\mathbb{1}_{(-\infty,a]}\phi_1 = \mathbb{1}_{(-\infty,a]}\mathcal{B}^{-1}\mathbb{1}_{(-\infty,a]} f = \mathbb{1}_{(-\infty,a]}\mathcal{B}^{-1} f = \mathbb{1}_{(-\infty,a]}\phi.$$

We thus compute

$$\begin{aligned} \operatorname{Re}\left\langle \mathbb{1}_{(-\infty,a]}\phi, \mathcal{B}\phi \right\rangle_{L_{2,\nu}(\mathbb{R};H)} &= \operatorname{Re}\left\langle \mathbb{1}_{(-\infty,a]}\phi_1, f \right\rangle_{L_{2,\nu}(\mathbb{R};H)} = \operatorname{Re}\left\langle \phi_1, \mathbb{1}_{(-\infty,a]} f \right\rangle_{L_{2,\nu}(\mathbb{R};H)} \\ &= \operatorname{Re}\left\langle \phi_1, \mathcal{B}\phi_1 \right\rangle_{L_{2,\nu}(\mathbb{R};H)} \geqslant c \left\langle \phi_1, \phi_1 \right\rangle_{L_{2,\nu}(\mathbb{R};H)} \\ &\geqslant c \left\| \mathbb{1}_{(-\infty,a]}\phi_1 \right\|^2_{L_{2,\nu}(\mathbb{R};H)} = c \left\| \mathbb{1}_{(-\infty,a]}\phi \right\|^2_{L_{2,\nu}(\mathbb{R};H)} \\ &= c \left\langle \mathbb{1}_{(-\infty,a]}\phi, \phi \right\rangle_{L_{2,\nu}(\mathbb{R};H)}, \end{aligned}$$

where in the last estimate we used that multiplication by $\mathbb{1}_{(-\infty,a]}$ is a contraction. □

Lemma 16.2.4 *Let* $\mathcal{B}\colon \operatorname{dom}(\mathcal{B}) \subseteq L_{2,\nu}(\mathbb{R}; H) \to L_{2,\nu}(\mathbb{R}; H)$ *be linear. Let* $\lambda, \mu \in \rho(\mathcal{B})$ *be contained in the same connected component of* $\rho(\mathcal{B})$*. Assume that* $(\mu - \mathcal{B})^{-1}$ *is causal. Then* $(\lambda - \mathcal{B})^{-1}$ *is causal.*

Proof Let Z be the connected component of $\rho(\mathcal{B})$ shared by both μ and λ. Define

$$M := \left\{ \eta \in Z\,;\ \forall a \in \mathbb{R}\colon \mathbb{1}_{(-\infty,a]}(m)(\eta - \mathcal{B})^{-1}\mathbb{1}_{(-\infty,a]}(m) = \mathbb{1}_{(-\infty,a]}(m)(\eta - \mathcal{B})^{-1} \right\}$$

Then, $\mu \in M$. Next, we show that M is open and closed in Z. For this, let $\eta_0 \in M$. By Proposition 2.4.1, we have $B(\eta_0, r) \subseteq \rho(\mathcal{B})$ with $r := 1/\|(\eta_0 - \mathcal{B})^{-1}\|$. As $B(\eta_0, r)$ is connected, we infer $B(\eta_0, r) \subseteq Z$. Furthermore, from Proposition 2.4.1, we infer for $\eta \in B(\eta_0, r)$ that

$$(\eta - \mathcal{B})^{-1} = \sum_{k=0}^{\infty} (\eta_0 - \eta)^k ((\eta_0 - \mathcal{B})^{-1})^{k+1}.$$

Hence, since $\eta_0 \in M$, we obtain for all $a \in \mathbb{R}$,

$$\begin{aligned}
\mathbb{1}_{(-\infty,a]}(m)(\eta - \mathcal{B})^{-1} &= \mathbb{1}_{(-\infty,a]}(m) \sum_{k=0}^{\infty} (\eta_0 - \eta)^k ((\eta_0 - \mathcal{B})^{-1})^{k+1} \\
&= \sum_{k=0}^{\infty} (\eta_0 - \eta)^k \mathbb{1}_{(-\infty,a]}(m) ((\eta_0 - \mathcal{B})^{-1})^{k+1} \\
&= \sum_{k=0}^{\infty} (\eta_0 - \eta)^k \mathbb{1}_{(-\infty,a]}(m) ((\eta_0 - \mathcal{B})^{-1})^{k+1} \mathbb{1}_{(-\infty,a]}(m) \\
&= \mathbb{1}_{(-\infty,a]}(m) \sum_{k=0}^{\infty} (\eta_0 - \eta)^k ((\eta_0 - \mathcal{B})^{-1})^{k+1} \mathbb{1}_{(-\infty,a]}(m) \\
&= \mathbb{1}_{(-\infty,a]}(m)(\eta - \mathcal{B})^{-1} \mathbb{1}_{(-\infty,a]}(m).
\end{aligned}$$

Thus, $B(\eta_0, r) \subseteq M$ and M is open in Z. Next, let $(\eta_n)_n$ be a sequence in M, convergent to some $\eta \in Z$. For $n \in \mathbb{N}$ the equality

$$\mathbb{1}_{(-\infty,a]}(m)(\eta_n - \mathcal{B})^{-1} = \mathbb{1}_{(-\infty,a]}(m)(\eta_n - \mathcal{B})^{-1} \mathbb{1}_{(-\infty,a]}(m) \quad (a \in \mathbb{R})$$

as well as the continuity of $(\cdot - \mathcal{B})^{-1}$ imply that $\eta \in M$. Hence, M is closed. We infer $M = Z$ from the connectedness of Z and, thus, $\lambda \in M$. □

Lemma 16.2.5 *Let $\nu \in \mathbb{R}$ and $\mathcal{M} \in L(L_{2,\nu}(\mathbb{R}; H))$ be causal. If there exists $c > 0$ such that*

$$\operatorname{Re} \langle \phi, \mathcal{M}\phi \rangle_{L_{2,\nu}(\mathbb{R}; H)} \geqslant c \langle \phi, \phi \rangle_{L_{2,\nu}(\mathbb{R}; H)} \quad (\phi \in L_{2,\nu}(\mathbb{R}; H)),$$

then $\mathcal{M}^{-1}$ is causal.

Proof We have $0 \in \rho(\mathcal{M})$ by Proposition 6.2.3(b). In particular, we obtain for all $a \in \mathbb{R}$ and $\phi \in L_{2,\nu}(\mathbb{R}; H)$, using causality of $\mathcal{M}$, that

$$\begin{aligned}
\operatorname{Re}\left\langle \mathbb{1}_{(-\infty,a]}\phi, \mathcal{M}\phi\right\rangle_{L_{2,\nu}(\mathbb{R};H)} &= \operatorname{Re}\left\langle \mathbb{1}_{(-\infty,a]}\phi, \mathbb{1}_{(-\infty,a]}\mathcal{M}\phi\right\rangle_{L_{2,\nu}(\mathbb{R};H)} \\
&= \operatorname{Re}\left\langle \mathbb{1}_{(-\infty,a]}\phi, \mathbb{1}_{(-\infty,a]}\mathcal{M}\mathbb{1}_{(-\infty,a]}\phi\right\rangle_{L_{2,\nu}(\mathbb{R};H)} \\
&= \operatorname{Re}\left\langle \mathbb{1}_{(-\infty,a]}\phi, \mathcal{M}\mathbb{1}_{(-\infty,a]}\phi\right\rangle_{L_{2,\nu}(\mathbb{R};H)} \\
&\geqslant c\left\langle \mathbb{1}_{(-\infty,a]}\phi, \mathbb{1}_{(-\infty,a]}\phi\right\rangle_{L_{2,\nu}(\mathbb{R};H)} \\
&= c\left\langle \mathbb{1}_{(-\infty,a]}\phi, \phi\right\rangle_{L_{2,\nu}(\mathbb{R};H)},
\end{aligned}$$

which yields causality of $\mathcal{M}^{-1}$ by Proposition 16.2.3 applied to $\mathcal{B} = \mathcal{M}$. □

Lemma 16.2.6 *Let $\mathcal{M}, \mathcal{N}, \mathcal{M}' \in L(L_{2,\nu}(\mathbb{R}; H))$. Assume*

$$\mathcal{M}\partial_{t,\nu} \subseteq \partial_{t,\nu}\mathcal{M} - \mathcal{M}'$$

and

$$\operatorname{Re}\left\langle \phi, (\partial_{t,\nu}\mathcal{M} + \mathcal{N})\phi\right\rangle \geqslant c\left\langle \phi, \phi\right\rangle \quad (\phi \in \operatorname{dom}(\partial_{t,\nu})).$$

Then

$$Z := \left\{\eta \in [0, \infty)\ ;\ \left(\partial_{t,\nu}(\mathcal{M} + \eta) + \mathcal{N}\right)^{-1} \textit{causal}\right\}$$

is closed.

Proof As it was mentioned before, the proof of $0 \in \rho\left(\partial_{t,\nu}(\mathcal{M} + \eta) + \mathcal{N}\right)$ for $\eta \in [0, \infty)$ is postponed to Exercise 16.4. For all $\eta \in [0, \infty)$ and $\phi \in \operatorname{dom}(\partial_{t,\nu})$ we have

$$\operatorname{Re}\left\langle \phi, (\partial_{t,\nu}(\mathcal{M} + \eta) + \mathcal{N})\phi\right\rangle \geqslant c\left\langle \phi, \phi\right\rangle \quad (\phi \in \operatorname{dom}(\partial_{t,\nu})).$$

Note that this inequality to hold for all $\phi \in \operatorname{dom}(\partial_{t,\nu})$ is sufficient for it to hold for all $\phi \in \operatorname{dom}(\partial_{t,\nu}(\mathcal{M} + \eta))$. Indeed, this is a consequence of $\operatorname{dom}(\partial_{t,\nu})$ being a core for $\partial_{t,\nu}(\mathcal{M} + \eta)$, which is easily seen (see also Lemma 16.3.3). Hence, by Proposition 16.2.3, $\eta \in Z$ if and only if

$$\operatorname{Re}\left\langle \mathbb{1}_{(-\infty,a]}\phi, (\partial_{t,\nu}(\mathcal{M} + \eta) + \mathcal{N})\phi\right\rangle \geqslant c\left\langle \mathbb{1}_{(-\infty,a]}\phi, \phi\right\rangle \quad (\phi \in \operatorname{dom}(\partial_{t,\nu})).$$

Before we show closedness of Z, we shortly recall that integration by parts yields for all $a \in \mathbb{R}$

$$\operatorname{Re}\left\langle \mathbb{1}_{(-\infty,a]}\phi, \partial_{t,\nu}\phi\right\rangle = \frac{1}{2}\left\|\phi(a)\right\|^2 \mathrm{e}^{-2\nu a} + \nu\left\langle \mathbb{1}_{(-\infty,a]}\phi, \phi\right\rangle \quad (\phi \in \operatorname{dom}(\partial_{t,\nu})).$$

In order to show that Z is closed, let $(\eta_n)_n$ be a sequence in Z, convergent to some $\eta \in [0, \infty)$. Then we compute for all $a \in \mathbb{R}$ and $\phi \in \operatorname{dom}(\partial_{t,\nu})$ and $n \in \mathbb{N}$

$$\begin{aligned}&\operatorname{Re}\left\langle \mathbb{1}_{(-\infty,a]}\phi, \left(\partial_{t,\nu}(\mathcal{M}+\eta)+\mathcal{N}\right)\phi\right\rangle \\ &= \operatorname{Re}\left\langle \mathbb{1}_{(-\infty,a]}\phi, \left(\partial_{t,\nu}(\mathcal{M}+\eta_n)+\mathcal{N}\right)\phi\right\rangle + \operatorname{Re}\left\langle \mathbb{1}_{(-\infty,a]}\phi, \partial_{t,\nu}(\eta-\eta_n)\phi\right\rangle \\ &\geqslant c\left\langle \mathbb{1}_{(-\infty,a]}\phi, \phi\right\rangle + \frac{1}{2}(\eta-\eta_n)\left\|\phi(a)\right\|^2 \exp(-2\nu a) + (\eta-\eta_n)\nu\left\langle \mathbb{1}_{(-\infty,a]}\phi, \phi\right\rangle.\end{aligned}$$

Letting $n \to \infty$, we infer

$$\operatorname{Re}\left\langle \mathbb{1}_{(-\infty,a]}\phi, \left(\partial_{t,\nu}(\mathcal{M}+\eta)+\mathcal{N}\right)\phi\right\rangle \geqslant c\left\langle \mathbb{1}_{(-\infty,a]}\phi, \phi\right\rangle$$

for $\phi \in \operatorname{dom}(\partial_{t,\nu})$. Hence, $\eta \in Z$. □

Proof of Theorem 16.2.1 Keeping Exercise 16.4 in mind, we only need to show that the solution operator $(\partial_{t,\nu}\mathcal{M}+\mathcal{N})^{-1}$ is causal.

By Lemma 16.2.6, it suffices to show that for all $\eta > 0$,

$$(\partial_{t,\nu}(\mathcal{M}+\eta)+\mathcal{N})^{-1}$$

is causal. Hence, we may assume that $0 \in \rho(\mathcal{M})$ and, using Lemma 16.2.5, that $\mathcal{M}^{-1}$ is causal. In this situation, it remains to show that

$$(\partial_{t,\nu}\mathcal{M}+\mathcal{N})^{-1} = \mathcal{M}^{-1}(\partial_{t,\nu}+\mathcal{N}\mathcal{M}^{-1})^{-1}$$

is causal. As $\mathcal{M}^{-1}$ is causal, it furthermore suffices to show causality of

$$(\partial_{t,\nu}+\mathcal{K})^{-1}$$

where $\mathcal{K} := \mathcal{N}\mathcal{M}^{-1}$ is causal. Using $\operatorname{Re}\mathcal{M} \geqslant 0$ and the inequality assumed for $\partial_{t,\nu}\mathcal{M}+\mathcal{N}$, we conclude that $(\partial_{t,\nu}+\mu+\mathcal{K})$ is continuously invertible for all $\mu \geqslant 0$. Since $\partial_{t,\nu}^{-1}$ is causal, Lemma 16.2.4 yields that $(\partial_{t,\nu}+\mu)^{-1}$ is causal. From $\operatorname{Re}(\partial_{t,\nu}+\mu) \geqslant \nu+\mu$ it follows that $\left\|(\partial_{t,\nu}+\mu)^{-1}\right\| \leqslant 1/(\nu+\mu)$. Hence, we find $\mu > 0$ such that $\left\|(\partial_{t,\nu}+\mu)^{-1}\mathcal{K}\right\| < 1$. Thus,

$$\begin{aligned}(\partial_{t,\nu}+\mu+\mathcal{K})^{-1} &= \left(1+(\partial_{t,\nu}+\mu)^{-1}\mathcal{K}\right)^{-1}(\partial_{t,\nu}+\mu)^{-1} \\ &= \sum_{k=0}^{\infty}(-1)^k\left((\partial_{t,\nu}+\mu)^{-1}\mathcal{K}\right)^k(\partial_{t,\nu}+\mu)^{-1}\end{aligned}$$

is causal as a composition of causal operators. Finally, Lemma 16.2.4 implies causality of $(\partial_{t,\nu}+\mathcal{K})^{-1}$ as desired. □

16.3 Non-Autonomous Picard's Theorem—The PDE Case

Let H be a Hilbert space. In Sect. 4.2, we have already discussed the notion of uniformly Lipschitz continuous mappings. Here we concentrate on linear uniformly Lipschitz continuous mappings, which we call *evolutionary* as a short hand:

Definition Let $\nu_0 \in \mathbb{R}$. A mapping

$$\mathcal{M}\colon S_c(\mathbb{R}; H) \to \bigcap_{\nu \geqslant \nu_0} L_{2,\nu}(\mathbb{R}; H)$$

is called *evolutionary (at ν_0)* if it is linear and uniformly Lipschitz continuous (at ν_0); that is, for all $\nu \geqslant \nu_0$, the mapping $\mathcal{M}\colon S_c(\mathbb{R}; H) \subseteq L_{2,\nu}(\mathbb{R}; H) \to L_{2,\nu}(\mathbb{R}; H)$ is linear and continuous. Moreover, its continuous extension to the whole of $L_{2,\nu}(\mathbb{R}; H)$, denoted by $\mathcal{M}^\nu$, satisfies $\sup_{\nu \geqslant \nu_0} \|\mathcal{M}^\nu\| < \infty$.

The set of all evolutionary mappings is defined as

$$S_{\mathrm{ev}}(H, \nu_0) := \left\{ \mathcal{M}\colon S_c(\mathbb{R}; H) \to \bigcap_{\nu \geqslant \nu_0} L_{2,\nu}(\mathbb{R}; H)\,;\ \mathcal{M} \text{ evolutionary at } \nu_0 \right\}.$$

We have seen that material law operators are evolutionary (see Theorem 5.3.6 and the concluding lines of the proof). In the non-autonomous version of Picard's theorem (Theorem 6.2.1), evolutionary mappings will replace the notion of material law operators. Hence, we allow for an explicit time-dependence in the coefficients.

Recall from Lemma 4.2.5(a), that $\mathcal{M}^\nu$ is causal and independent of ν in the sense of Lemma 4.2.5(c).

The non-autonomous version of Picard's theorem now reads as follows.

Theorem 16.3.1 *Let $\mu \in \mathbb{R}$, $\mathcal{M}, \mathcal{M}', \mathcal{N} \in S_{\mathrm{ev}}(H, \mu)$, $\operatorname{Re} \mathcal{M}^\nu \geqslant 0$ for all $\nu \geqslant \mu$ and $A\colon \operatorname{dom}(A) \subseteq H \to H$ be closed and densely defined. Assume that there exists $c > 0$ such that the following conditions are satifsfied:*

(a) $\mathcal{M}^\mu \partial_{t,\mu} \subseteq \partial_{t,\mu} \mathcal{M}^\mu - \left(\mathcal{M}'\right)^\mu$,

(b) *for all $\nu \geqslant \mu$ and $\phi \in \operatorname{dom}(\partial_{t,\nu})$ we have*

$$\operatorname{Re}\left\langle \phi, \left(\partial_{t,\nu}\mathcal{M}^\nu + \mathcal{N}^\nu\right)\phi\right\rangle_{L_{2,\nu}(\mathbb{R}; H)} \geqslant c\, \langle \phi, \phi\rangle_{L_{2,\nu}(\mathbb{R}; H)},$$

(c) *for all $x \in \operatorname{dom}(A)$ and $y \in \operatorname{dom}(A^*)$ we have*

$$\operatorname{Re}\, \langle x, Ax\rangle_H \geqslant 0 \text{ and } \operatorname{Re}\left\langle y, A^* y\right\rangle_H \geqslant 0.$$

Then for all $\nu \geqslant \max\{\mu, 0\}$, $\nu \neq 0$, the operator

$$\partial_{t,\nu}\mathcal{M}^\nu + \mathcal{N}^\nu + A\colon H_\nu^1(\mathbb{R}; H) \cap L_{2,\nu}\big(\mathbb{R}; \operatorname{dom}(A)\big) \subseteq L_{2,\nu}(\mathbb{R}; H) \to L_{2,\nu}(\mathbb{R}; H)$$

is closable and its closure is continuously invertible. Moreover, with $\mathcal{S}_\nu \in L(L_{2,\nu}(\mathbb{R}; H))$ being the inverse of this closure, $\|\mathcal{S}_\nu\|_{L(L_{2,\nu}(\mathbb{R};H))} \leqslant 1/c$, $\mathcal{S}_\nu$ is eventually independent of ν and $\mathcal{S}_\nu$ is causal.

Remark 16.3.2

(a) It is a consequence of Theorem 16.3.1 that the mapping

$$\mathcal{S}\colon S_c(\mathbb{R}; H) \to \bigcap_{\nu \geqslant \mu} L_{2,\nu}(\mathbb{R}; H)$$

$$f \mapsto \overline{\left(\partial_{t,\mu}\mathcal{M}^\mu + \mathcal{N}^\mu + A\right)}^{-1} f$$

is evolutionary.

(b) It will follow from the techniques used in the proof of Theorem 16.3.1, that a similar results holds without the assumption of evolutionarity for the operator coefficients. We refer to the formulation in Exercise 16.5 and ask the reader to provide a proof for this.

The proof of the non-autonomous version of Picard's theorem requires some preparations. Being still a linear theory, the well-posedness result is—similar to the autonomous version of Picard's theorem—based on Proposition 6.3.1. Furthermore, we need some results on the interaction of the time derivative and the non-autonomous coefficients. Thus, for the next lemma, we introduce the commutator

$$[A, B] := AB - BA$$

for two linear operators A and B on its natural domain

$$\operatorname{dom}(AB) \cap \operatorname{dom}(BA).$$

Lemma 16.3.3 *Let $\nu \in \mathbb{R}$, $\mathcal{M}, \mathcal{M}', \mathcal{N} \in S_{\mathrm{ev}}(H, \nu)$. For $\varepsilon > 0$ small enough, denote $S_\varepsilon := (1 + \varepsilon\partial_{t,\nu})^{-1}$.*

(a) *If $\mathcal{M}^\nu \partial_{t,\nu} \subseteq \partial_{t,\nu}\mathcal{M}^\nu - (\mathcal{M}')^\nu$, then for all $\varepsilon > 0$ we have*

$$\overline{[\partial_{t,\nu}\mathcal{M}^\nu, S_\varepsilon]} = \varepsilon\partial_{t,\nu}S_\varepsilon(\mathcal{M}')^\nu S_\varepsilon \in L(L_{2,\nu}(\mathbb{R}; H)).$$

In this case, we also have that $\overline{[\partial_{t,\nu}\mathcal{M}^\nu, S_\varepsilon]} \to 0$ in the strong operator topology of $L(L_{2,\nu}(\mathbb{R}; H))$.

(b) *We have that $[\mathcal{N}, S_\varepsilon] \to 0$ as $\varepsilon \to 0$ in the strong operator topology of $L(L_{2,\nu}(\mathbb{R}; H))$.*

Proof

(a) Let $\varepsilon > 0$ and $\phi \in \operatorname{dom}(\partial_{t,\nu})$. Then

$$\begin{aligned}\overline{[\partial_{t,\nu}\mathcal{M}^\nu, S_\varepsilon]}\phi &= \partial_{t,\nu}(\mathcal{M}^\nu S_\varepsilon - S_\varepsilon \mathcal{M}^\nu)\phi \\ &= \partial_{t,\nu} S_\varepsilon((1+\varepsilon\partial_{t,\nu})\mathcal{M}^\nu - \mathcal{M}^\nu(1+\varepsilon\partial_{t,\nu}))S_\varepsilon\phi \\ &= \varepsilon\partial_{t,\nu} S_\varepsilon(\mathcal{M}')^\nu S_\varepsilon\phi,\end{aligned}$$

which shows the first equality. Since $S_\varepsilon \to 1$ as $\varepsilon \to 0$ in the strong operator topology and $\varepsilon\partial_{t,\nu}S_\varepsilon = (1-S_\varepsilon) \to 0$ as $\varepsilon \to 0$ in the strong operator topology, we infer the convergence statement in (a).

(b) This statement follows from $S_\varepsilon \to 1$ in the strong operator topology. □

Lemma 16.3.4 *Let $\mu \in \mathbb{R}$, $\mathcal{M}, \mathcal{M}', \mathcal{N} \in S_{\mathrm{ev}}(H,\mu)$ and $A\colon \operatorname{dom}(A) \subseteq H \to H$ be closed and densely defined. Assume $\mathcal{M}^\mu\partial_{t,\mu} \subseteq \partial_{t,\mu}\mathcal{M}^\mu - \left(\mathcal{M}'\right)^\mu$. Then for all $\nu \geqslant \mu$*

$$(\partial_{t,\nu}\mathcal{M}^\nu + \mathcal{N}^\nu + A)^* = \overline{(\partial_{t,\nu}\mathcal{M}^\nu + \mathcal{N}^\nu)^* + A^*} = \overline{(\mathcal{M}^\nu)^*\partial_{t,\nu}^* + (\mathcal{N}^\nu)^* + A^*}.$$

Proof Let $\nu \geqslant \mu$. It is not difficult to see that $\mathcal{M}^\mu\partial_{t,\mu} \subseteq \partial_{t,\mu}\mathcal{M}^\mu - \left(\mathcal{M}'\right)^\mu$ implies $\mathcal{M}^\nu\partial_{t,\nu} \subseteq \partial_{t,\nu}\mathcal{M}^\nu - \left(\mathcal{M}'\right)^\nu$, see Exercise 16.2.

Let $g \in \operatorname{dom}\left((\partial_{t,\nu}\mathcal{M}^\nu + \mathcal{N}^\nu + A)^*\right)$. For $\varepsilon > 0$ small enough, we define $S_\varepsilon := (1+\varepsilon\partial_{t,\nu})^{-1}$ as well as $g_\varepsilon := S_\varepsilon^* g$. For $u \in \operatorname{dom}(\partial_{t,\nu}\mathcal{M}^\nu + \mathcal{N}^\nu + A)$ we compute

$$\begin{aligned}&\left\langle(\partial_{t,\nu}\mathcal{M}^\nu + \mathcal{N}^\nu + A)u, g_\varepsilon\right\rangle \\ &= \left\langle S_\varepsilon(\partial_{t,\nu}\mathcal{M}^\nu + \mathcal{N}^\nu + A)u, g\right\rangle \\ &= \left\langle(\partial_{t,\nu}\mathcal{M}^\nu + \mathcal{N}^\nu + A)S_\varepsilon u, g\right\rangle - \left\langle[\partial_{t,\nu}\mathcal{M}^\nu, S_\varepsilon]u + [\mathcal{N}^\nu, S_\varepsilon]u, g\right\rangle.\end{aligned} \tag{16.2}$$

We read off that $g_\varepsilon \in \operatorname{dom}\left((\partial_{t,\nu}\mathcal{M}^\nu + \mathcal{N}^\nu + A)^*\right)$ and

$$\begin{aligned}&(\partial_{t,\nu}\mathcal{M}^\nu + \mathcal{N}^\nu + A)^* g_\varepsilon \\ &\quad = S_\varepsilon^*(\partial_{t,\nu}\mathcal{M}^\nu + \mathcal{N}^\nu + A)^* g - [\partial_{t,\nu}\mathcal{M}^\nu, S_\varepsilon]^* g - [\mathcal{N}^\nu, S_\varepsilon]^* g.\end{aligned}$$

By Lemma 9.3.3, we infer that $g_\varepsilon \to g$ weakly as $\varepsilon \to 0$. Similarly, we obtain

$$S_\varepsilon^*(\partial_{t,\nu}\mathcal{M}^\nu - \mathcal{N}^\nu + A)^* g + [\partial_{t,\nu}\mathcal{M}^\nu, S_\varepsilon]^* g - [\mathcal{N}^\nu, S_\varepsilon]^* g \to (\partial_{t,\nu}\mathcal{M}^\nu + \mathcal{N}^\nu + A)^* g$$

weakly as $\varepsilon \to 0$. Next, we show that $g_\varepsilon \in \operatorname{dom}(A)$ for all $\varepsilon > 0$. For this, we realise that $g_\varepsilon \in \operatorname{dom}(\partial_{t,\nu}^*) = \operatorname{dom}(\partial_{t,\nu})$ and, thus, revisiting (16.2), we infer

$$\begin{aligned}\langle Au, g_\varepsilon\rangle &= -\left\langle(\partial_{t,\nu}\mathcal{M}^\nu + \mathcal{N}^\nu)u, g_\varepsilon\right\rangle + \left\langle(\partial_{t,\nu}\mathcal{M}^\nu + \mathcal{N}^\nu + A)S_\varepsilon u, g\right\rangle \\ &\quad - \left\langle[\partial_{t,\nu}\mathcal{M}^\nu, S_\varepsilon]u, g\right\rangle - \left\langle[\mathcal{N}^\nu, S_\varepsilon]u, g\right\rangle\end{aligned}$$

$$\begin{aligned}&= -\left\langle u, ((\mathcal{M}^\nu)^*\partial_{t,\nu}^* + (\mathcal{N}^\nu)^*)g_\varepsilon\right\rangle + \left\langle u, S_\varepsilon^*(\partial_{t,\nu}\mathcal{M}^\nu + \mathcal{N}^\nu + A)^*g\right\rangle \\ &\quad - \left\langle u, [\partial_{t,\nu}\mathcal{M}^\nu, S_\varepsilon]^*g + [\mathcal{N}^\nu, S_\varepsilon]^*g\right\rangle.\end{aligned}$$

Since $H_\nu^1(\mathbb{R}; H) \cap L_{2,\nu}\big(\mathbb{R}; \operatorname{dom}(A)\big)$ is dense in $L_{2,\nu}\big(\mathbb{R}; \operatorname{dom}(A)\big)$, we read off that $g_\varepsilon \in \operatorname{dom}(A^*)$. Thus, since $g_\varepsilon \in \operatorname{dom}(\partial_{t,\nu}^*)$ anyway, we obtain by the first statements in Theorem 2.3.2 and Theorem 2.3.4 that

$$(\partial_{t,\nu}\mathcal{M}^\nu + \mathcal{N}^\nu + A)^*g_\varepsilon = (\mathcal{M}^\nu)^*\partial_{t,\nu}^*g_\varepsilon + (\mathcal{N}^\nu)^*g_\varepsilon + A^*g_\varepsilon,$$

which together with the above convergence result shows the assertion. □

Lemma 16.3.5 *Let $\mu, \nu \in \mathbb{R}$, $\mu \geqslant \nu$. Let $\mathcal{S}_\nu \in L(L_{2,\nu}(\mathbb{R}; H))$ as well as $\mathcal{S}_\mu \in L(L_{2,\mu}(\mathbb{R}; H))$ be causal and $D \subseteq L_{2,\nu}(\mathbb{R}; H) \cap L_{2,\mu}(\mathbb{R}; H)$ dense in $L_{2,\mu}(\mathbb{R}; H)$ such that $\mathcal{S}_\nu = \mathcal{S}_\mu$ on D. Then $\mathcal{S}_\nu = \mathcal{S}_\mu$ on $L_{2,\nu}(\mathbb{R}; H) \cap L_{2,\mu}(\mathbb{R}; H)$.*

Proof Let $f \in L_{2,\nu}(\mathbb{R}; H) \cap L_{2,\mu}(\mathbb{R}; H)$. By density of D, we may find a sequence $(f_n)_n$ in D such that $f_n \to f$ in $L_{2,\mu}(\mathbb{R}; H)$. Let $a \in \mathbb{R}$. Then $\mathbb{1}_{(-\infty,a]}f_n \to \mathbb{1}_{(-\infty,a]}f$ in $L_{2,\nu}(\mathbb{R}; H) \cap L_{2,\mu}(\mathbb{R}; H)$. Since both $\mathcal{S}_\mu$ and $\mathcal{S}_\nu$ are causal, we infer for $n \in \mathbb{N}$ that

$$\mathbb{1}_{(-\infty,a]}\mathcal{S}^\mu\mathbb{1}_{(-\infty,a]}f_n = \mathbb{1}_{(-\infty,a]}\mathcal{S}^\mu f_n = \mathbb{1}_{(-\infty,a]}\mathcal{S}^\nu f_n = \mathbb{1}_{(-\infty,a]}\mathcal{S}^\nu\mathbb{1}_{(-\infty,a]}f_n.$$

Letting $n \to \infty$, we deduce that both the left-hand side as well as the right-hand side converge in $L_{2,\mathrm{loc}}(\mathbb{R}; H)$. Consequently, we infer, using causality again that

$$\mathbb{1}_{(-\infty,a]}\mathcal{S}^\mu f = \mathbb{1}_{(-\infty,a]}\mathcal{S}^\mu\mathbb{1}_{(-\infty,a]}f = \mathbb{1}_{(-\infty,a]}\mathcal{S}^\nu\mathbb{1}_{(-\infty,a]}f = \mathbb{1}_{(-\infty,a]}\mathcal{S}^\nu f.$$

This equality holds for all $a \in \mathbb{R}$, thus $\mathcal{S}^\mu f = \mathcal{S}^\nu f$ and the assertion follows. □

The following lemma is proved in the (easy) Exercise 16.7.

Lemma 16.3.6 *Let H_0, H_1 be Hilbert spaces. Let $B\colon \operatorname{dom}(B) \subseteq H_0 \to H_1$ be closed and densely defined. Let V be a Hilbert space such that $V \hookrightarrow \operatorname{dom}(B)$ continuously and densely. If $D \subseteq V$ is a dense subspace, then D is a core for B.*

Proof of Theorem 16.3.1 Define $\widetilde{B} := \partial_{t,\nu}\mathcal{M}^\nu + \mathcal{N}^\nu + A$ with $\operatorname{dom}(\widetilde{B}) = H_\nu^1(\mathbb{R}; H) \cap L_{2,\nu}\big(\mathbb{R}; \operatorname{dom}(A)\big)$. By the last equality in Lemma 16.3.4, we have $\operatorname{dom}(\widetilde{B}^*) \supseteq H_\nu^1(\mathbb{R}; H) \cap L_{2,\nu}\big(\mathbb{R}; \operatorname{dom}(A^*)\big)$. Hence, $\widetilde{B}^*$ is densely defined and, therefore, by Lemma 2.2.7, $\widetilde{B}$ is closable. Next, we want to apply Proposition 6.3.1 to $B := \overline{\widetilde{B}}$. For this, we let $\phi \in \operatorname{dom}(\widetilde{B})$ and compute

$$\begin{aligned}\operatorname{Re}\langle\phi, B\phi\rangle &= \operatorname{Re}\left\langle\phi, (\partial_{t,\nu}\mathcal{M}^\nu + \mathcal{N}^\nu + A)\phi\right\rangle \\ &\geqslant c\,\langle\phi, \phi\rangle + \operatorname{Re}\langle\phi, A\phi\rangle \geqslant c\,\langle\phi, \phi\rangle.\end{aligned}$$

Since $\operatorname{dom}(\widetilde{B})$ is a core for B, we deduce

$$\operatorname{Re}\langle\phi, B\phi\rangle \geqslant c\langle\phi, \phi\rangle \quad (\phi \in \operatorname{dom}(B)).$$

Using Lemma 16.3.4, we obtain $D := \operatorname{dom}\big(\big(\partial_{t,\nu}\mathcal{M}^\nu + \mathcal{N}^\nu\big)^*\big) \cap L_{2,\nu}\big(\mathbb{R}; \operatorname{dom}(A^*)\big)$ is a core for B^*. Using Theorem 16.2.1, we estimate for all $\psi \in D$ that

$$\operatorname{Re}\big\langle\psi, B^*\psi\big\rangle = \operatorname{Re}\big\langle\psi, \big(\partial_{t,\nu}\mathcal{M}^\nu + \mathcal{N}^\nu\big)^*\psi + A^*\psi\big\rangle \geqslant c\langle\psi, \psi\rangle.$$

Hence,

$$\operatorname{Re}\big\langle\psi, B^*\psi\big\rangle \geqslant c\langle\psi, \psi\rangle \quad (\psi \in \operatorname{dom}(B^*)).$$

Thus, Proposition 6.3.1 applies and we deduce that $0 \in \rho(B)$ and $\big\|B^{-1}\big\| \leqslant 1/c$.

Next, since $(\partial_{t,\nu}\mathcal{M}^\nu + \mathcal{N}^\nu)^{-1}$ is causal by Theorem 16.2.1, using Proposition 16.2.3 for $\phi \in H_\nu^1(\mathbb{R}; H) \cap L_{2,\nu}\big(\mathbb{R}; \operatorname{dom}(A)\big) = \operatorname{dom}(\widetilde{B})$ we obtain for $a \in \mathbb{R}$ that

$$\begin{aligned}\operatorname{Re}\big\langle\mathbb{1}_{(-\infty,a]}\phi, B\phi\big\rangle &= \operatorname{Re}\big\langle\mathbb{1}_{(-\infty,a]}\phi, (\partial_{t,\nu}\mathcal{M}^\nu + \mathcal{N}^\nu + A)\phi\big\rangle \\ &= \operatorname{Re}\big\langle\mathbb{1}_{(-\infty,a]}\phi, (\partial_{t,\nu}\mathcal{M}^\nu + \mathcal{N}^\nu)\phi\big\rangle\phi + \operatorname{Re}\big\langle\mathbb{1}_{(-\infty,a]}\phi, \mathbb{1}_{(-\infty,a]}A\phi\big\rangle \\ &\geqslant c\big\langle\mathbb{1}_{(-\infty,a]}\phi, \phi\big\rangle + \operatorname{Re}\big\langle\mathbb{1}_{(-\infty,a]}\phi, A\mathbb{1}_{(-\infty,a]}\phi\big\rangle \geqslant c\big\langle\mathbb{1}_{(-\infty,a]}\phi, \phi\big\rangle.\end{aligned}$$

The inequality $\operatorname{Re}\big\langle\mathbb{1}_{(-\infty,a]}\phi, B\phi\big\rangle \geqslant c\big\langle\mathbb{1}_{(-\infty,a]}\phi, \phi\big\rangle$ carries over to all $\phi \in \operatorname{dom}(B)$ using that $\operatorname{dom}(\widetilde{B})$ is, by definition, a core for B. Again appealing to Proposition 16.2.3 we obtain that B^{-1} is causal. Finally, in order to show that $\mathcal{S}_\nu$ is eventually independent of ν, we want to apply Lemma 16.3.5. Since we have shown that for all $\nu \geqslant \eta \geqslant \mu$, the operators $\mathcal{S}_\nu$ and $\mathcal{S}_\eta$ are continuous and causal, it remains to construct a set $U \subseteq L_{2,\nu}(\mathbb{R}; H) \cap L_{2,\eta}(\mathbb{R}; H)$ dense in $L_{2,\nu}(\mathbb{R}; H)$ such that $\mathcal{S}_\nu = \mathcal{S}_\eta$ on U. We put

$$U := (\partial_{t,\nu}\mathcal{M}^\nu + \mathcal{N}^\nu + A)\big[C_c^\infty\big(\mathbb{R}; \operatorname{dom}(A)\big)\big],$$

which is evidently a subset of $L_{2,\nu}(\mathbb{R}; H)$. Observe that $C_c^\infty\big(\mathbb{R}; \operatorname{dom}(A)\big) \subseteq L_{2,\eta}(\mathbb{R}; H) \cap L_{2,\nu}(\mathbb{R}; H)$. Moreover, $\mathcal{M}^\nu = \mathcal{M}^\eta$ as well as $\mathcal{N}^\nu = \mathcal{N}^\eta$ on $L_{2,\eta}(\mathbb{R}; H) \cap L_{2,\nu}(\mathbb{R}; H)$. Thus, both $\mathcal{M}^\nu$ and $\mathcal{N}^\nu$ leave $L_{2,\eta}(\mathbb{R}; H) \cap L_{2,\nu}(\mathbb{R}; H)$ invariant, by Lemma 4.2.5. Hence, since $A\big[C_c^\infty\big(\mathbb{R}; \operatorname{dom}(A)\big)\big] \subseteq C_c^\infty(\mathbb{R}; H)$, we infer that $U \subseteq L_{2,\eta}(\mathbb{R}; H) \cap L_{2,\nu}(\mathbb{R}; H)$.

Finally, by Lemma 9.4.1, $C_c^\infty\big(\mathbb{R}; \operatorname{dom}(A)\big)$ is dense in $L_{2,\nu}\big(\mathbb{R}; \operatorname{dom}(A)\big) \cap H_\nu^1(\mathbb{R}; H)$. We now apply Lemma 16.3.6 to $C_c^\infty\big(\mathbb{R}; \operatorname{dom}(A)\big) \subseteq V := L_{2,\nu}(\mathbb{R}; \operatorname{dom}(A)) \cap H(\mathbb{R}; H)$ and B to get that $C_c^\infty\big(\mathbb{R}; \operatorname{dom}(A)\big)$ is a core for B. Since B is surjective, this implies that $U = B\big[C_c^\infty\big(\mathbb{R}; \operatorname{dom}(A)\big)\big] \subseteq L_{2,\nu}(\mathbb{R}; H)$ is dense which yields the assertion. □

16.4 Comments

Traditionally, non-autonomous equations have been dealt with—similar to the autonomous case—by mimicking techniques and results from non-autonomous ordinary differential equations. In consequence, the fundamental solution is the central object of attention, which finds itself in the concept of so-called evolution families $(U(t,s))_{t\geqslant s}$ or propagators, see e.g. [53, 112]. Similar to the autonomous case, one is interested in the initial value problem

$$\begin{cases} u'(t) + A(t)u(t) = 0, & t > 0, \\ u(0) = u_0, \end{cases}$$

for a given parameter dependent operator family $(A(t))_t$ of *unbounded* operators. The solution is then given by $u(t) = U(t,0)u_0$. In applications, for instance to parabolic equations, $A(t) = -\operatorname{div} a(t) \operatorname{grad}$.

One is then interested in whether $(A(t))_t$ gives rise to an evolution family. There, the main issue is to understand the behaviour of the possibly different domains of $A(t)$ for any given t. Focussing on inhomogeneous problems rather than initial value problems, we again are changing the perspective in the case of evolutionary equations. The presented time-space perspective entirely dispenses with the possible domain issues and requires only mild regularity conditions of the coefficients. In particular, as it has been demonstrated for the heat equation in Sect. 16.1, we merely require boundedness and measurability for a, whereas for Maxwell's equations we need Lipschitz continuity for the coefficients ε and μ.

The first result on the well-posedness of non-autonomous evolutionary equations has been found in [92]. In this source, the focus was on multiplication operators as coefficients and Lipschitz continuity of the operator coefficients with respect to time was assumed. The method of proof has been used to generalise this to the commutator assumption presented here, see [137, 138]. Theorem 16.3.1 also has a nonlinear analogue. This can be found in [122]. For an autonomous well-posedness result for nonlinear evolutionary inclusions we also refer to Chap. 17.

Exercises

Exercise 16.1 Let $V\colon \mathbb{R} \to \mathbb{R}$ be Lipschitz continuous.

(a) Let $\phi \in C_c^\infty(\mathbb{R})$. Show that $\phi V \in H_\nu^1(\mathbb{R})$ with bounded derivative. Show that there exists a bounded measurable function V' such that $V(t) - V(0) = \int_0^t V'(\tau)\mathrm{d}\tau$.
(b) Let V be bounded. Show that $V(\mathrm{m})$ is evolutionary at 0 and that

$$V(\mathrm{m})^\nu \partial_{t,\nu} \subseteq \partial_{t,\nu} V(\mathrm{m})^\nu - V'(\mathrm{m})^\nu.$$

(c) In the situation of (b), show that for $\phi \in \operatorname{dom}(\partial_{t,\nu})$, we have

$$\operatorname{Re}\left\langle \phi, \partial_{t,\nu} V(\mathrm{m})\phi \right\rangle = \nu \left\langle \phi, V(\mathrm{m})\phi \right\rangle + \frac{1}{2}\left\langle \phi, V'(\mathrm{m})\phi \right\rangle.$$

Exercise 16.2 Let H be a Hilbert space, $\mu \in \mathbb{R}$. Let $\mathcal{M}, \mathcal{M}' \in S_{\mathrm{ev}}(H, \mu)$. Assume that

$$\mathcal{M}^{\mu}\partial_{t,\mu} \subseteq \partial_{t,\mu}\mathcal{M}^{\mu} - (\mathcal{M}')^{\mu}.$$

Show that then for all $\nu \geqslant \mu$ we have

$$\mathcal{M}^{\nu}\partial_{t,\nu} \subseteq \partial_{t,\nu}\mathcal{M}^{\nu} - (\mathcal{M}')^{\nu}.$$

Exercise 16.3 Let H be a Hilbert space, $\nu, c > 0$, $M \in \mathcal{M}(H, \nu)$. Assume that

$$\operatorname{Re} zM(z) \geqslant c.$$

Show that then

$$\operatorname{Re}\left\langle \partial_{t,\nu} M(\partial_{t,\nu})\phi, \mathbb{1}_{(-\infty,a]}\phi \right\rangle \geqslant c \left\| \mathbb{1}_{(-\infty,a]}\phi \right\|^2$$

for all $\phi \in \operatorname{dom}(\partial_{t,\nu})$ and $a \in \mathbb{R}$.

Exercise 16.4 In the situation of Theorem 16.2.1, show that $0 \in \rho(\partial_{t,\nu}\mathcal{M} + \mathcal{N})$ and $\left\| (\partial_{t,\nu}\mathcal{M} + \mathcal{N})^{-1} \right\| \leqslant 1/c$.
Hint: Show $\operatorname{Re}\left(\partial_{t,\nu}\mathcal{M} + \mathcal{N}\right)^* \geqslant c$ first.

Exercise 16.5 Prove the following 'non-causal' version of Theorem 16.3.1: Let H a Hilbert space, $\nu \in \mathbb{R}$. Let $\mathcal{M}, \mathcal{M}', \mathcal{N} \in L(L_{2,\nu}(\mathbb{R}; H))$ and $A\colon \operatorname{dom}(A) \subseteq H \to H$ be closed and densely defined. Assume that there exists $c > 0$ such that the following conditions are satisfsied:

(a) $\mathcal{M}\partial_{t,\nu} \subseteq \partial_{t,\nu}\mathcal{M} - \mathcal{M}'$,
(b) for all $\phi \in \operatorname{dom}(\partial_{t,\nu})$ we have

$$\operatorname{Re}\left\langle \phi, \left(\partial_{t,\nu}\mathcal{M} + \mathcal{N}\right)\phi \right\rangle_{L_{2,\nu}(\mathbb{R};H)} \geqslant c \left\langle \phi, \phi \right\rangle_{L_{2,\nu}(\mathbb{R};H)},$$

(c) for all $x \in \operatorname{dom}(A)$ and $y \in \operatorname{dom}(A^*)$ we have

$$\operatorname{Re}\left\langle x, Ax \right\rangle_H \geqslant 0 \text{ and } \operatorname{Re}\left\langle y, A^*y \right\rangle_H \geqslant 0.$$

Then

$$\partial_{t,\nu}\mathcal{M} + \mathcal{N} + A\colon H_\nu^1(\mathbb{R}; H) \cap L_{2,\nu}\big(\mathbb{R}; \operatorname{dom}(A)\big) \subseteq L_{2,\nu}(\mathbb{R}; H) \to L_{2,\nu}(\mathbb{R}; H)$$

is closable and its closure is continuously invertible. Denoting the respective inverse by $\mathcal{S}$, we have $\|\mathcal{S}\|_{L(L_{2,\nu}(\mathbb{R};H))} \leqslant 1/c$.

Exercise 16.6 Without using Theorem 16.3.1 or Exercise 16.5 show that if $M \in \mathcal{M}(H,\nu)$ and $\mathcal{N} \in \mathcal{S}_{\mathrm{ev}}(H,\nu)$ satisfy

$$\operatorname{Re}\left\langle \phi, (\partial_{t,\nu} M(\partial_{t,\nu}) + \mathcal{N}^{\nu})\phi \right\rangle \geqslant c \langle \phi, \phi \rangle \quad (\phi \in \operatorname{dom}(\partial_{t,\nu}))$$

for some $c > 0$, then $0 \in \rho\left(\overline{\partial_{t,\nu} M(\partial_{t,\nu}) + \mathcal{N}^{\nu} + A}\right)$, for all skew-selfadjoint $A\colon \operatorname{dom}(A) \subseteq H \to H$.
Hint: Compute the adjoint of $\partial_{t,\nu} M(\partial_{t,\nu}) + \mathcal{N}^{\nu} + A$ with the help of Theorem 6.2.1 and Theorem 2.3.2.

Exercise 16.7 Prove Lemma 16.3.6.

References

53. T. Kato, Integration of the equation of evolution in a Banach space. J. Math. Soc. Jpn. **5**(2), 208–234 (1953)
92. R. Picard et al., On non-autonomous evolutionary problems. J. Evol. Equ. **13**, 751–776 (2013)
112. H. Tanabe, *Equations of Evolution*, vol. 6. Monographs and Studies in Mathematics. Translated from the Japanese by N. Mugibayashi and H. Haneda (Pitman (Advanced Publishing Program), Boston, MA, London, 1979)
122. S. Trostorff, Well-posedness for a general class of differential inclusions. J. Differ. Equ. **268**, 6489–6516 (2020)
137. M. Waurick, On non-autonomous integro-differential-algebraic evolutionary problems. Math. Methods Appl. Sci. **38**(4), 665–676 (2015)
138. M. Waurick, On the continuous dependence on the coefficients of evolutionary equations. Habilitation. Technische Universität Dresden, 2016 http://arxiv.org/abs/1606.07731

Chapter 17
Evolutionary Inclusions

This chapter is devoted to the study of *evolutionary inclusions*. In contrast to evolutionary equations, we will replace the skew-selfadjoint operator A by a so-called maximal monotone relation $A \subseteq H \times H$ in the Hilbert space H. The resulting problem is then no longer an equation, but just an inclusion; that is, we consider problems of the form

$$(u, f) \in \overline{\partial_{t,\nu} M(\partial_{t,\nu}) + A}, \tag{17.1}$$

where $f \in L_{2,\nu}(\mathbb{R}; H)$ is given and $u \in L_{2,\nu}(\mathbb{R}; H)$ is to be determined. This generalisation allows the treatment of certain non-linear problems, since we will not require any linearity for the relation A. Moreover, the property that A is just a relation and not neccessarily an operator can be used to treat hysteresis phenomena, which for instance occur in the theory of elasticity and electro-magnetism.

We begin to define the notion of maximal monotone relations in the first part of this chapter. In particular, we introduce the notion of the so-called Yosida approximation of A and provide a useful perturbation result for maximal monotone relations, which will be the key argument for proving the well-posedness of (17.1). For this, we prove the celebrated Theorem of Minty, which characterises the maximal monotone relations by a range condition. The second section is devoted to the main result of this chapter, namely the well-posedness of (17.1), which generalises Picard's theorem (see Theorem 6.2.1) to a broader class of problems. In the concluding section we consider Maxwell's equations in a polarisable medium as an application.

C. Seifert et al., *Evolutionary Equations*, Operator Theory: Advances and Applications 287, https://doi.org/10.1007/978-3-030-89397-2_17

17.1 Maximal Monotone Relations and the Theorem of Minty

Definition Let $A \subseteq H \times H$. We call A *monotone* if

$$\forall (u, v), (x, y) \in A : \operatorname{Re} \langle u - x, v - y \rangle \geqslant 0.$$

Moreover, we call A *maximal monotone* if A is monotone and for each monotone relation $B \subseteq H \times H$ with $A \subseteq B$ it follows that $A = B$.

Remark 17.1.1 Let $A \subseteq H \times H$ be a monotone relation.

(a) It is clear that A is maximal monotone if and only if for each $x, y \in H$ with

$$\forall (u, v) \in A : \operatorname{Re} \langle u - x, v - y \rangle \geqslant 0$$

it follows that $(x, y) \in A$.

(b) From (a) it follows that A is *demiclosed*; i.e., for each sequence $((x_n, y_n))_{n\in\mathbb{N}}$ in A with $x_n \to x$ in H and $y_n \to y$ weakly or $x_n \to x$ weakly and $y_n \to y$ in H for some $x, y \in H$ as $n \to \infty$ it follows that $(x, y) \in A$ (note that in both cases we have $\langle u - x_n, v - y_n \rangle \to \langle u - x, v - y \rangle$ for each $(u, v) \in A$).

We start to present some first properties of monotone and maximal monotone relations.

Proposition 17.1.2 *Let $A \subseteq H \times H$ be monotone and $\lambda > 0$. Then the following statements hold:*

(a) *The inverse relation $(1 + \lambda A)^{-1}$ is a Lipschitz-continuous mapping, which satisfies $\left\|(1 + \lambda A)^{-1}\right\|_{\mathrm{Lip}} \leqslant 1$.*

(b) *If $1 + \lambda A$ is onto, then A is maximal monotone.*

Proof For showing (a), we assume that $(f, u), (g, x) \in (1 + \lambda A)^{-1}$ for some $f, g, u, x \in H$. Then we find $v, y \in H$ such that $(u, v), (x, y) \in A$ and $u + \lambda v = f$ as well as $x + \lambda y = g$. The monotonicity of A then yields

$$\|u - x\|^2 = \operatorname{Re} \langle f - g - \lambda(v - y), u - x \rangle \leqslant \operatorname{Re} \langle f - g, u - x \rangle \leqslant \|f - g\| \, \|u - x\| .$$

If now $f = g$, then $u = x$. Hence, $(1 + \lambda A)^{-1}$ is a mapping and the inequality proves its Lipschitz-continuity with $\left\|(1 + \lambda A)^{-1}\right\|_{\mathrm{Lip}} \leqslant 1$.

To prove (b), let $B \subseteq H \times H$ be monotone with $A \subseteq B$ and let $(x, y) \in B$. Since $1 + \lambda A$ is onto, we find $(u, v) \in A \subseteq B$ such that $u + \lambda v = x + \lambda y$. Since $(1 + \lambda B)^{-1}$ is a mapping by (a), we infer that

$$x = (1 + \lambda B)^{-1}(x + \lambda y) = (1 + \lambda B)^{-1}(u + \lambda v) = u$$

and hence, also $v = y$, which proves that $(x, y) \in A$ and thus, $A = B$. □

Example 17.1.3 Let $B\colon \operatorname{dom}(B) \subseteq H \to H$ be a densely defined, closed linear operator. Assume $\operatorname{Re}\langle u, Bu\rangle \geqslant 0$ and $\operatorname{Re}\langle v, B^*v\rangle \geqslant 0$ for all $u \in \operatorname{dom}(B)$ and $v \in \operatorname{dom}(B^*)$. Then B is maximal monotone. Indeed, the monotonicity follows from the linearity of B and by Proposition 6.3.1 the operator $1 + B$ is continuously invertible, hence onto. Thus, the maximal monotonicity follows by Proposition 17.1.2(b). In particular, every skew-selfadjoint operator is maximal monotone. Moreover, if $M\colon \operatorname{dom}(M) \subseteq \mathbb{C} \to L(H)$ is a material law such that there exist $c > 0$, $\nu_0 \geqslant s_b(M)$ with

$$\operatorname{Re}\langle zM(z)\phi, \phi\rangle \geqslant c\,\|\phi\|^2 \quad (\phi \in H, z \in \mathbb{C}_{\operatorname{Re}\geqslant\nu_0}),$$

then $\partial_{t,\nu}M(\partial_{t,\nu}) - c$ is maximal monotone for each $\nu \geqslant \nu_0$.

Our first goal is to show that the implication in Proposition 17.1.2(b) is actually an equivalence. This is Minty's theorem. For this, we start to introduce subgradients of convex, proper, lower semi-continuous mappings, which form the probably most prominent example of maximal monotone relations.

Definition Let $f\colon H \to (-\infty, \infty]$. We call f

(a) *convex* if for all $x, y \in H$, $\lambda \in (0, 1)$ we have

$$f(\lambda x + (1-\lambda)y) \leqslant \lambda f(x) + (1-\lambda)f(y).$$

(b) *proper* if there exists $x \in H$ with $f(x) < \infty$.
(c) *lower semi-continuous (l.s.c.)* if for each $c \in \mathbb{R}$ the sublevel set

$$[f \leqslant c] = \{x \in H\,;\ f(x) \leqslant c\}$$

is closed.
(d) *coercive* if for each $c \in \mathbb{R}$ the sublevel set $[f \leqslant c]$ is bounded.

Remark 17.1.4 If $f\colon H \to (-\infty, \infty]$ is convex, the sublevel sets $[f \leqslant c]$ are convex for each $c \in \mathbb{R}$. Hence, if f is convex, l.s.c. and coercive, the sets $[f \leqslant c]$ are weakly sequentially compact (or, by the Eberlein–Šmulian theorem [50, theorem 13.1], equivalently, weakly compact) for each $c \in \mathbb{R}$. Indeed, if $(x_n)_{n\in\mathbb{N}}$ is a sequence in $[f \leqslant c]$ for some $c \in \mathbb{R}$, then it is bounded and thus, posseses a weakly convergent subsequence with weak limit $x \in H$. Since $[f \leqslant c]$ is closed and convex, Mazur's theorem [50, Corollary 2.11] yields that it is weakly closed and thus, $x \in [f \leqslant c]$ proving the claim.

Definition Let $f\colon H \to (-\infty, \infty]$ be convex. We define the *subgradient* of f by

$$\partial f := \{(x, y) \in H \times H\,;\ \forall u \in H:\ f(u) \geqslant f(x) + \operatorname{Re}\langle y, u - x\rangle\}.$$

Remark 17.1.5 Note that $u \mapsto f(x) + \operatorname{Re}\langle y, u - x\rangle$ is an affine function touching the graph of f in x. Thus, the subgradient is the set of all pairs $(x, y) \in H$ such that there exists an affine function with slope y touching the graph of f in x. It is not hard to show that if f is differentiable in x, then $(x, y) \in \partial f$ if and only if $y = f'(x)$ (see Exercise 17.1). Thus, the subgradient of f provides a generalisation of the derivative for arbitrary convex functions.

Proposition 17.1.6 *Let $f\colon H \to (-\infty, \infty]$ be convex and proper. Then the following statements hold:*

(a) *If $(x, y) \in \partial f$, then $f(x) < \infty$. Moreover, the subgradient ∂f is monotone.*
(b) *If f is l.s.c. and coercive, then there exists $x \in H$ such that $f(x) = \inf_{u\in H} f(u)$.*
(c) *Let $\alpha \geqslant 0$, $x, y \in H$ and $g\colon H \to (-\infty, \infty]$ with $g(u) := \frac{\alpha}{2}\|u - y\|^2 + f(u)$ for $u \in H$. Then $g(x) = \inf_{u\in H} g(u)$ if and only if $(x, \alpha(y - x)) \in \partial f$.*
(d) *Let $\alpha > 0$ and $y \in H$. If f is l.s.c., then $g\colon H \to (-\infty, \infty]$ with $g(u) := \frac{\alpha}{2}\|u - y\|^2 + f(u)$ for $u \in H$ is convex, proper, l.s.c and coercive. In particular $1 + \alpha\partial f$ is onto and hence, ∂f is maximal monotone.*

Proof

(a) If $(x, y) \in \partial f$ we have $f(u) \geqslant f(x) + \operatorname{Re}\langle y, u - x\rangle$ for each $u \in H$. Since f is proper, we find $u \in H$ such that $f(u) < \infty$ and hence, also $f(x) < \infty$. Let now $(u, v), (x, y) \in \partial f$. Then we have $f(u) \geqslant f(x) + \operatorname{Re}\langle y, u - x\rangle$ and $f(x) \geqslant f(u) + \operatorname{Re}\langle v, x - u\rangle = f(u) - \operatorname{Re}\langle v, u - x\rangle$. Summing up both expressions (note that $f(x), f(u) < \infty$ by what we have shown before), we infer

$$\operatorname{Re}\langle y - v, u - x\rangle \leqslant 0,$$

which shows the monotonicity.
(b) Let $(x_n)_{n\in\mathbb{N}}$ in H with $f(x_n) \to \inf_{u\in H} f(u) =: d$. Note that $d \in \mathbb{R}$, since f is proper. Without loss of generality, we can assume that $x_n \in [f \leqslant d + 1]$ for each $n \in \mathbb{N}$ and by Remark 17.1.4 we can assume that $x_n \to x$ weakly as $n \to \infty$ for some $x \in H$. Let $\varepsilon > 0$. Since $x_n \in [f \leqslant d + \varepsilon]$ for sufficiently large $n \in \mathbb{N}$, we derive $x \in [f \leqslant d + \varepsilon]$ again by Remark 17.1.4 and so, $f(x) \leqslant d + \varepsilon$ for each $\varepsilon > 0$, showing the claim.
(c) Assume that $g(x) = \inf_{u\in H} g(u)$ and let $u \in H$. Since f is proper, so is g and thus, we have $g(x) < \infty$, which in turn gives $f(x) < \infty$. Let $\lambda \in (0, 1]$ and set $w := \lambda u + (1 - \lambda)x$. Then the convexity of f yields

$$\begin{aligned}\lambda\left(f(u) - f(x)\right) &\geqslant f(w) - f(x)\\ &= g(w) - g(x) + \frac{\alpha}{2}(\|x - y\|^2 - \|w - y\|^2)\\ &\geqslant \frac{\alpha}{2}(\|x - y\|^2 - \|w - y\|^2)\end{aligned}$$

$$= \frac{\alpha}{2}\big(\|x - y\|^2 - \|\lambda(u - x) + x - y\|^2 \big)$$
$$= \frac{\alpha}{2}\big(-2\lambda \operatorname{Re}\langle u - x, x - y\rangle - \lambda^2 \|u - x\|^2 \big).$$

Dividing the latter expression by λ and taking the limit $\lambda \to 0$, we infer

$$-\alpha \operatorname{Re}\langle u - x, x - y\rangle \leqslant f(u) - f(x),$$

which proves $(x, \alpha(y - x)) \in \partial f$.
Assume now that $(x, \alpha(y - x)) \in \partial f$. For each $u \in H$ we have

$$\|x - y\|^2 - 2\operatorname{Re}\langle y - x, u - x\rangle = \|y - x - (u - x)\|^2 - \|u - x\|^2 \leqslant \|u - y\|^2$$

and thus,

$$f(u) \geqslant f(x) + \operatorname{Re}\langle \alpha(y - x), u - x\rangle \geqslant f(x) + \frac{\alpha}{2}\big(\|x - y\|^2 - \|u - y\|^2 \big),$$

which shows the claim.

(d) We first show that there exists an affine function $h\colon H \to \mathbb{R}$ with $h \leqslant f$. For this, we consider the epigraph of f given by

$$\operatorname{epi} f := \{(x, \beta) \in H \times \mathbb{R};\ f(x) \leqslant \beta\}.$$

Since f is convex and l.s.c., one easily verifies that epi f is convex and closed. Moreover, since f is proper, epi $f \neq \varnothing$. Let now $z \in H$ with $f(z) < \infty$ and $\eta < f(z)$. Then $(z, \eta) \in (H \times \mathbb{R}) \setminus \operatorname{epi} f$ and by the Hahn–Banach theorem we find $w \in H$ and $\gamma \in \mathbb{R}$ such that

$$\operatorname{Re}\langle w, z\rangle + \gamma\eta < \operatorname{Re}\langle w, x\rangle + \gamma\beta$$

for all $(x, \beta) \in \operatorname{epi} f$. In particular

$$\operatorname{Re}\langle w, z\rangle + \gamma\eta < \operatorname{Re}\langle w, x\rangle + \gamma f(x)$$

for each $x \in H$ and since this holds also for $x = z$, we infer $\gamma > 0$. Choosing $h(x) := \frac{1}{\gamma}\operatorname{Re}\langle w, z - x\rangle + \eta$ for $x \in H$, we have found the asserted affine function.

Using this, we have that

$$g(u) \geqslant \frac{\alpha}{2}\|u - y\|^2 + h(u) \quad (u \in H)$$

and since the right-hand side tends to ∞ as $\|u\| \to \infty$, we derive that g is coercive. Moreover, g is convex, proper and l.s.c. (see Exercise 17.2) and thus, there exists $x \in H$ with $g(x) = \inf_{u\in H} g(u)$ by (b). By (c), $(x, \alpha(y - x)) \in \partial f$ and thus, $(x, y) \in 1 + \alpha\partial f$. Since $y \in H$ was arbitrary, $1 + \alpha\partial f$ is onto and so, ∂f is maximal monotone by (a) and Proposition 17.1.2(b). □

We can now prove Minty's theorem.

Theorem 17.1.7 (Minty) *Let $A \subseteq H \times H$ maximal monotone. Then $1+\lambda A$ is onto for all $\lambda > 0$.*

Proof Since λA is maximal monotone for each $\lambda > 0$, it suffices to prove the statement for $\lambda = 1$. Moreover, since $A - (0, f)$ is maximal monotone for each $f \in H$, it suffices to show $0 \in \operatorname{ran}(1+A)$. For this, define $f_A \colon H \times H \to (-\infty, \infty]$ by (note that $A \neq \varnothing$ by maximal monotonicity)

$$f_A(u, v) := \sup \{\operatorname{Re}\langle u, y\rangle + \operatorname{Re}\langle v, x\rangle - \operatorname{Re}\langle x, y\rangle \;;\; (x, y) \in A\}.$$

As a supremum of affine functions, we see that f_A is convex and l.s.c. Moreover, we have that

$$\begin{aligned} f_A(u, v) &= -\inf\{-\operatorname{Re}\langle u, y\rangle - \operatorname{Re}\langle v, x\rangle + \operatorname{Re}\langle x, y\rangle \;;\; (x, y) \in A\} \\ &= -\inf\{\operatorname{Re}\langle x - u, y - v\rangle \;;\; (x, y) \in A\} + \operatorname{Re}\langle u, v\rangle \end{aligned}$$

for each $u, v \in H$ and since A is maximal monotone, we get by using Remark 17.1.1

$$\begin{aligned} \inf\{\operatorname{Re}\langle x - u, y - v\rangle \;;\; (x, y) \in A\} \geqslant 0 &\Leftrightarrow (u, v) \in A \\ &\Leftrightarrow \inf\{\operatorname{Re}\langle x - u, y - v\rangle \;;\; (x, y) \in A\} = 0 \end{aligned}$$

and so

$$\inf\{\operatorname{Re}\langle x - u, y - v\rangle \;;\; (x, y) \in A\} \leqslant 0 \quad (u, v \in H).$$

In particular, we get $f_A(u, v) \geqslant \operatorname{Re}\langle u, v\rangle$ for each $u, v \in H$ and $f_A(u, v) = \operatorname{Re}\langle u, v\rangle$ if and only if $(u, v) \in A$. Thus, f_A is proper since $A \neq \varnothing$. By Proposition 17.1.6(d) we obtain that $0 \in \operatorname{ran}(1 + \partial f_A)$ and thus, we find $(u_0, v_0) \in H \times H$ with $((u_0, v_0), (-u_0, -v_0)) \in \partial f_A$. Hence, by definition of ∂f_A,

$$\begin{aligned} f_A(u, v) &\geqslant f_A(u_0, v_0) + \operatorname{Re}\langle(-u_0, -v_0), (u - u_0, v - v_0)\rangle \\ &= f_A(u_0, v_0) + \|u_0\|^2 + \|v_0\|^2 - \operatorname{Re}\langle u_0, u\rangle - \operatorname{Re}\langle v_0, v\rangle \end{aligned}$$

for all $(u, v) \in H \times H$. In particular, using that $f_A(u, v) = \operatorname{Re}\langle u, v\rangle$ for $(u, v) \in A$ we get

$$0 \geqslant f_A(u_0, v_0) + \|u_0\|^2 + \|v_0\|^2 - \operatorname{Re}\langle u_0, u\rangle - \operatorname{Re}\langle v_0, v\rangle - \operatorname{Re}\langle u, v\rangle \quad ((u, v) \in A).$$

Taking the supremum over all $(u, v) \in A$, we infer

$$\begin{aligned}0 &\geqslant f_A(u_0, v_0) + \|u_0\|^2 + \|v_0\|^2 + f_A(-u_0, -v_0),\\ &\geqslant \operatorname{Re}\langle u_0, v_0\rangle + \|u_0\|^2 + \|v_0\|^2 + \operatorname{Re}\langle -u_0, -v_0\rangle = \|u_0 + v_0\|^2\end{aligned}$$

Thus, $u_0 + v_0 = 0$ and instead of inequalities, we actually have equalities in the expression above. Thus, $f_A(u_0, v_0) = \operatorname{Re}\langle u_0, v_0\rangle$ and so, $(u_0, v_0) \in A$. From $u_0 + v_0 = 0$ it thus follows that $0 \in \operatorname{ran}(1 + A)$. □

Next, we show how to extend maximal monotone relations on a Hilbert space H to the Bochner–Lebesgue space $L_2(\mu; H)$ for a σ-finite measure space $(\Omega, \mathcal{A}, \mu)$. The condition $(0, 0) \in A$ can be dropped if $\mu(\Omega) < \infty$.

Corollary 17.1.8 *Let* $A \subseteq H \times H$ *maximal monotone with* $(0, 0) \in A$. *Moreover, let* $(\Omega, \mathcal{A}, \mu)$ *be a* σ*-finite measure space and define*

$$A_{L_2(\mu;H)} := \{(f, g) \in L_2(\mu; H) \times L_2(\mu; H)\,;\ (f(t), g(t)) \in A \quad (t \in \Omega \text{ a.e.})\}\,.$$

Then $A_{L_2(\mu;H)}$ *is maximal monotone.*

Proof The monotonicity of $A_{L_2(\mu;H)}$ is clear. For showing the maximal monotonicity we prove that $1 + A_{L_2(\mu;H)}$ is onto (see Proposition 17.1.2(b)). For this, let $h \in L_2(\mu; H)$ and set $f(t) := (1 + A)^{-1}(h(t))$ for each $t \in \Omega$. Note that f is well-defined by Theorem 17.1.7. Since $(1 + A)^{-1}$ is continuous by Proposition 17.1.2(a) and h is Bochner-measurable, f is also Bochner-measurable. Moreover, using that $(0, 0) \in 1 + A$ and $\left\|(1 + A)^{-1}\right\|_{\mathrm{Lip}} \leqslant 1$, we compute

$$\int_\Omega \|f(t)\|^2 \,\mathrm{d}\mu(t) \leqslant \int_\Omega \|h(t)\|^2 \,\mathrm{d}\mu(t) < \infty$$

and so, $f \in L_2(\mu; H)$. Thus, $h - f \in L_2(\mu; H)$, which yields $(f, h - f) \in A_{L_2(\mu;H)}$ and so, $h \in \operatorname{ran}(1 + A_{L_2(\mu;H)})$. □

17.2 The Yosida Approximation and Perturbation Results

We now have all concepts at hand to introduce the Yosida approximation for a maximal monotone relation.

Definition Let $A \subseteq H \times H$ be maximal monotone and $\lambda > 0$. We define

$$A_\lambda := \lambda^{-1}\left(1 - (1 + \lambda A)^{-1}\right).$$

The family $(A_\lambda)_{\lambda>0}$ is called *Yosida approximation of* A.

Since for a maximal monotone relation $A \subseteq H \times H$ the resolvent $(1+\lambda A)^{-1}$ is actually a Lipschitz-continuous mapping (by Proposition 17.1.2(a)), whose domain is H (by Theorem 17.1.7), the same holds for A_λ. We collect some useful properties of the Yosida approximation.

Proposition 17.2.1 *Let $A \subseteq H \times H$ maximal monotone and $\lambda > 0$. Then the following statements hold:*

(a) *For all $x \in H$ we have $\big((1+\lambda A)^{-1}(x), A_\lambda(x)\big) \in A$.*
(b) *A_λ is monotone and $\|A_\lambda\|_{\mathrm{Lip}} \leqslant \frac{1}{\lambda}$.*

Proof

(a) For all $x \in H$ we have that $\big((1+\lambda A)^{-1}(x), x\big) \in 1+\lambda A$, and therefore, $\big((1+\lambda A)^{-1}(x), A_\lambda(x)\big) \in A$.
(b) Let $x, y \in H$. Then we compute

$$\begin{aligned}
&\lambda \operatorname{Re}\langle A_\lambda(x) - A_\lambda(y), x - y\rangle \\
&\quad = \|x-y\|^2 - \operatorname{Re}\Big\langle (1+\lambda A)^{-1}(x) - (1+\lambda A)^{-1}(y), x - y\Big\rangle \\
&\quad \geqslant \|x-y\|^2 - \Big\|(1+\lambda A)^{-1}(x) - (1+\lambda A)^{-1}(y)\Big\| \, \|x-y\| \\
&\quad \geqslant 0
\end{aligned}$$

by Proposition 17.1.2(a) and hence, A_λ is monotone. Moreover,

$$\begin{aligned}
&\operatorname{Re}\langle A_\lambda(x) - A_\lambda(y), x - y\rangle \\
&\quad = \operatorname{Re}\Big\langle A_\lambda(x) - A_\lambda(y), (1+\lambda A)^{-1}(x) - (1+\lambda A)^{-1}(y)\Big\rangle \\
&\qquad + \lambda \, \|A_\lambda(x) - A_\lambda(y)\|^2 \\
&\quad \geqslant \lambda \, \|A_\lambda(x) - A_\lambda(y)\|^2 ,
\end{aligned}$$

where we have used (a) and the monotonicity of A. The Cauchy–Schwarz inequality now yields $\|A_\lambda\|_{\mathrm{Lip}} \leqslant \frac{1}{\lambda}$. □

We state a result on the strong convergence of the resolvents of a maximal monotone relation, which we already have used in previous sections for the resolvent of $\partial_{t,\nu}$. For the projection $P_C(x)$ of $x \in H$ onto a non-empty closed convex set $C \subseteq H$, recall Exercise 4.4 and that $y = P_C(x)$ if and only if $y \in C$ and

$$\operatorname{Re}\langle x - y, u - y\rangle_H \leqslant 0 \quad (u \in C).$$

Proposition 17.2.2 *Let $A \subseteq H \times H$ be maximal monotone. Then $\overline{\operatorname{dom}(A)}$ is convex and for all $x \in H$ we have $(1+\lambda A)^{-1}(x) \to P_{\overline{\operatorname{dom}(A)}}(x)$ as $\lambda \to 0+$, where $P_{\overline{\operatorname{dom}(A)}}$ denotes the projection onto $\overline{\operatorname{dom}(A)}$.*

Proof We set $C := \overline{\operatorname{conv} \operatorname{dom}(A)}$. Then C is closed and convex. Next, we prove that $(1+\lambda A)^{-1}(x) \to P_C(x)$ as $\lambda \to 0+$ for all $x \in H$. So let $x \in H$ and set $x_\lambda := (1+\lambda A)^{-1}(x)$ for each $\lambda > 0$. Then we have $A_\lambda(x) = \frac{1}{\lambda}(x - x_\lambda)$ and hence, using Proposition 17.2.1(a) and the monotonicity of A, we infer $\operatorname{Re}\left\langle x_\lambda - u, \frac{1}{\lambda}(x - x_\lambda) - v\right\rangle \geqslant 0$ for each $(u, v) \in A$. Consequently, we obtain

$$\|x_\lambda\|^2 \leqslant \operatorname{Re}\langle x_\lambda - u, x\rangle + \operatorname{Re}\langle x_\lambda, u\rangle - \lambda \operatorname{Re}\langle x_\lambda - u, v\rangle \quad ((u, v) \in A). \qquad (17.2)$$

In particular, we see that $(x_\lambda)_{\lambda>0}$ is bounded as $\lambda \to 0$ and so, for each nullsequence we find a subsequence $(\lambda_n)_n$ with $\lambda_n \to 0$ such that $x_{\lambda_n} \to z$ weakly for some $z \in H$. By (17.2) it follows that

$$\|z\|^2 \leqslant \operatorname{Re}\langle z - u, x\rangle + \operatorname{Re}\langle z, u\rangle \quad (u \in \operatorname{dom}(A)).$$

It is easy to see that this inequality carries over to each $u \in C$ and thus $\operatorname{Re}\langle z - u, z - x\rangle \leqslant 0$ for each $u \in C$ which proves $z = P_C(x)$ and hence, $x_{\lambda_n} \to P_C(x)$ weakly. Next we prove that the convergence also holds in the norm topology. From (17.2) we see that

$$\limsup_{n\to\infty} \left\|x_{\lambda_n}\right\|^2 \leqslant \operatorname{Re}\langle P_C(x) - u, x\rangle + \operatorname{Re}\langle P_C(x), u\rangle \quad (u \in \operatorname{dom}(A))$$

and again, this inequality stays true for each $u \in C$. In particular, choosing $u = P_C(x)$ we infer $\limsup_{n\to\infty} \left\|x_{\lambda_n}\right\|^2 \leqslant \|P_C(x)\|^2$, which together with the weak convergence, yields the convergence in norm (see Exercise 17.3). A subsequence argument (cf. Exercise 14.3) reveals $x_\lambda \to P_C(x)$ in H as $\lambda \to 0$.
It remains to show that $\overline{\operatorname{dom}(A)}$ is convex. By what we have shown above, we have $(1+\lambda A)^{-1}(x) \to x$ as $\lambda \to 0$ for each $x \in C$ and since $(1+\lambda A)^{-1}(x) \in \operatorname{dom}(A)$ for each $\lambda > 0$, we infer $x \in \overline{\operatorname{dom}(A)}$. Thus, $C \subseteq \overline{\operatorname{dom}(A)}$ and since the other inclusion holds trivially the proof is completed. □

We conclude this section with some perturbation results.

Lemma 17.2.3 *Let $A \subseteq H \times H$ be maximal monotone and $C \colon H \to H$ Lipschitz-continuous and monotone. Then $A + C$ is maximal monotone.*

Proof The monotonicity of $A + C$ is clear. If C is constant, then the maximality of $A + C$ is obvious. If C is non-constant we choose $0 < \lambda < \frac{1}{\|C\|_{\mathrm{Lip}}}$. Then for all $f \in H$ the mapping

$$u \mapsto (1+\lambda A)^{-1}(f - \lambda C(u))$$

defines a strict contraction (use Proposition 17.1.2(a) and $\operatorname{dom}((1+\lambda A)^{-1}) = H$ by Theorem 17.1.7) and thus, posseses a fixed point $x \in H$, which then satisfies $(x, f) \in 1+\lambda(A+C)$. Thus, $A+C$ is maximal monotone by Proposition 17.1.2(b). □

We note that the latter lemma particularily applies to $C = B_\lambda$ for a maximal monotone relation $B \subseteq H \times H$ and $\lambda > 0$ by Proposition 17.2.1(b).

Proposition 17.2.4 *Let $A, B \subseteq H \times H$ be two maximal monotone relations, $c > 0$ and $f \in H$. For $\lambda > 0$ we set*

$$x_\lambda := (c + A + B_\lambda)^{-1}(f).$$

Then $f \in \operatorname{ran}(c + A + B)$ if and only if $\sup_{\lambda>0} \|B_\lambda(x_\lambda)\| < \infty$ and in the latter case $x_\lambda \to x$ as $\lambda \to 0$ with $(x, f) \in c + A + B$, which identifies x uniquely.

Proof Note that x_λ is well-defined for $\lambda > 0$ by Lemma 17.2.3, Theorem 17.1.7 and Proposition 17.1.2.

For all $\lambda > 0$ we find $y_\lambda \in H$ such that $(x_\lambda, y_\lambda) \in A$ and $cx_\lambda + y_\lambda + B_\lambda(x_\lambda) = f$.

We first assume that there exist $x, y, z \in H$ such that $(x, y) \in A$, $(x, z) \in B$ and $cx + y + z = f$. Thus, we have

$$c(x - x_\lambda) = y_\lambda + B_\lambda(x_\lambda) - y - z,$$

which gives

$$\begin{aligned}
0 \leqslant c\,\|x_\lambda - x\|^2 &= \operatorname{Re}\langle y - y_\lambda, x_\lambda - x\rangle + \operatorname{Re}\langle z - B_\lambda(x_\lambda), x_\lambda - x\rangle \\
&\leqslant \operatorname{Re}\langle z - B_\lambda(x_\lambda), x_\lambda - x\rangle \\
&= \operatorname{Re}\left\langle z - B_\lambda(x_\lambda), (1+\lambda B)^{-1}(x_\lambda) - x\right\rangle + \operatorname{Re}\langle z - B_\lambda(x_\lambda), \lambda B_\lambda(x_\lambda)\rangle \\
&\leqslant \operatorname{Re}\langle z - B_\lambda(x_\lambda), \lambda B_\lambda(x_\lambda)\rangle
\end{aligned}$$

where we have used the monotonicity of A in the second line and the monotonicity of B as well as Proposition 17.2.1(a) in the last line. The latter implies

$$\|B_\lambda(x_\lambda)\|^2 \leqslant \operatorname{Re}\langle z, B_\lambda(x_\lambda)\rangle\,,$$

and the claim follows by the Cauchy–Schwarz inequality.

Assume now that $K := \sup_{\lambda>0} \|B_\lambda(x_\lambda)\| < \infty$ and let $\mu, \lambda > 0$. As above, we compute

$$\begin{aligned}
c\,\|x_\lambda - x_\mu\|^2 &= \operatorname{Re}\left\langle y_\mu - y_\lambda, x_\lambda - x_\mu\right\rangle + \operatorname{Re}\left\langle B_\mu(x_\mu) - B_\lambda(x_\lambda), x_\lambda - x_\mu\right\rangle \\
&\leqslant \operatorname{Re}\left\langle B_\mu(x_\mu) - B_\lambda(x_\lambda), x_\lambda - x_\mu\right\rangle
\end{aligned}$$

$$
\begin{aligned}
&= \operatorname{Re}\Big\langle B_\mu(x_\mu) - B_\lambda(x_\lambda), (1+\lambda B)^{-1}(x_\lambda) - (1+\mu B)^{-1}(x_\mu)\Big\rangle \\
&\quad + \operatorname{Re}\big\langle B_\mu(x_\mu) - B_\lambda(x_\lambda), \lambda B_\lambda(x_\lambda) - \mu B_\mu(x_\mu)\big\rangle \\
&\leqslant \operatorname{Re}\big\langle B_\mu(x_\mu) - B_\lambda(x_\lambda), \lambda B_\lambda(x_\lambda) - \mu B_\mu(x_\mu)\big\rangle \\
&\leqslant 2(\lambda+\mu)K^2.
\end{aligned}
$$

Thus, for a nullsequence $(\lambda_n)_{n\in\mathbb{N}}$ in $(0,\infty)$ we infer that $(x_{\lambda_n})_{n\in\mathbb{N}}$ is a Cauchy sequence whose limit we denote by x. Since $(B_{\lambda_n}(x_{\lambda_n}))_{n\in\mathbb{N}}$ is bounded, we can assume, by passing to a suitable subsequence, that $B_{\lambda_n}(x_{\lambda_n}) \to z$ weakly for some $z \in H$. Then

$$
\left\|(1+\lambda_n B)^{-1}(x_{\lambda_n}) - x\right\| \leqslant \left\|x_{\lambda_n} - x\right\| + \left\|\lambda_n B_{\lambda_n}(x_{\lambda_n})\right\| \to 0 \quad (n\to\infty)
$$

and since $((1+\lambda_n B)^{-1}(x_{\lambda_n}), B_{\lambda_n}(x_{\lambda_n})) \in B$ for each $n \in \mathbb{N}$ by Proposition 17.2.1(a), the demi-closedness of B (see Remark 17.1.1) reveals $(x,z) \in B$. Moreover,

$$
y_{\lambda_n} = f - B_{\lambda_n}(x_{\lambda_n}) - c x_{\lambda_n} \to f - z - cx =: y \quad (n\to\infty)
$$

weakly and hence, by the demi-closedness of A, we infer $(x,y) \in A$, which completes the proof of the asserted equivalence. By a subsequence argument (cf. Exercise 14.3) we obtain the asserted convergence (note that $x = (c+A+B)^{-1}(f)$ is uniquely determined by f). □

To treat the example in Sect. 17.4 we need another perturbation result, for which we need to introduce the notion of local boundedness of a relation.

Definition Let $A \subseteq H \times H$ and $x \in \operatorname{dom}(A)$. Then A is called *locally bounded at* x if there exists $\delta > 0$ such that

$$
A[B(x,\delta)] = \{y \in H\,;\, \exists z \in B(x,\delta) : (z,y) \in A\}
$$

is bounded.

Proposition 17.2.5 *Let* $A \subseteq H \times H$ *be maximal monotone such that* $\operatorname{int}\operatorname{conv}\operatorname{dom}(A) \neq \varnothing$. *Then* $\operatorname{int}\operatorname{dom}(A) = \operatorname{int}\operatorname{conv}\operatorname{dom}(A) = \operatorname{int}\overline{\operatorname{dom}(A)}$ *and* A *is locally bounded at each point* $x \in \operatorname{int}\operatorname{dom}(A)$.

In order to prove this proposition, we need the following lemma.

Lemma 17.2.6 *Let* $(D_n)_{n\in\mathbb{N}}$ *be a sequence of subsets of* H *with* $D_n \subseteq D_{n+1}$ *for each* $n \in \mathbb{N}$ *and* $D := \bigcup_{n\in\mathbb{N}} D_n$. *If* $\operatorname{int}\operatorname{conv} D \neq \varnothing$, *then* $\operatorname{int}\operatorname{conv} D = \bigcup_{n\in\mathbb{N}} \operatorname{int}\overline{\operatorname{conv} D_n}$.

Proof Set $C := \operatorname{int}\operatorname{conv} D$. By Exercise 17.4 we have $\overline{C} = \overline{\operatorname{conv} D}$. Since $(D_n)_{n\in\mathbb{N}}$ is increasing we have $\operatorname{conv} D = \bigcup_{n\in\mathbb{N}} \operatorname{conv} D_n$ and hence, $C \subseteq \bigcup_{n\in\mathbb{N}} \overline{\operatorname{conv} D_n} \subseteq \overline{C}$. Since C is a Baire space by Exercise 17.5, we find $n_0 \in \mathbb{N}$ such that $\operatorname{int}\overline{\operatorname{conv} D_{n_0}} \neq \varnothing$ and hence, $\operatorname{int}\overline{\operatorname{conv} D_n} \neq \varnothing$ for each $n \geqslant n_0$. Hence, $\overline{\operatorname{conv} D_n} = \overline{\operatorname{int}\overline{\operatorname{conv} D_n}}$ for each $n \geqslant n_0$ by Exercise 17.4. Thus,

$$\overline{C} = \overline{\bigcup_{n\in\mathbb{N}} \overline{\operatorname{conv} D_n}} = \overline{\bigcup_{n\in\mathbb{N}} \overline{\operatorname{int}\overline{\operatorname{conv} D_n}}} = \overline{\bigcup_{n\in\mathbb{N}} \operatorname{int}\overline{\operatorname{conv} D_n}}.$$

Finally, since $\bigcup_{n\in\mathbb{N}} \operatorname{int}\overline{\operatorname{conv} D_n}$ is open and convex, we infer $C = \bigcup_{n\in\mathbb{N}} \operatorname{int}\overline{\operatorname{conv} D_n}$ by Exercise 17.4. □

Proof of Proposition 17.2.5 We first show that A is locally bounded at each point in $\operatorname{int}\operatorname{conv}\operatorname{dom}(A)$. For this, we set

$$A_n := \{(x, y) \in A\,;\ \|x\|, \|y\| \leqslant n\} \quad (n \in \mathbb{N}).$$

Then $\operatorname{dom}(A) = \bigcup_{n\in\mathbb{N}} \operatorname{dom}(A_n)$ and $\operatorname{dom}(A_n) \subseteq \operatorname{dom}(A_{n+1})$ for each $n \in \mathbb{N}$. Since $\operatorname{int}\operatorname{conv}\operatorname{dom}(A) \neq \varnothing$, Lemma 17.2.6 gives $\operatorname{int}\operatorname{conv}\operatorname{dom}(A) = \bigcup_{n\in\mathbb{N}} \operatorname{int}\overline{\operatorname{conv}\operatorname{dom}(A_n)}$. Thus, it suffices to show that A is locally bounded at each $x \in \operatorname{int}\overline{\operatorname{conv}\operatorname{dom}(A_n)}$ for each $n \in \mathbb{N}$. So, let $x \in \operatorname{int}\overline{\operatorname{conv}\operatorname{dom}(A_n)}$ for some $n \in \mathbb{N}$. Then we find $\delta > 0$ such that $B[x, \delta] \subseteq \overline{\operatorname{conv}\operatorname{dom}(A_n)}$. We show that $A[B(x, \frac{\delta}{2})]$ is bounded. So, let $(u, v) \in A$ with $\|u - x\| < \frac{\delta}{2}$ and note that $u \in \overline{\operatorname{conv}\operatorname{dom}(A_n)} \subseteq B[0, n]$. Then for each $(a, b) \in A_n$ we have $\operatorname{Re}\langle u - a, v - b\rangle \geqslant 0$ and thus

$$\begin{aligned}\operatorname{Re}\langle a - u, v\rangle &= \operatorname{Re}\langle a - u, v - b\rangle + \operatorname{Re}\langle a - u, b\rangle \\ &\leqslant \operatorname{Re}\langle a - u, b\rangle \leqslant 2n^2 \quad (a \in \operatorname{dom}(A_n)).\end{aligned}$$

Clearly, this inequality carries over to each $a \in \overline{\operatorname{conv}\operatorname{dom}(A_n)}$. If $v \neq 0$ we choose $a := \frac{\delta}{2\|v\|} v + u \in B[u, \frac{\delta}{2}] \subseteq B[x, \delta] \subseteq \overline{\operatorname{conv}\operatorname{dom}(A_n)}$, and obtain

$$\|v\| \leqslant \frac{4n^2}{\delta},$$

which shows the boundedness of $A[B(x, \frac{\delta}{2})]$.
To complete the proof we need to show that $\operatorname{int}\operatorname{dom}(A) = \operatorname{int}\operatorname{conv}\operatorname{dom}(A) = \operatorname{int}\overline{\operatorname{dom}(A)}$. First we note that $\overline{\operatorname{dom}(A)}$ is convex by Proposition 17.2.2 and hence, $\overline{\operatorname{conv}\operatorname{dom}(A)} = \overline{\operatorname{dom}(A)}$. Now Exercise 17.4(b) gives

$$\operatorname{int}\overline{\operatorname{dom}(A)} = \operatorname{int}\overline{\operatorname{conv}\operatorname{dom}(A)} = \operatorname{int}\operatorname{conv}\operatorname{dom}(A).$$

To show the missing equality it suffices to prove that $\operatorname{int}\operatorname{conv}\operatorname{dom}(A) \subseteq \operatorname{dom}(A)$. So, let $x \in \operatorname{int}\operatorname{conv}\operatorname{dom}(A)$. Then $x \in \overline{\operatorname{dom}(A)}$ and hence, we find a sequence

$((x_n, y_n))_{n\in\mathbb{N}}$ in A with $x_n \to x$. Since A is locally bounded at x, the sequence $(y_n)_{n\in\mathbb{N}}$ is bounded and hence, we can assume without loss of generality that $y_n \to y$ weakly for some $y \in H$. The demi-closedness of A (see Remark 17.1.1) yields $(x, y) \in A$ and thus, $x \in \operatorname{dom}(A)$. □

Now we can prove the following perturbation result.

Theorem 17.2.7 *Let $A, B \subseteq H \times H$ be maximal monotone, $\big(\operatorname{int}\operatorname{dom}(A)\big) \cap \operatorname{dom}(B) \neq \varnothing$. Then $A + B$ is maximal monotone.*

Proof By shifting A and B, we can assume without loss of generality that $(0, 0) \in A \cap B$ and $0 \in (\operatorname{int}\operatorname{dom}(A)) \cap \operatorname{dom}(B)$. We need to prove that $\operatorname{ran}(1+A+B) = H$. So, let $y \in H$ and set

$$x_\lambda := (1 + A + B_\lambda)^{-1}(y) \quad (\lambda > 0).$$

Since $(0, 0) \in A \cap B_\lambda$ and $\left\|(1 + A + B_\lambda)^{-1}\right\|_{\mathrm{Lip}} \leqslant 1$, we infer that $\|x_\lambda\| \leqslant \|y\|$ for each $\lambda > 0$. For showing $y \in \operatorname{ran}(1 + A + B)$ we need to prove that $\sup_{\lambda>0} \|B_\lambda(x_\lambda)\| < \infty$ by Proposition 17.2.4. By definition we find $y_\lambda \in H$ such that $(x_\lambda, y_\lambda) \in A$ and $y = x_\lambda + y_\lambda + B_\lambda(x_\lambda)$ for each $\lambda > 0$. Since A is locally bounded at $0 \in \operatorname{int}\operatorname{dom}(A)$ by Proposition 17.2.5 we find $R, \delta > 0$ with $B(0, \delta) \subseteq \operatorname{dom}(A)$ and

$$\forall (u, v) \in A : \|u\| < \delta \Rightarrow \|v\| \leqslant R.$$

For $\lambda > 0$ we define $u_\lambda := \frac{\delta}{2\|y_\lambda\|} y_\lambda$ if $y_\lambda \neq 0$ and $u_\lambda := 0$ if $y_\lambda = 0$. Then $\|u_\lambda\| \leqslant \frac{\delta}{2} < \delta$ and thus, $u_\lambda \in \operatorname{dom}(A)$. Hence, there exist $v_\lambda \in H$ with $(u_\lambda, v_\lambda) \in A$ and $\|v_\lambda\| \leqslant R$ for each $\lambda > 0$. The monotonicity of A then yields

$$\begin{aligned}
0 &\leqslant \operatorname{Re}\langle y_\lambda - v_\lambda, x_\lambda - u_\lambda\rangle \\
&= \operatorname{Re}\langle y_\lambda, x_\lambda\rangle - \operatorname{Re}\langle v_\lambda, x_\lambda\rangle - \operatorname{Re}\langle y_\lambda, u_\lambda\rangle + \operatorname{Re}\langle v_\lambda, u_\lambda\rangle \\
&\leqslant \operatorname{Re}\langle y - x_\lambda - B_\lambda(x_\lambda), x_\lambda\rangle - \operatorname{Re}\langle y_\lambda, u_\lambda\rangle + R\|y\| + \frac{\delta}{2}R \\
&\leqslant \operatorname{Re}\langle y, x_\lambda\rangle - \operatorname{Re}\langle y_\lambda, u_\lambda\rangle + R\|y\| + \frac{\delta}{2}R \\
&\leqslant \|y\|^2 - \operatorname{Re}\langle y_\lambda, u_\lambda\rangle + R\|y\| + \frac{\delta}{2}R,
\end{aligned}$$

where we have used the monotonicity of B_λ and $B_\lambda(0) = 0$ in the fourth line. Hence, we obtain

$$\frac{\delta}{2}\|y_\lambda\| = \operatorname{Re}\langle y_\lambda, u_\lambda\rangle \leqslant \|y\|^2 + R\|y\| + \frac{\delta}{2}R,$$

which shows that $(y_\lambda)_{\lambda>0}$ is bounded and thus, also $\sup_{\lambda>0} \|B_\lambda(x_\lambda)\| < \infty$. □

17.3 A Solution Theory for Evolutionary Inclusions

In this section we provide a solution theory for evolutionary inclusions by generalising Picard's theorem (see Theorem 6.2.1) to the following situation.

Throughout, we assume that $A \subseteq H \times H$ is a maximal monotone relation with $(0, 0) \in A$. Moreover, let $M\colon \operatorname{dom}(M) \subseteq \mathbb{C} \to L(H)$ be a material law satisfying the usual positive definiteness constraint

$$\exists \nu_0 \geqslant s_b(M), c > 0 \forall z \in \mathbb{C}_{\mathrm{Re} \geqslant \nu_0}, \phi \in H : \mathrm{Re}\langle \phi, zM(z)\phi \rangle \geqslant c \|\phi\|^2 .$$

Then for $\nu \geqslant \max\{\nu_0, 0\}$, $\nu \neq 0$, we consider *evolutionary inclusions* of the form

$$(u, f) \in \overline{\partial_{t,\nu} M(\partial_{t,\nu}) + A_{L_{2,\nu}(\mathbb{R};H)}}, \tag{17.3}$$

where $A_{L_{2,\nu}(\mathbb{R};H)}$ is defined as in Corollary 17.1.8. The solution theory for this kind of problems is as follows.

Theorem 17.3.1 *Let $\nu \geqslant \max\{\nu_0, 0\}$, $\nu \neq 0$. Then the inverse relation $S_\nu := \left(\overline{\partial_{t,\nu} M(\partial_{t,\nu}) + A_{L_{2,\nu}(\mathbb{R};H)}}\right)^{-1}$ is a Lipschitz-continuous mapping, $\operatorname{dom}(S_\nu) = L_{2,\nu}(\mathbb{R}; H)$ and $\|S_\nu\|_{\mathrm{Lip}} \leqslant \frac{1}{c}$. Moreover, the solution mapping S_ν is causal and independent of ν in the sense that $S_\nu(f) = S_\mu(f)$ for each $f \in L_{2,\nu}(\mathbb{R}; H) \cap L_{2,\mu}(\mathbb{R}; H)$ and $\mu \geqslant \nu \geqslant \max\{\nu_0, 0\}$, $\nu \neq 0$.*

In order to prove this theorem, we need some prerequisites. We start with an estimate, which will give us the uniqueness of the solution as well as the causality of the solution mapping S_ν.

Proposition 17.3.2 *Let $\nu \geqslant \max\{\nu_0, 0\}$, $\nu \neq 0$, and*

$$(u, f), (x, g) \in \overline{\partial_{t,\nu} M(\partial_{t,\nu}) + A_{L_{2,\nu}(\mathbb{R};H)}}.$$

Then for all $a \in \mathbb{R}$

$$\left\| \mathbb{1}_{(-\infty,a]}(u - x) \right\|_{L_{2,\nu}} \leqslant \frac{1}{c} \left\| \mathbb{1}_{(-\infty,a]}(f - g) \right\|_{L_{2,\nu}} .$$

Proof By definition, we find sequences $((u_n, f_n))_{n\in\mathbb{N}}$ and $((x_n, g_n))_{n\in\mathbb{N}}$ in $\partial_{t,\nu} M(\partial_{t,\nu}) + A_{L_{2,\nu}(\mathbb{R};H)}$ such that $u_n \to u, x_n \to x, f_n \to f$ and $g_n \to g$ as $n \to \infty$. In particular, for each $n \in \mathbb{N}$ we find $v_n, y_n \in L_{2,\nu}(\mathbb{R}; H)$ such that $(u_n, v_n), (x_n, y_n) \in A_{L_{2,\nu}(\mathbb{R};H)}$ and

$$\partial_{t,\nu} M(\partial_{t,\nu}) u_n + v_n = f_n,$$

$$\partial_{t,\nu} M(\partial_{t,\nu}) x_n + y_n = g_n.$$

Since $(0,0) \in A$, we infer $(\mathbb{1}_{(-\infty,a]}u_n, \mathbb{1}_{(-\infty,a]}v_n), (\mathbb{1}_{(-\infty,a]}x_n, \mathbb{1}_{(-\infty,a]}y_n) \in A_{L_{2,\nu}(\mathbb{R};H)}$ and hence, we may estimate

$$\begin{aligned}
&\operatorname{Re}\left\langle \mathbb{1}_{(-\infty,a]}(f_n - g_n), u_n - x_n \right\rangle \\
&\quad = \operatorname{Re}\left\langle \mathbb{1}_{(-\infty,a]}\partial_{t,\nu}M(\partial_{t,\nu})(u_n - x_n), u_n - x_n \right\rangle \\
&\qquad + \operatorname{Re}\left\langle \mathbb{1}_{(-\infty,a]}v_n - \mathbb{1}_{(-\infty,a]}y_n, \mathbb{1}_{(-\infty,a]}u_n - \mathbb{1}_{(-\infty,a]}x_n \right\rangle \\
&\quad \geqslant \operatorname{Re}\left\langle \mathbb{1}_{(-\infty,a]}\partial_{t,\nu}M(\partial_{t,\nu})(u_n - x_n), u_n - x_n \right\rangle,
\end{aligned}$$

where we used Corollary 17.1.8. Moreover, since $z \mapsto (zM(z))^{-1}$ is a material law, $(\partial_{t,\nu}M(\partial_{t,\nu}))^{-1}$ is causal. By Proposition 16.2.3, for $\phi \in \operatorname{dom}(\partial_{t,\nu}M(\partial_{t,\nu}))$ we have $\operatorname{Re}\left\langle \mathbb{1}_{(-\infty,a]}\partial_{t,\nu}M(\partial_{t,\nu})\phi, \phi \right\rangle \geqslant c \left\| \mathbb{1}_{(-\infty,a]}\phi \right\|^2$. Thus, we end up with

$$\operatorname{Re}\left\langle \mathbb{1}_{(-\infty,a]}(f_n - g_n), u_n - x_n \right\rangle \geqslant c \left\| \mathbb{1}_{(-\infty,a]}(u_n - x_n) \right\|^2,$$

which yields

$$\left\| \mathbb{1}_{(-\infty,a]}(u_n - x_n) \right\| \leqslant \frac{1}{c} \left\| \mathbb{1}_{(-\infty,a]}(f_n - g_n) \right\|.$$

Letting $n \to \infty$, we derive the assertion. □

Next, we address the existence of a solution for (17.3) for suitable right-hand sides f. For this, we provide another useful characterisation for the weak differentiability of a function in $L_{2,\nu}(\mathbb{R}; H)$.

Lemma 17.3.3 *Let $\nu \in \mathbb{R}$, $u \in L_{2,\nu}(\mathbb{R}; H)$. Then $u \in \operatorname{dom}(\partial_{t,\nu})$ if and only if $\sup_{0<h\leqslant h_0} \frac{1}{h} \|\tau_h u - u\| < \infty$ for some $h_0 > 0$. In either case*

$$\frac{1}{h}(\tau_h u - u) \to \partial_{t,\nu} u \quad (h \to 0)$$

in $L_{2,\nu}(\mathbb{R}; H)$.

Proof For $h > 0$ we consider the operator $D_h \colon L_{2,\nu}(\mathbb{R}; H) \to L_{2,\nu}(\mathbb{R}; H)$ given by $D_h v = \frac{1}{h}(\tau_h v - v)$. If $v \in C_c^1(\mathbb{R}; H)$ we estimate

$$\begin{aligned}
\|D_h v\|^2 &= \int_{\mathbb{R}} \frac{1}{h^2} \|v(t+h) - v(t)\|^2 \mathrm{e}^{-2\nu t}\,\mathrm{d}t = \int_{\mathbb{R}} \frac{1}{h^2} \left\| \int_0^h v'(t+s)\,\mathrm{d}s \right\|^2 \mathrm{e}^{-2\nu t}\,\mathrm{d}t \\
&\leqslant \int_{\mathbb{R}} \frac{1}{h} \int_0^h \left\|v'(t+s)\right\|^2 \mathrm{d}s\, \mathrm{e}^{-2\nu t}\,\mathrm{d}t = \frac{1}{h} \int_0^h \int_{\mathbb{R}} \left\|v'(t+s)\right\|^2 \mathrm{e}^{-2\nu t}\,\mathrm{d}t\,\mathrm{d}s \\
&\leqslant \mathrm{e}^{2\nu h} \left\|v'\right\|^2.
\end{aligned}$$

By density of $C_c^1(\mathbb{R}; H)$ in $H_\nu^1(\mathbb{R}; H)$ we infer that

$$\sup_{0\leqslant h\leqslant 1} \|D_h\|_{L(H_\nu^1(\mathbb{R};H), L_{2,\nu}(\mathbb{R};H))} \leqslant \mathrm{e}^{\nu}.$$

Moreover, for $v \in C_c^1(\mathbb{R}; H)$ it is clear that $D_h v \to v'$ in $L_{2,\nu}(\mathbb{R}; H)$ as $h \to 0$ by dominated convergence. Since $(D_h)_{0 \leqslant h \leqslant 1}$ is uniformly bounded, the convergence carries over to elements in $H_\nu^1(\mathbb{R}; H)$, which proves the first asserted implication and the convergence statement.

Assume now that $\sup_{0<h\leqslant h_0} \frac{1}{h} \|\tau_h u - u\| < \infty$ for some $h_0 > 0$. Choosing a suitable sequence $(h_n)_{n\in\mathbb{N}}$ in $(0, h_0]$ with $h_n \to 0$ as $n \to \infty$, we can assume that $\frac{1}{h_n}(\tau_{h_n} u - u) \to v$ weakly for some $v \in L_{2,\nu}(\mathbb{R}; H)$. Then we compute for each $\phi \in C_c^\infty(\mathbb{R}; H)$

$$
\begin{aligned}
\langle v, \phi \rangle &= \lim_{n\to\infty} \int_{\mathbb{R}} \frac{1}{h_n} \langle u(t + h_n) - u(t), \phi(t) \rangle \, \mathrm{e}^{-2\nu t} \, \mathrm{d}t \\
&= \lim_{n\to\infty} \int_{\mathbb{R}} \frac{1}{h_n} \left\langle u(t), \phi(t - h_n)\mathrm{e}^{2\nu h_n} - \phi(t) \right\rangle \mathrm{e}^{-2\nu t} \, \mathrm{d}t \\
&= \int_{\mathbb{R}} \left\langle u(t), -\phi'(t) + 2\nu\phi(t) \right\rangle \mathrm{e}^{-2\nu t} \, \mathrm{d}t = \left\langle u, \partial_{t,\nu}^* \phi \right\rangle,
\end{aligned}
$$

which—as $C_c^\infty(\mathbb{R}; H)$ is a core for $\partial_{t,\nu}^*$ (see Proposition 3.2.4 and Corollary 3.2.6)—shows $u \in \operatorname{dom}(\partial_{t,\nu}^{**}) = \operatorname{dom}(\partial_{t,\nu})$. □

Proposition 17.3.4 *Let $\nu \geqslant \nu_0$ and $f \in \operatorname{dom}(\partial_{t,\nu})$. Then there exists $u \in \operatorname{dom}(\partial_{t,\nu})$ such that*

$$
(u, f) \in \partial_{t,\nu} M(\partial_{t,\nu}) + A_{L_{2,\nu}(\mathbb{R};H)}.
$$

Proof We recall that $B := \partial_{t,\nu} M(\partial_{t,\nu}) - c$ is maximal monotone by Example 17.1.3. Let $\lambda > 0$ and set

$$
u_\lambda := \left(c + B + \left(A_{L_{2,\nu}(\mathbb{R};H)}\right)_\lambda\right)^{-1}(f) = \left(\partial_{t,\nu} M(\partial_{t,\nu}) + \left(A_{L_{2,\nu}(\mathbb{R};H)}\right)_\lambda\right)^{-1}(f).
$$

We remark that $\left(A_{L_{2,\nu}(\mathbb{R};H)}\right)_\lambda = \left(A_\lambda\right)_{L_{2,\nu}(\mathbb{R};H)}$ (see Exercise 17.6). Hence, we have $\tau_h \left(A_{L_{2,\nu}(\mathbb{R};H)}\right)_\lambda = \left(A_{L_{2,\nu}(\mathbb{R};H)}\right)_\lambda \tau_h$ for each $h > 0$. Thus, we obtain

$$
\tau_h u_\lambda = \left(\partial_{t,\nu} M(\partial_{t,\nu}) + \left(A_{L_{2,\nu}(\mathbb{R};H)}\right)_\lambda\right)^{-1}(\tau_h f)
$$

and so, due to the monotonicity of B and $\left(A_{L_{2,\nu}(\mathbb{R};H)}\right)_\lambda$,

$$
\|\tau_h u_\lambda - u_\lambda\| \leqslant \frac{1}{c} \|\tau_h f - f\| .
$$

Dividing both sides by h and using Lemma 17.3.3, we infer that $u_\lambda \in \operatorname{dom}(\partial_{t,\nu})$ and

$$
\left\|\partial_{t,\nu} u_\lambda\right\| = \lim_{h\to 0} \frac{1}{h} \|\tau_h u_\lambda - u_\lambda\| \leqslant \frac{1}{c} \sup_{0<h\leqslant 1} \frac{1}{h} \|\tau_h f - f\| =: K
$$

and hence,

$$\sup_{\lambda>0} \left\| \left(A_{L_{2,\nu}(\mathbb{R};H)}\right)_{\lambda}(u_{\lambda}) \right\| = \sup_{\lambda>0} \left\| f - \partial_{t,\nu} M(\partial_{t,\nu}) u_{\lambda} \right\| \leqslant \|f\| + K \left\| M(\partial_{t,\nu}) \right\|.$$

Proposition 17.2.4 implies $u_\lambda \to u$ as $\lambda \to 0$ and $(u, f) \in \partial_{t,\nu} M(\partial_{t,\nu}) + A_{L_{2,\nu}(\mathbb{R};H)}$. Moreover, since $(\partial_{t,\nu} u_\lambda)_{\lambda>0}$ is uniformly bounded, we can choose a suitable nullsequence $(\lambda_n)_{n\in\mathbb{N}}$ in $(0, \infty)$ such that $\partial_{t,\nu} u_{\lambda_n} \to v$ weakly for some $v \in L_{2,\nu}(\mathbb{R}; H)$. Since $\partial_{t,\nu}$ is closed and hence, weakly closed (either use $\partial_{t,\nu}^{**} = \partial_{t,\nu}$ or Mazur's theorem [50, Corollary 2.11]) again), we infer that $u \in \operatorname{dom}(\partial_{t,\nu})$. □

We are now in the position to prove Theorem 17.3.1.

Proof of Theorem 17.3.1 Let $\nu \geqslant \nu_0$. Since $\partial_{t,\nu} M(\partial_{t,\nu}) - c$ is monotone (Example 17.1.3), the relation $\partial_{t,\nu} M(\partial_{t,\nu}) + A_{L_{2,\nu}(\mathbb{R};H)} - c$ is monotone and thus, $(\partial_{t,\nu} M(\partial_{t,\nu}) + A_{L_{2,\nu}(\mathbb{R};H)})^{-1}$ defines a Lipschitz-continuous mapping with smallest Lipschitz-constant less than or equal to $\frac{1}{c}$. Since this mapping is densely defined by Proposition 17.3.4, it follows that $S_\nu = \overline{\left(\partial_{t,\nu} M(\partial_{t,\nu}) + A_{L_{2,\nu}(\mathbb{R};H)}\right)^{-1}}$ is Lipschitz-continuous with $\|S_\nu\|_{\mathrm{Lip}} \leqslant \frac{1}{c}$ and $\operatorname{dom}(S_\nu) = L_{2,\nu}(\mathbb{R}; H)$. Moreover, S_ν is causal, since for $f, g \in L_{2,\nu}(\mathbb{R}; H)$ with $\mathbb{1}_{(-\infty,a]} f = \mathbb{1}_{(-\infty,a]} g$ for some $a \in \mathbb{R}$ it follows that $\mathbb{1}_{(-\infty,a]} S_\nu(f) = \mathbb{1}_{(-\infty,a]} S_\nu(g)$ by Proposition 17.3.2. Thus, the only thing left to be shown is the independence of the parameter ν. So, let $f \in L_{2,\nu}(\mathbb{R}; H) \cap L_{2,\mu}(\mathbb{R}; H)$ for some $\nu_0 \leqslant \nu \leqslant \mu$. Then we find a sequence $(\phi_n)_{n\in\mathbb{N}}$ in $C_c^1(\mathbb{R}; H)$ with $\phi_n \to f$ in both $L_{2,\nu}(\mathbb{R}; H)$ and $L_{2,\mu}(\mathbb{R}; H)$. We set $u_n := S_\nu(\phi_n) \in L_{2,\nu}(\mathbb{R}; H)$ and since $0 = S_\nu(0)$, we derive that $\inf \operatorname{spt} u_n \geqslant \inf \operatorname{spt} \phi_n > -\infty$ by Proposition 17.3.2. Thus, $u_n \in L_{2,\mu}(\mathbb{R}; H)$ and since $u_n \in \operatorname{dom}(\partial_{t,\nu})$ by Proposition 17.3.4 and $\operatorname{spt} \partial_{t,\nu} u_n \subseteq \operatorname{spt} u_n$, we infer that also $\partial_{t,\nu} u_n \in L_{2,\mu}(\mathbb{R}; H)$, which shows $u_n \in \operatorname{dom}(\partial_{t,\mu})$ and $\partial_{t,\mu} u_n = \partial_{t,\nu} u_n$ by Exercise 11.1. By Theorem 5.3.6 it follows that

$$\begin{aligned} \partial_{t,\nu} M(\partial_{t,\nu}) u_n &= M(\partial_{t,\nu}) \partial_{t,\nu} u_n = M(\partial_{t,\nu}) \partial_{t,\mu} u_n \\ &= M(\partial_{t,\mu}) \partial_{t,\mu} u_n = \partial_{t,\mu} M(\partial_{t,\mu}) u_n. \end{aligned}$$

Since we have $(u_n, \phi_n - \partial_{t,\nu} M(\partial_{t,\nu}) u_n) \in A_{L_{2,\nu}(\mathbb{R};H)}$ it follows that $(u_n, \phi_n - \partial_{t,\mu} M(\partial_{t,\mu}) u_n) \in A_{L_{2,\mu}(\mathbb{R};H)}$ by the definition of $A_{L_{2,\mu}(\mathbb{R};H)}$ and thus, $u_n = S_\mu(\phi_n)$. Letting $n \to \infty$, we finally derive $S_\mu(f) = S_\nu(f)$. □

17.4 Maxwell's Equations in Polarisable Media

We recall Maxwell's equations from Chap. 6. Let $\Omega \subseteq \mathbb{R}^3$ open. Then the electric field E and the magnetic induction B are linked via Faraday's law

$$\partial_{t,\nu} B + \operatorname{curl}_0 E = 0,$$

where we assume the electric boundary condition for E. Moreover, the electric displacement D, the current j_c and the magnetic field H are linked via Ampère's law

$$\partial_{t,\nu} D + j_c - \operatorname{curl} H = j_0,$$

where j_0 is a given external current. Classically, D and E as well as B and H are linked by the constitutive relations

$$D = \varepsilon E, \text{ and } B = \mu H,$$

where $\varepsilon, \mu \in L(L_2(\Omega)^3)$ model the dielectricity and magnetic permeability, respectively. In a non-polarisable medium, we would additionally assume Ohm's law that links j_c and E by $j_c = \sigma E$ with $\sigma \in L(L_2(\Omega)^3)$. In polarisable media however, this relation is replaced as follows

$$\begin{aligned} \|E\| < E_0 &\Rightarrow j_c = \sigma E \\ \|E\| = E_0 &\Rightarrow \exists \lambda \geqslant 0 : \; j_c = (\sigma + \lambda) E, \end{aligned} \tag{17.4}$$

where $E_0 > 0$ is the called the threshold of ionisation of the underlying medium. The above relation is used to model the following phenomenon: Assume that the medium is not or weakly electrically conductive (i.e., σ is very small) but if the electric field is strong enough (i.e., reaching the threshold E_0), the medium polarises and allows for a current flow proportional to the electric field. Such phenomena occur for instance in certain gases between two capacitor plates, where the gas becomes a conductor if the electric field is strong enough.

Our first goal is to formulate (17.4) in terms of a binary relation. For this, we set

$$B := \left\{ (u, v) \in L_2(\Omega)^3 \times L_2(\Omega)^3 ; \; \|u\| \leqslant E_0, \; \operatorname{Re} \langle u, v \rangle = E_0 \|v\| \right\}.$$

Lemma 17.4.1 *Let $u, v \in L_2(\Omega)^3$. Then $(u, v) \in B$ if and only if*

$$(\|u\| \leqslant E_0) \text{ and } (\|u\| < E_0 \Rightarrow v = 0) \text{ and } (\|u\| = E_0 \Rightarrow \exists \lambda \geqslant 0 : \; v = \lambda u).$$

Proof Assume first that $(u, v) \in B$. Then $\|u\| \leqslant E_0$ by definition. Moreover,

$$E_0 \|v\| = \operatorname{Re} \langle u, v \rangle \leqslant \|u\| \, \|v\|$$

and hence, if $\|u\| < E_0$ it follows that $v = 0$. Moreover, if $\|u\| = E_0$ we have equality and thus, u and v are linearly dependent; that is, we find $\lambda_1, \lambda_2 \in \mathbb{C}$ with $\lambda_1 \lambda_2 \neq 0$ such that $\lambda_1 u + \lambda_2 v = 0$. Note that $\lambda_2 \neq 0$ since $u \neq 0$ and hence, we get

$v = \lambda u$ with $\lambda := -\frac{\lambda_1}{\lambda_2}$. We then have

$$0 \leqslant |\lambda| E_0^2 = \|v\| \, E_0 = \operatorname{Re} \langle u, v \rangle = \operatorname{Re} \lambda \, \|u\|^2 = \operatorname{Re} \lambda \, E_0^2,$$

which shows $0 \leqslant \operatorname{Re} \lambda = |\lambda|$ and thus, $\lambda \geqslant 0$. The other implication is trivial. □

The latter lemma shows that (E, j_c) satisfies (17.4) if and only if $(E, j_c - \sigma E) \in B$, or equivalently $(E, j_c) \in \sigma + B$. Thus, we may reformulate Maxwell's equations in a polarisable medium Ω as follows

$$\left(\begin{pmatrix} E \\ H \end{pmatrix}, \begin{pmatrix} j_0 \\ 0 \end{pmatrix} \right) \in \partial_{t,\nu} \begin{pmatrix} \varepsilon & 0 \\ 0 & \mu \end{pmatrix} + \begin{pmatrix} \sigma & 0 \\ 0 & 0 \end{pmatrix} + \begin{pmatrix} B & -\operatorname{curl} \\ \operatorname{curl}_0 & 0 \end{pmatrix}.$$

To apply our solution theory in Theorem 17.3.1, we need to ensure that

$$A := \begin{pmatrix} B & -\operatorname{curl} \\ \operatorname{curl}_0 & 0 \end{pmatrix} = \begin{pmatrix} B & 0 \\ 0 & 0 \end{pmatrix} + \begin{pmatrix} 0 & -\operatorname{curl} \\ \operatorname{curl}_0 & 0 \end{pmatrix} \tag{17.5}$$

defines a maximal monotone relation on $L_2(\Omega)^6 \times L_2(\Omega)^6$. This will be done by the perturbation result presented in Theorem 17.2.7. We start by showing the maximal monotonicity of B.

Lemma 17.4.2 *We define the function* $I \colon L_2(\Omega)^3 \to (-\infty, \infty]$ *by*

$$I(u) = \begin{cases} 0 & \text{if } \|u\| \leqslant E_0 \\ \infty & \text{otherwise.} \end{cases}$$

Then I is convex, proper and l.s.c. Moreover, $B = \partial I$. In particular, B is maximal monotone.

Proof This is part of Exercise 17.7. □

Proposition 17.4.3 *The relation A given by (17.5) is maximal monotone with $(0, 0) \in A$.*

Proof Since B is maximal monotone by Lemma 17.4.2, it is easy to see that $\begin{pmatrix} B & 0 \\ 0 & 0 \end{pmatrix}$ is maximal monotone, too. Moreover, by definition we see that $0 \in \operatorname{int} \operatorname{dom}(B)$ and thus, $0 \in \operatorname{int} \operatorname{dom} \begin{pmatrix} B & 0 \\ 0 & 0 \end{pmatrix} = \operatorname{int} \operatorname{dom}(B) \times L_2(\Omega)^3$. Since clearly $0 \in \operatorname{dom} \begin{pmatrix} 0 & -\operatorname{curl} \\ \operatorname{curl}_0 & 0 \end{pmatrix}$ and $\begin{pmatrix} 0 & -\operatorname{curl} \\ \operatorname{curl}_0 & 0 \end{pmatrix}$ is maximal monotone (see Example 17.1.3), the assertion follows from Theorem 17.2.7. □

Theorem 17.4.4 *Let $\varepsilon, \mu, \sigma \in L(L_2(\Omega)^3)$ with ε, μ selfadjoint. Moreover, assume there exist $\nu_0, c > 0$ such that*

$$\nu\varepsilon + \operatorname{Re}\sigma \geqslant c \text{ and } \mu \geqslant c \quad (\nu \geqslant \nu_0).$$

Then for each $\nu \geqslant \nu_0$ we have that

$$S_\nu := \left(\overline{\partial_{t,\nu}\begin{pmatrix}\varepsilon & 0\\ 0 & \mu\end{pmatrix} + \begin{pmatrix}\sigma & 0\\ 0 & 0\end{pmatrix} + \begin{pmatrix}B & -\operatorname{curl}\\ \operatorname{curl}_0 & 0\end{pmatrix}_{L_{2,\nu}(\mathbb{R};L_2(\Omega)^6)}}\right)^{-1}$$

is a Lipschitz-continuous mapping with $\operatorname{dom}(S_\nu) = L_{2,\nu}(\mathbb{R}; L_2(\Omega)^6)$ and $\|S_\nu\|_{\mathrm{Lip}} \leqslant \frac{1}{c}$. Moreover, S_ν is causal and independent of ν in the sense that $S_\nu(f) = S_\eta(f)$ whenever $\nu, \eta \geqslant \nu_0$ and $f \in L_{2,\nu}(\mathbb{R}; L_2(\Omega)^6) \cap L_{2,\eta}(\mathbb{R}; L_2(\Omega)^6)$.

Proof This follows from Theorem 17.3.1 applied to $M(z) := \begin{pmatrix}\varepsilon & 0\\ 0 & \mu\end{pmatrix} + z^{-1}\begin{pmatrix}\sigma & 0\\ 0 & 0\end{pmatrix}$ and A as in (17.5). □

17.5 Comments

The concept of maximal monotone relations in Hilbert spaces was first introduced by Minty in 1960 for the study of networks [66] and became a well-studied subject also with generalisations to the Banach space case. For this topic we refer to the monographs [16] and [49, Chapter 3]. The concept of subgradients is older and it was found out by Rockafellar [99] that subgradients are maximal monotone. Indeed, one can show that subgradients are precisely the cyclically maximal monotone relations (see e.g. [16, Theoreme 2.5]).

The Theorem of Minty was proved in 1962, [65] and generalised to the case of reflexive Banach spaces by Rockafellar in 1970 [100]. The proof presented here follows [106] and was kindly communicated by Ralph Chill and Hendrik Vogt.

The classical way to approach differential inclusions of the form $(u, f) \in \partial_t + A$ where A is maximal monotone uses the theory of nonlinear semigroups of contractions, introduced by Komura in the Hilbert space case, [56] and generalised to the Banach space case by Crandall and Pazy, [24]. The results on evolutionary inclusions presented in this chapter are based on [117, 118] and were further generalised to non-autonomous problems in [122, 126].

The model for Maxwell's equations in polarisable media can be found in [36, Chapter VII]. We note that in this reference, condition (17.4) is replaced by

$$|E| < E_0 \Rightarrow j_c = \sigma E$$
$$|E| = E_0 \Rightarrow \exists\lambda \geqslant 0 : \ j_c = (\sigma + \lambda)E,$$

which should hold almost everywhere. To solve this problem, one cannot apply Theorem 17.2.7, since 0 is not an interior point of the domain of the corresponding relation and thus, a weaker notion of solution is needed to tackle this problem, see [36, Theorem 8.1].

Exercises

Exercise 17.1 Let $f: H \to (-\infty, \infty]$ be convex, proper and l.s.c. Moreover, assume that f is differentiable in $x \in H$ (in particular, $f < \infty$ in a neighbourhood of x). Show that $(x, y) \in \partial f$ if and only if $y = f'(x)$.

Exercise 17.2 Let $f, g: H \to (-\infty, \infty]$. Prove that

(a) $f + g$ is convex if f and g are convex.
(b) $f + g$ is l.s.c. if f and g are l.s.c.

Exercise 17.3 Let H be a Hilbert space, $(x_n)_{n\in\mathbb{N}}$ in H and $x \in H$. Show, that $x_n \to x$ if and only if $x_n \to x$ weakly and $\limsup_{n\to\infty} \|x_n\| \leqslant \|x\|$.

Exercise 17.4 Let X be a normed space (or, more generally, a topological vector space) and $C \subseteq X$ convex. Prove the following statements:

(a) If $x \in \operatorname{int} C$ and $y \in \overline{C}$, then $(1-t)x + ty \in \operatorname{int} C$ for each $t \in [0, 1)$.
(b) If $\operatorname{int} C \neq \varnothing$, then $\overline{C} = \overline{\operatorname{int} C}$ and $\operatorname{int} \overline{C} = \operatorname{int} C$.
(c) If C is open and $K \subseteq X$ is open with $\overline{K} \subseteq \overline{C}$. Then $K \subseteq C$.

Hint: For (a) take an open set $U \subseteq X$ with $0 \in U$ such that $x + U - U \subseteq C$ and show $(1-t)x + ty + (1-t)U \subseteq C$.

Exercise 17.5 Let X be a topological space and $U \subseteq X$ open. We equip U with the trace topology. Prove the following statements:

(a) For $A \subseteq U$ we have $\overline{A}^U = \overline{A}^X \cap U$ and $\operatorname{int}_U A = \operatorname{int}_X A$.
(b) If $A \subseteq U$ is closed in U and $\operatorname{int}_U A = \varnothing$, then $\operatorname{int}_X \overline{A}^X = \varnothing$.
(c) If X is a Baire space, then U is a Baire space.

Recall, that a topological space X is a *Baire space* if for each sequence $(A_n)_{n\in\mathbb{N}}$ of closed sets with $\operatorname{int} A_n = \varnothing$ it follows that $\operatorname{int} \bigcup_{n\in\mathbb{N}} A_n = \varnothing$ or, equivalently, if for each sequence $(U_n)_{n\in\mathbb{N}}$ of open and dense sets it follows that $\bigcap_{n\in\mathbb{N}} U_n$ is dense.

Exercise 17.6 Let $A \subseteq H \times H$ be maximal monotone.

(a) Let $\mu, \lambda > 0$. Show that $(A_\lambda)_\mu = A_{\lambda+\mu}$.
(b) Let $(0, 0) \in A$ and $(\Omega, \mathcal{A}, \mu)$ a σ-finite measure space. Prove that $(A_\lambda)_{L_2(\mu)} = (A_{L_2(\mu)})_\lambda$ for each $\lambda > 0$.

Exercise 17.7 Let H be a Hilbert space and $C \subseteq H$ non-empty, convex and closed. Moreover, define $I_C\colon H \to (-\infty, \infty]$ by

$$I_C(x) := \begin{cases} 0 & \text{if } x \in C, \\ \infty & \text{otherwise.} \end{cases}$$

Show that I_C is convex, proper and l.s.c. and show

$$(x, y) \in \partial I_C \Leftrightarrow x \in C, \forall u \in C : \operatorname{Re} \langle y, u - x \rangle \leqslant 0.$$

Moreover, prove Lemma 17.4.2.

References

16. H. Brézis, *Opérateurs maximaux monotones et semi-groupes de contractions dans les espaces de Hilbert*, vol. 5 (Elsevier, Amsterdam, 1973)
24. M.G. Crandall, A. Pazy, Semi-groups of nonlinear contractions and dissipative sets. J. Funct. Anal. **3**, 376–418 (1969). English.
36. G. Duvaut, J.L. Lions, *Inequalities in Mechanics and Physics. Translated from the French by C.W. John*, vol. 219 (Springer, Berlin, 1976)
49. S. Hu, N.S. Papageorgiou, *Handbook of Multivalued Analysis. Vol. I*, Vol. 419. Mathematics and Its Applications. Theory (Kluwer Academic Publishers, Dordrecht, 1997)
50. J. Voigt, *A Course on Topological Vector Spaces* (Birkhäuser-Verlag, Cham, Switzerland, 2020)
56. Y. Komura, Nonlinear semigroups in Hilbert space. J. Math. Soc. Jpn. **19**, 493–507 (1967)
65. G.J. Minty, Monotone (nonlinear) operators in Hilbert space. Duke Math. J. **29**, 341–346 (1962)
66. G.J. Minty, Monotone networks. Proc. R. Soc. Lond. Ser. A **257**, 194–212 (1960)
99. R.T. Rockafellar, On the maximal monotonicity of subdifferential mappings. Pac. J. Math. **33**, 209–216 (1970)
100. R.T. Rockafellar, On the maximality of sums of nonlinear monotone operators. Trans. Am. Math. Soc. **149**, 75–88 (1970)
106. S. Simons, C. Zalinescu, A new proof for Rockafellar's characterization of maximal monotone operators. Proc. Am. Math. Soc. **132**(10), 2969–2972 (2004)
117. S. Trostorff, An alternative approach to well-posedness of a class of differential inclusions in Hilbert spaces. Nonlinear Anal. **75**(15), 5851–5865 (2012)
118. S. Trostorff, Autonomous evolutionary inclusions with applications to problems with nonlinear boundary conditions. Int. J. Pure Appl. Math. **85**(2), 303–338 (2013)
122. S. Trostorff, Well-posedness for a general class of differential inclusions. J. Differ. Equ. **268**, 6489–6516 (2020)
126. S. Trostorff, M. Wehowski, Well-posedness of non-autonomous evolutionary inclusions. Nonlinear Anal. Theory Methods Appl. Ser. A Theory Methods **101**, 47–65 (2014)

BY

Appendix A
Derivations of Main Equations

In this appendix we will derive the main equations studied in this book from a mere Physics' point of view. We will start with the heat equation and then turn to Maxwell's equations. After that, we derive the equations for linear elasticity and finally deduce the wave equation from elasticity theory.

A.1 Heat Equation

The heat equation describes the energy transport between materials due to a difference in temperature, where the transport evolves from high temperature to low temperature. Let $\Omega \subseteq \mathbb{R}^d$ be open. Let $\theta \colon \mathbb{R} \times \Omega \to \mathbb{R}$ be the heat distribution. As a physical principle, we ask for conservation of total energy. For a Borel subset $V \subseteq \Omega$ with smooth boundary let $Q_V \colon \mathbb{R} \to \mathbb{R}$ given by $Q_V(t) := \int_V \theta(t,x)\,\mathrm{d}x$ be the time-dependent heat content (i.e., the energy) in V. Then for a system without external heat sources, changes of Q_V can only result in heat fluxes along the boundary of V. Let $q \colon \mathbb{R} \times \Omega \to \mathbb{R}^d$ be the heat flux, which can be interpreted as a density. Then

$$\partial_t Q_V(t) = -\int_{\partial V} q(t,x) \cdot \nu(x)\,\mathrm{d}S(x),$$

where ν is the outward unit normal on ∂V. By Gauss' divergence theorem, we thus have

$$\partial_t Q_V(t) = -\int_V \operatorname{div} q(t,x)\,\mathrm{d}x.$$

C. Seifert et al., *Evolutionary Equations*, Operator Theory: Advances and Applications 287, https://doi.org/10.1007/978-3-030-89397-2

On the other hand, interchanging the time derivative and integration, we observe

$$\partial_t Q_V(t) = \int_V \partial_t \theta(t, x)\, \mathrm{d}x.$$

Hence,

$$\int_V \big(\partial_t \theta(t, x) + \operatorname{div} q(t, x)\big)\, \mathrm{d}x = 0.$$

Since $V \subseteq \Omega$ was arbitrary, we conclude the continuity equation

$$\partial_t \theta + \operatorname{div} q = 0.$$

In presence of an external heat source $Q\colon \mathbb{R} \times \Omega \to \mathbb{R}$, the continuity equation turns into the heat flux balance

$$\partial_t \theta + \operatorname{div} q = Q.$$

In order to incorporate that the energy transport runs from regions of high temperature to regions of low temperature, we make use of Fourier's law stating that the heat flux at time t and position x is determined by the gradient of the temperature at t and x; that is,

$$q(t, x) = -a(x) \operatorname{grad} \theta(t, x),$$

where $a\colon \Omega \to \mathbb{R}^{d \times d}$ is the heat conductivity, and we may assume that $a(x)$ is invertible for all $x \in \Omega$. We thus arrive at the heat equation

$$\begin{aligned} \partial_t \theta + \operatorname{div} q &= Q, \\ a^{-1} q + \operatorname{grad} \theta &= 0, \end{aligned}$$

or, put differently,

$$\partial_t \theta - \operatorname{div}(a \operatorname{grad} \theta) = Q.$$

A.2 Maxwell's Equations

Maxwell's equations are the governing equations in electrodynamics and describe the evolution of the electromagnetic fields. Let $\Omega \subseteq \mathbb{R}^3$ be a *domain*; that is, open and connected. The physical quantities of interest in Maxwell's equations in vacuum are the time-dependent electric field $E\colon \mathbb{R} \times \Omega \to \mathbb{R}^3$ and magnetic induction $B\colon \mathbb{R} \times \Omega \to \mathbb{R}^3$ on Ω, since they can be observed via their action as a force.

Given two point charges q, q' at distinct points $x, x' \in \Omega$, respectively, the Coulomb force

$$F = q\frac{1}{4\pi\varepsilon_0}q'\frac{x - x'}{\|x - x'\|^3}$$

can be observed, where ε_0 is the dielectric constant in vacuum. More precisely, F is the force on the point charge q at x induced by the point charge q' at x':

q F q'
x x'

The electrical field at time t and position x induced by q' at x' is then given by

$$E(t, x) = \frac{1}{4\pi\varepsilon_0}q'\frac{x - x'}{\|x - x'\|^3}$$

such that it acts locally on the point charge q at x via the Coulomb force

$$F = qE(t, x).$$

Let us generalise from point charges, formally given by $q'\delta_{x'}$, to charge densities. Let $\rho\colon \mathbb{R} \times \Omega \to \mathbb{R}$ be the time-dependent charge density. Then the electric field at time t and position x is given by

$$E(t, x) = \frac{1}{4\pi\varepsilon_0}\int_\Omega \rho(t, x')\frac{x - x'}{\|x - x'\|^3}\,\mathrm{d}x'.$$

By Exercise A.1 we can rewrite this as

$$\begin{aligned} E(t, x) &= -\frac{1}{4\pi\varepsilon_0}\int_\Omega \operatorname{grad}\frac{\rho(t, x')}{\|x - x'\|}\,\mathrm{d}x' = -\operatorname{grad}\Big(\frac{1}{4\pi\varepsilon_0}\int_\Omega \frac{\rho(t, x')}{\|x - x'\|}\,\mathrm{d}x'\Big) \\ &= -\operatorname{grad}\Phi(t, x), \end{aligned}$$

where $\Phi\colon \mathbb{R} \times \Omega \to \mathbb{R}$ given by $\Phi(t, x) := \frac{1}{4\pi\varepsilon_0}\int_\Omega \frac{\rho(t,x')}{\|x-x'\|}\,\mathrm{d}x'$ is the electric potential.

Analogously, the magnetic induction acts as a force as follows. We first consider two closed non-intersecting curves C and C' in Ω decribing two wires and let I and I' be (constant) currents on C and C', respectively. Then the force between these two wires is given by

$$F = I\frac{\mu_0}{4\pi}I'\int_C\int_{C'}\Big(\frac{x - x'}{\|x - x'\|^3} \times \mathrm{d}x'\Big) \times \mathrm{d}x,$$

where μ_0 is the permeability in vacuum.

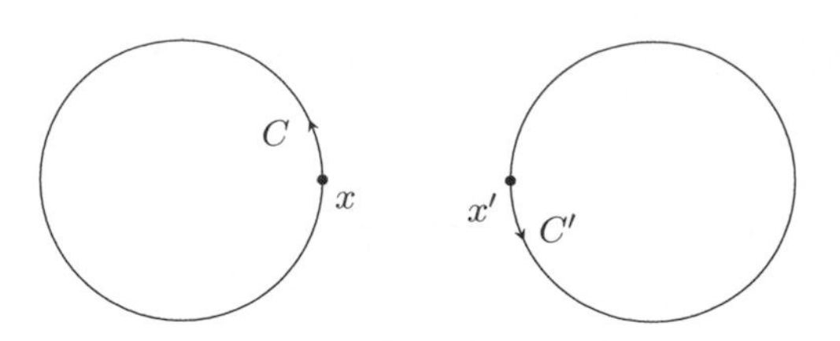

Thus, the magnetic induction at time t induced by the wire C' acting at a point x from C is given by

$$B(t, x) = -\frac{\mu_0}{4\pi} I' \int_{C'} \frac{x - x'}{\|x - x'\|^3} \times \mathrm{d}x',$$

such that it acts via the force

$$F = I \int_C B(t, x) \times \mathrm{d}x.$$

Let us generalise from constant currents on one-dimensional curves C' to current densities. Let $j : \mathbb{R} \times \Omega \to \mathbb{R}^3$ be the time-dependent current density. Then the magnetic induction at time t and position x is given by

$$B(t, x) = -\frac{\mu_0}{4\pi} \int_\Omega \frac{x - x'}{\|x - x'\|^3} \times j(t, x')\, \mathrm{d}x'.$$

By Exercise A.1 we can rewrite this as

$$\begin{aligned} B(t, x) &= \frac{\mu_0}{4\pi} \int_\Omega \operatorname{grad} \frac{1}{\|x - x'\|} \times j(t, x')\, \mathrm{d}x' \\ &= \operatorname{curl} \left(\frac{\mu_0}{4\pi} \int_\Omega \frac{j(t, x')}{\|x - x'\|} \mathrm{d}x' \right) = \operatorname{curl} A(t, x), \end{aligned}$$

where $A : \mathbb{R} \times \Omega \to \mathbb{R}^3$ given by $A(t, x) := \frac{\mu_0}{4\pi} \int_\Omega \frac{j(t,x')}{\|x-x'\|} \mathrm{d}x'$ is the vector potential.

We now relate the charge density ρ and the current density j. As a physical principle, we ask for conservation of total charge. For a Borel subset $V \subseteq \Omega$ with smooth boundary let $Q_V : \mathbb{R} \to \mathbb{R}$ given by $Q_V(t) := \int_V \rho(t, x)\, \mathrm{d}x$ be the time-dependent total charge in V. Then changes of Q_V can only result in currents along the boundary of V; that is,

$$\partial_t Q_V(t) = -\int_{\partial V} j(t, x) \cdot \nu(x)\, \mathrm{d}S(x),$$

where ν is the outward unit normal on ∂V. By Gauss' divergence theorem, we thus have

$$\partial_t Q_V(t) = -\int_V \operatorname{div} j(t, x)\, \mathrm{d}x.$$

On the other hand, interchanging the (time) differentiation and integration, we observe

$$\partial_t Q_V(t) = \int_V \partial_t \rho(t, x)\, \mathrm{d}x.$$

Hence,

$$\int_V \big(\partial_t \rho(t, x) + \operatorname{div} j(t, x)\big)\, \mathrm{d}x = 0.$$

Since $V \subseteq \Omega$ was arbitrary, we conclude the continuity equation

$$\partial_t \rho + \operatorname{div} j = 0.$$

We now derive the two fundamental equations, namely Faraday's law and Ampère's law. We start with Faraday's law. Let $\Sigma \subseteq \Omega$ be a two-dimensional submanifold with boundary curve $\partial\Sigma$ which we may think of as a wire.

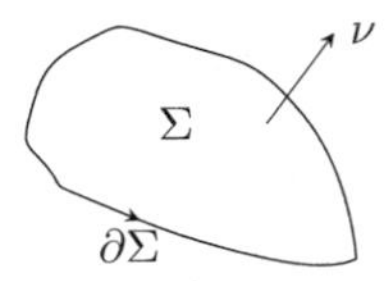

Then a changing magnetic field through Σ induces a voltage along $\partial\Sigma$ as

$$U(t) = -\int_\Sigma \partial_t B(t, x) \cdot \nu(x)\, \mathrm{d}S(x).$$

Since voltages result from electric fields, we also have

$$U(t) = \int_{\partial\Sigma} E(t, x)\, \mathrm{d}x = \int_\Sigma \operatorname{curl} E(t, x) \cdot \nu(x)\, \mathrm{d}S(x),$$

where we invoked Stoke's theorem and ν is again the unit normal on Σ (oriented accordingly to a parametrisation of $\partial\Sigma$). Thus,

$$\int_\Sigma \big(\partial_t B(t, x) + \operatorname{curl} E(t, x)\big) \cdot \nu(x)\, \mathrm{d}S(x) = 0.$$

Since $\Sigma \subseteq \Omega$ was arbitrary, we conclude Faraday's law

$$\partial_t B = -\operatorname{curl} E.$$

We now derive Ampère's law by considering $\operatorname{curl} B = \operatorname{curl}\operatorname{curl} A = \operatorname{grad}\operatorname{div} A - \Delta A$, where $\Delta A = (\Delta A_1, \Delta A_2, \Delta A_3)$ and $\Delta A_j = \operatorname{div}\operatorname{grad} A_j$ for $j \in \{1, 2, 3\}$. We calculate by Exercise A.1

$$\begin{aligned}\operatorname{div} A(t,x) &= \frac{\mu_0}{4\pi}\int_\Omega \operatorname{div}\frac{j(t,x')}{\|x-x'\|}\,\mathrm{d}x' = \frac{\mu_0}{4\pi}\int_\Omega \Big(-\operatorname{grad}_{x'}\frac{1}{\|x-x'\|}\Big)\cdot j(t,x')\,\mathrm{d}x' \\ &= \frac{\mu_0}{4\pi}\int_\Omega \frac{\operatorname{div} j(t,x')}{\|x-x'\|}\,\mathrm{d}x'.\end{aligned}$$

By the continuity equation, we further obtain

$$\begin{aligned}\operatorname{div} A(t,x) &= -\frac{\mu_0}{4\pi}\int_\Omega \frac{\partial_t \rho(t,x')}{\|x-x'\|}\,\mathrm{d}x' \\ &= -\frac{\mu_0}{4\pi}\partial_t\int_\Omega \frac{\rho(t,x')}{\|x-x'\|}\,\mathrm{d}x' = -\varepsilon_0\mu_0\partial_t\Phi(t,x).\end{aligned}$$

Thus,

$$\begin{aligned}\operatorname{grad}\operatorname{div} A(t,x) &= -\varepsilon_0\mu_0 \operatorname{grad}\partial_t\Phi(t,x) \\ &= -\varepsilon_0\mu_0\partial_t \operatorname{grad}\Phi(t,x) = \varepsilon_0\mu_0\partial_t E(t,x).\end{aligned}$$

Moreover, by Exercise A.2 (assuming that $j(t,\cdot)$ can be smoothly extended to $\mathbb{R}^3$),

$$\begin{aligned}\Delta A(t,x) &= \frac{\mu_0}{4\pi}\int \Delta\frac{1}{\|x-x'\|} j(t,x')\,\mathrm{d}x' = \frac{\mu_0}{4\pi}\int \Big(\Delta_{x'}\frac{1}{\|x-x'\|}\Big) j(t,x')\,\mathrm{d}x' \\ &= \frac{\mu_0}{4\pi}\int \frac{1}{\|x-x'\|}\Delta j(t,x')\,\mathrm{d}x' = -\mu_0 j(t,x).\end{aligned}$$

We conclude Ampère's law

$$\operatorname{curl} B = \varepsilon_0\mu_0\partial_t E + \mu_0 j.$$

So far we only considered the equations in vacuum. In materials two additional effects, polarisation and magnetisation, occur due to the interaction of the fields with the medium. Let $P\colon \mathbb{R}\times\Omega \to \mathbb{R}^3$ be the polarisation; that is, the averaged electrical dipole moments. Further, let $M\colon \mathbb{R}\times\Omega\to\mathbb{R}^3$ be the magnetisation; that is, the averaged magnetic dipole moments. Then the current density gets two additional terms $j_P, j_M\colon \mathbb{R}\times\Omega\to\mathbb{R}^3$, where $j_P = \partial_t P$ and $j_M = \operatorname{curl} M$. Thus, $j = j_c + j_P + j_M$ where j_c corresponds to the free charged carriers or free current (as the current density in vacuum) and $j_P + j_M$ forms the bound currents. In order to take these two effects into account, we define the electric displacement $D\colon \mathbb{R}\times\Omega\to\mathbb{R}^3$ by $D := \varepsilon_0 E + P$ and the magnetic field $H\colon \mathbb{R}\times\Omega\to\mathbb{R}^3$ by $H := \frac{1}{\mu_0}B - M$, such that $B = \mu_0 H + M$. Then one typically expands P and M in terms of E and H. We only consider linear models; that is, $P = \chi_e E$ and $M = \chi_m H$ with electric and magnetic susceptibility $\chi_e, \chi_m \colon \Omega \to \mathbb{R}^{3\times 3}$, respectively. Then $D = \varepsilon E$ where $\varepsilon = \varepsilon_0(1+\chi_e) \colon \Omega \to \mathbb{R}^{3\times 3}$ is the dielectricity and $B = \mu H$ where $\mu = \mu_0(1+\chi_m) \colon \Omega \to \mathbb{R}^{3\times 3}$ is the magnetic permeability. Polarisation and magnetisation have no effect on Faraday's law, but on Ampère's law, which now states

$$\operatorname{curl} H = \partial_t \varepsilon E + j_c.$$

In case of an external current $j_0 \colon \mathbb{R} \times \Omega \to \mathbb{R}^3$, we observe

$$\operatorname{curl} H = \partial_t \varepsilon E + j_c - j_0.$$

Finally, Ohm's law couples the free current j_c with the electric field E by $j_c = \sigma E$, where $\sigma \colon \Omega \to \mathbb{R}^{3\times 3}$ is the electric conductivity, so that we obtain

$$\operatorname{curl} H = \partial_t \varepsilon E + \sigma E - j_0.$$

We thus arrive at Maxwell's equations

$$\begin{aligned} \partial_t \varepsilon E + \sigma E - \operatorname{curl} H &= j_0, \\ \partial_t \mu H + \operatorname{curl} E &= 0. \end{aligned}$$

A.3 Linear Elasticity

The theory of elasticity is devoted to the study of distortion of bodies due to forces, which is reversible in the sense that the body will return to its original state when the force is removed. In order to reasonably neglect thermodynamical effects we assume that the deformation occurs slowly to obtain thermodynamical equilibrium and the temperature of the body is constant. Also, we assume that the behaviour of the material does not depend on memory effects, so hysteresis is excluded. Moreover, we exclude rigid body moves (i.e., translations and rotations) due to the forces.

Let $\Omega \subseteq \mathbb{R}^d$ be a domain which models the body. Then the displacement field $u \colon \mathbb{R} \times \Omega \to \mathbb{R}^d$ describes the deformation vector of the body at time t and position x. For $x, y \in \Omega$ we write $x' = x + u(t,x)$ and $y' = y + u(t,y)$ for the new positions of x and y, respectively, after the deformation at time t.

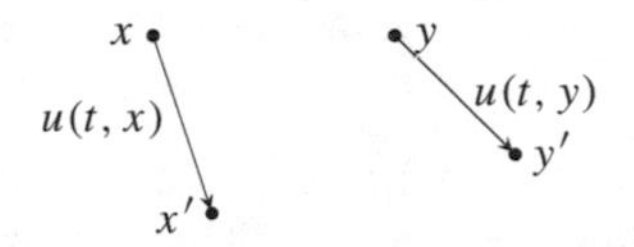

Then, assuming spatially smooth and slowly varying deformations u (i.e., small spatial derivatives of u), by a linearisation of $u(t,\cdot)$ we obtain

$$u(t,x) \approx u(t,y) + \partial_y u(t,y)(x-y)$$

for x close to y and therefore

$$\begin{aligned}
\left|x'-y'\right|^2 &= |x+u(t,x)-(y+u(t,y))|^2 \\
&= |x-y|^2 + 2\langle u(t,x)-u(t,y), x-y\rangle + |u(t,x)-u(t,y)|^2 \\
&\approx |x-y|^2 + 2\left\langle \partial_y u(t,y)(x-y), x-y\right\rangle,
\end{aligned}$$

where we neglected the quadratic term $|u(t,x)-u(t,y)|^2 \approx \left|\partial_y u(t,y)(x-y)\right|^2$.

Since

$$\begin{aligned}
\left\langle \partial_y u(t,y)(x-y), x-y\right\rangle &= \sum_{j,k=1}^{d} \partial_k u_j(t,y)(x_k-y_k)(x_j-y_j) \\
&= \sum_{j,k=1}^{d} \left(\frac{1}{2}\partial_k u_j(t,y) + \frac{1}{2}\partial_j u_k(t,y)\right)(x_k-y_k)(x_j-y_j) \\
&= \left\langle \frac{1}{2}\left(\partial_k u_j(t,y) + \partial_j u_k(t,y)\right)_{j,k\in\{1,\ldots,d\}}(x-y), x-y\right\rangle,
\end{aligned}$$

we may introduce the symmetrised gradient of u as $\operatorname{Grad} u\colon \mathbb{R}\times\Omega \to \mathbb{R}^d$ defined by $\operatorname{Grad} u(t,y) := \frac{1}{2}\left(\partial_k u_j(t,y) + \partial_j u_k(t,y)\right)_{j,k\in\{1,\ldots,d\}}$ to get

$$\left|x'-y'\right|^2 \approx |x-y|^2 + 2\langle \operatorname{Grad} u(t,y)(x-y), x-y\rangle.$$

Note that $\varepsilon(u)(t,y) := \operatorname{Grad} u(t,y)$ is called the strain tensor of u at t and y.

Due to the displacement u, there appear forces between the molecules of the material trying to push them back to their equilibrium state. These forces induced by the displacement u result from stresses along the boundary of Ω. Let $T := T_u\colon \mathbb{R}\times\Omega \to \mathbb{R}^{d\times d}_{\mathrm{sym}}$ be the stress tensor corresponding to the displacement u. Then the forces between the molecules are given by the divergence of T; that is, by $\operatorname{Div} T\colon \mathbb{R}\times\Omega\to\mathbb{R}^d$,

$$\operatorname{Div} T(t,x) := \left(\sum_{k=1}^{d} \partial_k T_{jk}(t,x)\right)_{j\in\{1,\ldots,d\}}.$$

In thermodynamics, the free energy $\mathcal{F}$ of a system describes the maximum amount of work that a system can perform. Thus, we may expand the free energy $\mathcal{F}_u$ of the deformed system in terms of the strain tensor $\varepsilon(u) = \operatorname{Grad} u$ around the undeformed system $\mathcal{F}_0$. Since changes of the free energy result from stresses, we observe $T = \frac{\partial \mathcal{F}_u}{\partial \varepsilon(u)}$. Since the stress tensor vanishes for deformation 0, there exists a so-called elasticity tensor $C\colon \Omega \to L\big(\mathbb{R}^{d\times d}_{\mathrm{sym}}, \mathbb{R}^{d\times d}_{\mathrm{sym}}\big)$ such that

$$\mathcal{F}_u = \mathcal{F}_0 + \frac{1}{2}\left\langle \varepsilon(u), C\varepsilon(u)\right\rangle.$$

Thus,

$$T = \frac{\partial \mathcal{F}_u}{\partial \varepsilon(u)} = C\varepsilon(u) = C\,\mathrm{Grad}\,u.$$

This is Hooke's law of linear elasticity. Using Hooke's law, we get

$$\mathrm{Div}\,T = \mathrm{Div}\,C\,\mathrm{Grad}\,u.$$

In order to obtain the governing equations for linear elasticity, we make use of Newton's law. Let $\rho\colon \mathbb{R}\times\Omega \to \mathbb{R}$ be the mass density of the body. Then Newton's law on conservation of momentum yields

$$\partial_t \rho \partial_t u = F,$$

where F describes the acting forces on the system. These forces decompose into the internal forces between the molecules due to the displacement u and we have seen that this is given by $\mathrm{Div}\,T$. Moreover, there may be external forces $f\colon \mathbb{R}\times\Omega \to \mathbb{R}^d$ (for example gravity). Thus, $F = \mathrm{Div}\,T + f$, and therefore

$$\partial_t \rho \partial_t u - \mathrm{Div}\,T = f.$$

Taking into account Hooke's law, we arrive at the governing equation of linear elasticity as

$$\partial_t \rho \partial_t u - \mathrm{Div}\,C\,\mathrm{Grad}\,u = f.$$

A.4 Scalar Wave Equation

The scalar wave equation can be derived from linear elasticity. Indeed, let $\Omega \subseteq \mathbb{R}^d$ be open and consider scalar displacements $u\colon \mathbb{R}\times\Omega \to \mathbb{R}$, so we only consider displacements in one particular direction. Also, we may assume constant mass density; that is, $\rho\colon \mathbb{R}\times\Omega \to \mathbb{R}$ is constant. Without loss of generality, we therefore set $\rho = 1$. Let $f\colon \mathbb{R}\times\Omega \to \mathbb{R}$ be an external force in direction of the displacements. Then from linear elasticity we obtain

$$\partial_t^2 u - \mathrm{div}\,T = f,$$

where $T : \mathbb{R} \times \Omega \to \mathbb{R}^d$ is the stress obtained by the displacements. If we further make use of Hooke's law $T = C \operatorname{grad} u$ with the elasticity tensor $C : \Omega \to L(\mathbb{R}^d, \mathbb{R}^d) = \mathbb{R}^{d \times d}$, we arrive at the scalar wave equation

$$\partial_t^2 u - \operatorname{div} C \operatorname{grad} u = f.$$

A.5 Comments

The physical derivations of the equations treated in this appendix are well-known and can be found in many textbooks. We refer to [74–76] for foundations on physics of electrodynamics, thermodynamics and statistical physics. The final form of Maxwell's equations appeared in [62], however they had been derived in his earlier works already. The vector form of Maxwell's equations appeared in the 1880s. The equations of linear elasticity stem from elastodynamics.

Exercises

Exercise A.1 Let $\Omega \subseteq \mathbb{R}^3$ be open, $x' \in \Omega$, $f : \Omega \setminus \{x'\} \to \mathbb{R}$ defined by $f(x) := \frac{1}{\|x-x'\|}$. Show that f is differentiable and $\operatorname{grad} f(x) = \frac{x-x'}{\|x-x'\|^3}$ for all $x \in \Omega \setminus \{x'\}$.

Exercise A.2 Let $K : \mathbb{R}^3 \setminus \{0\} \to \mathbb{R}$, $K(x) := -\frac{1}{4\pi}\frac{1}{\|x\|}$. Then $\Delta K(x) = \operatorname{div} \operatorname{grad} K(x) = 0$ for all $x \in \mathbb{R}^3 \setminus \{0\}$ and

$$-\int_{\mathbb{R}^3} K(x) \Delta\varphi(x) \, \mathrm{d}x = \varphi(0)$$

for all $\varphi \in C_c^\infty(\mathbb{R}^3)$.

References

62. J.C. Maxwell, VIII. A dynamical theory of the electromagnetic field. Phil. Trans. R. Soc. **155**, 459–512 (1865)
74. W. Nolting, *Theoretical Physics. 3. Electrodynamics* (Springer, Cham, 2016)
75. W. Nolting, *Theoretical Physics. 5. Thermodynamics* (Springer, Cham, 2017)
76. W. Nolting, *Theoretical Physics. 8. Statistical Physics* (Springer, Cham, 2018)

Bibliography

1. M. Abramowitz, I.A. Stegun, *Handbook of Mathematical Functions with Formulas, Graphs, and Mathematical Tables*, vol. 55. National Bureau of Standards Applied Mathematics Series. For sale by the Superintendent of Documents, U.S. Government Printing Office, Washington, D.C., 1964
2. M. Achache, E.M. Ouhabaz, 'Lions' maximal regularity problem with $H^{\frac{1}{2}}$-regularity in time'. J. Differ. Equ. **266**(6), 3654–3678 (2019)
3. F. Andreu et al., Finite propagation speed for limited flux diffusion equations. Arch. Ration. Mech. Anal. **182**(2), 269 (2006)
4. W. Arendt, C.J.K. Batty, Tauberian theorems and stability of one-parameter semigroups. Trans. Am. Math. Soc. **306**(2), 837–852 (1988)
5. W. Arendt et al., *Form Methods for Evolution Equations, and Applications*. 18th Internet Seminar, 2015
6. W. Arendt et al., *Vector-Valued Laplace Transforms and Cauchy Problems. 2nd ed.* (Birkhäuser, Basel, 2011)
7. R. Arens, Operational calculus of linear relations. Pac. J. Math. **11**, 9–23 (1961)
8. P. Auscher, M. Egert, On non-autonomous maximal regularity for elliptic operators in divergence form. Arch. Math. (Basel) **107**(3), 271–284 (2016)
9. C. Bardos, G. Lebeau, J. Rauch, Sharp sufficient conditions for the observation, control, and stabilization of waves from the boundary. SIAM J. Control Optim. **30**(5), 1024–1065 (1992)
10. A. Bátkai, S. Piazzera, *Semigroups for Delay Equations*, vol. 10. Research Notes in Mathematics (A K Peters, Ltd., Wellesley, MA, 2005)
11. A. Bensoussan, J.-L. Lions, G. Papanicolaou, *Asymptotic Analysis for Periodic Structures*, vol. 5. Studies in Mathematics and Its Applications (North-Holland Publishing Co., Amsterdam-New York, 1978)
12. S. Benzoni-Gavage, D. Serre, *Multidimensional Hyperbolic Partial Differential Equations*. Oxford Mathematical Monographs. First-Order Systems and Applications (The Clarendon Press, Oxford University Press, Oxford, 2007)
13. T. Berger, A. Ilchmann, S. Trenn, The quasi-Weierstrass form for regular matrix pencils. Linear Algebra Appl. **436**(10), 4052–4069 (2012)
14. T. Berger, C. Trunk, H. Winkler, Linear relations and the Kronecker canonical form. Linear Algebra Appl. **488**, 13–44 (2016)
15. A. Borichev, Y. Tomilov, Optimal polynomial decay of functions and operator semigroups. Math. Ann. **347**(2), 455–478 (2010)
16. H. Brézis, *Opérateurs maximaux monotones et semi-groupes de contractions dans les espaces de Hilbert*, vol. 5 (Elsevier, Amsterdam, 1973).

C. Seifert et al., *Evolutionary Equations*, Operator Theory: Advances and Applications 287, https://doi.org/10.1007/978-3-030-89397-2

17. M. Briane, J. Casado-Díaz, F. Murat, The div-curl lemma "trente ans après": an extension and an application to the G-convergence of unbounded monotone operators. J. Math. Pures Appl. (9) **91**(5), 476–494 (2009)
18. A. Buffa, M. Costabel, D. Sheen, On traces for $\mathbf{H}(\mathbf{curl}, \Omega)$ in Lipschitz domains. J. Math. Anal. Appl. **276**(2), 845–867 (2002)
19. P. Cannarsa, D. Sforza, Global solutions of abstract semilinear parabolic equations with memory terms. NoDEA Nonlinear Differ. Equ. Appl. **10**(4), 399–430 (2003)
20. K. Cherednichenko, M. Waurick, Resolvent estimates in homogenisation of periodic problems of fractional elasticity. J. Differ. Equ. **264**(6), 3811–3835 (2018)
21. D. Cioranescu, P. Donato, *An Introduction to Homogenization*, vol. 17. Oxford Lecture Series in Mathematics and Its Applications (The Clarendon Press, Oxford University Press, New York, 1999)
22. P. Cojuhari, A. Gheondea, Closed embeddings of Hilbert spaces. J. Math. Anal. Appl. **369**(1), 60–75 (2010)
23. S. Cooper, M. Waurick, Fibre homogenisation. J. Funct. Anal. **276**(11), 3363–3405 (2019)
24. M.G. Crandall, A. Pazy, Semi-groups of nonlinear contractions and dissipative sets. J. Funct. Anal. **3**, 376–418 (1969). English
25. R. Cross, *Multivalued Linear Operators*, vol. 213. Monographs and Textbooks in Pure and Applied Mathematics (Marcel Dekker, Inc., New York, 1998)
26. G. Da Prato, P. Grisvard, Sommes d'opérateurs linéaires et équations différentielles opérationnelles. J. Math. Pures Appl. (9) **54**(3), 305–387 (1975)
27. C.M. Dafermos, An abstract Volterra equation with applications to linear viscoelasticity. J. Differ. Equ. **7**, 554–569 (1970)
28. L. Dai, *Singular Control Systems*, vol. 118 (Springer, Berlin etc., 1989)
29. R. Datko, Uniform asymptotic stability of evolutionary processes in a Banach space. SIAM J. Math. Anal. **3**, 428–445 (1972)
30. D. Dier, R. Zacher, Non-autonomous maximal regularity in Hilbert spaces. J. Evol. Equ. **17**(3), 883–907 (2017)
31. J. Diestel, J.J. Uhl Jr., *Vector Measures*. With a foreword by B. J. Pettis, Mathematical Surveys, No. 15 (American Mathematical Society, Providence, RI, 1977)
32. W.F. Donoghue Jr., *Distributions and Fourier Transforms*, vol. 32. Pure and Applied Mathematics (Academic Press, New York, 1969)
33. G. Dore, L_p regularity for abstract differential equations. Funct. Anal. Relat. Top. 1991 **1540**, 25–38 (1993)
34. M. Dreher, R. Quintanilla, R. Racke, Ill-posed problems in thermomechanics. Appl. Math. Lett. **22**(9), 1374–1379 (2009)
35. P.L. Duren, *Theory of H^p Spaces* (Academic Press XII, New York and London, 1970), 258 p.
36. G. Duvaut, J.L. Lions, *Inequalities in Mechanics and Physics. Translated from the French by C.W. John*, vol. 219 (Springer, Berlin, 1976).
37. A.F.M. ter Elst, G. Gorden, M. Waurick, The Dirichlet-to-Neumann operator for divergence form problems. Ann. Mat. Pura Appl. (4) **198**(1), 177–203 (2019)
38. K.-J. Engel, R. Nagel, *One-Parameter Semigroups for Linear Evolution Equations*, vol. 194. Graduate Texts in Mathematics. With contributions by S. Brendle, M. Campiti, T. Hahn, G. Metafune, G. Nickel, D. Pallara, C. Perazzoli, A. Rhandi, S. Romanelli, R. Schnaubelt (Springer, New York, 2000)
39. L.C. Evans, *Partial Differential Equations*, vol. 19. Graduate Studies in Mathematics (American Mathematical Society, Providence, RI, 1998)
40. R.P. Feynman, R.B. Leighton, M. Sands, *The Feynman Lectures on Physics, Vol. 2: Mainly Electromagnetism and Matter* (Addison-Wesley Publishing Co., Reading, MA, London, 1964)
41. Y. Fourès, I. Segal, Causality and analyticity. Trans. Am. Math. Soc. **78**, 385–405 (1955)
42. S. Franz, M. Waurick, Resolvent estimates and numerical implementation for the homogenisation of one-dimensional periodic mixed type problems. Zeitschrift für Angewandte Mathematik und Mechanik **98**(7), 1284–1294 (2018)

43. K.O. Friedrichs, Symmetric hyperbolic linear differential equations. Commun. Pure Appl. Math. **7**, 345–392 (1954)
44. K.O. Friedrichs, Symmetric positive linear differential equations. Commun. Pure Appl. Math. **11**, 333–418 (1958)
45. D. Gilbarg, N.S. Trudinger, *Elliptic Partial Differential Equations of Second Order*. Second. Vol. 224. Grundlehren der Mathematischen Wissenschaften [Fundamental Principles of Mathematical Sciences] (Springer, Berlin, 1983)
46. M. Haase, *Functional Calculus*. 21st Internet Seminar. 2017/18
47. M. Haase, *The Functional Calculus for Sectorial Operators,* vol. 169 (Birkhäuser, Basel, 2006)
48. E. Hille, R.S. Phillips, *Functional Analysis and Semi-Groups*. American Mathematical Society Colloquium Publications, vol. 31. rev. edn. (American Mathematical Society, Providence, RI, 1957)
49. S. Hu, N.S. Papageorgiou, *Handbook of Multivalued Analysis. Vol. I*, vol. 419. Mathematics and Its Applications. Theory (Kluwer Academic Publishers, Dordrecht, 1997)
50. J. Voigt, *A Course on Topological Vector Spaces* (Birkhäuser-Verlag, Cham, Switzerland, 2020)
51. B. Jacob, J.R. Partington, Graphs, closability, and causality of linear time-invariant discrete-time systems. Int. J. Control **73**(11), 1051–1060 (2000)
52. A. Kalauch et al., A Hilbert space perspective on ordinary differential equations with memory term. J. Dyn. Differ. Equ. **26**(2), 369–399 (2014)
53. T. Kato, Integration of the equation of evolution in a Banach space. J. Math. Soc. Jpn. **5**(2), 208–234 (1953)
54. T. Kato, *Perturbation Theory for Linear Operators*. Classics in Mathematics. Reprint of the 1980 edition (Springer, Berlin, 1995)
55. D. Khusainov, M. Pokojovy, R. Racke, Strong and mild extrapolated L^2-solutions to the heat equation with constant delay. SIAM J. Math. Anal. **47**(1), 427–454 (2015)
56. Y. Komura, Nonlinear semigroups in Hilbert space. J. Math. Soc. Jpn. **19**, 493–507 (1967).
57. P. Kunkel, V. Mehrmann, *Differential-Algebraic Equations*. EMS Textbooks in Mathematics. Analysis and Numerical Solution (European Mathematical Society (EMS), Zürich, 2006)
58. S. Kwapien, Isomorphic characterizations of inner product spaces by orthogonal series with vector valued coefficients. Studia Mathematica **44**(6), 583–595 (1972)
59. R. Leis, Zur Theorie elektromagnetischer Schwingungen in anisotropen inhomogenen Medien. Math. Z. **106**, 213–224 (1968)
60. J.-L. Lions, E. Magenes, *Non-Homogeneous Boundary Value Problems and Applications. Vol. I.* Translated from the French by P. Kenneth, Die Grundlehren der mathematischen Wissenschaften, Band 181 (Springer, New York-Heidelberg, 1972)
61. Y.I. Lyubich, Q.P. Vũ, Asymptotic stability of linear differential equations in Banach spaces. Studia Math. **88**(1), 37–42 (1988)
62. J.C. Maxwell, VIII. A dynamical theory of the electromagnetic field. Phil. Trans. R. Soc. **155**, 459–512 (1865)
63. D.F. McGhee, R. Picard, A note on anisotropic, inhomogeneous, poro-elastic media. Math. Methods Appl. Sci. **33**(3), 313–322 (2010)
64. H. Minkowski, Die Grundgleichungen für die elektromagnetischen Vorgänge in bewegten Körpern. Math. Ann. **68**(4), 472–525 (1910)
65. G.J. Minty, Monotone (nonlinear) operators in Hilbert space. Duke Math. J. **29**, 341–346 (1962).
66. G.J. Minty, Monotone networks. Proc. R. Soc. Lond. Ser. A **257**, 194–212 (1960).
67. D. Morgenstern, Beträge zur nichtlinearen Funktionalanalysis. Ph.D. thesis. TU Berlin, 1952
68. S. Mukhopadyay et al., On some models in linear thermo-elasticity with rational material laws. Math. Mech. Solids **21**(9), 1149–1163 (2016)
69. M.A. Murad, J.H. Cushman, Multiscale flow and deformation in hydrophilic swelling porous media. Int. J. Eng. Sci. **34**(3), 313–338 (1996)

70. F. Murat, Compacité par compensation. Ann. Scuola Norm. Sup. Pisa Cl. Sci. (4) **5**(3), 489–507 (1978)
71. F. Murat, L. Tartar, H-convergence. *Topics in the Mathematical Modelling of Composite Materials*, vol. 31. Progr. Nonlinear Differential Equations Appl. (Birkhäuser Boston, Boston, MA, 1997), pp. 21–43
72. J. Nečas, *Direct Methods in the Theory of Elliptic Equations*. Springer Monographs in Mathematics. Translated from the 1967 French original by Gerard Tronel and Alois Kufner, Editorial coordination and preface by Šárka Nečasová and a contribution by Christian G. Simader (Springer, Heidelberg, 2012)
73. B. Nolte, S. Kempfle, I. Schäfer, Does a real material behave fractionally? Applications of fractional differential operators to the damped structure borne sound in viscoelastic solids. J. Comput. Acoust. **11**(03), 451–489 (2003). eprint: https://doi.org/10.1142/S0218396X03002024
74. W. Nolting, *Theoretical Physics. 3. Electrodynamics* (Springer, Cham, 2016)
75. W. Nolting, *Theoretical Physics. 5. Thermodynamics* (Springer, Cham, 2017)
76. W. Nolting, *Theoretical Physics. 8. Statistical Physics* (Springer, Cham, 2018)
77. R.S. Palais, *Seminar on the Atiyah-Singer Index Theorem.* With contributions by M.F. Atiyah, A. Borel, E.E. Floyd, R.T. Seeley, W. Shih, R. Solovay. Annals of Mathematics Studies, No. 57 (Princeton University Press, Princeton, NJ, 1965)
78. R.E. Paley, N. Wiener, *Fourier Transforms in the Complex Domain.* (Am. Math. Soc. Colloq. Publ. 19) (Am. Math. Soc. VIII, New York, 1934)
79. D. Pauly, R. Picard, S. Trostorff, M. Waurick, On a class of degenerate abstract parabolic problems and applications to some Eddy current models. J. Funct. Anal. **280**(7), 108847 (2021)
80. D. Pauly, A global div-curl-lemma for mixed boundary conditions in weak Lipschitz domains and a corresponding generalized $\mathrm{A}_0^* - \mathrm{A}_1$-lemma in Hilbert spaces. Analysis (Berlin) **39**(2), 33–58 (2019)
81. A. Pazy, *Semigroups of Linear Operators and Applications to Partial Differential Equations*, vol. 44. Applied Mathematical Sciences (Springer, New York, 1983)
82. R. Picard, A structural observation for linear material laws in classical mathematical physics. Math. Methods Appl. Sci. **32**, 1768–1803 (2009)
83. R. Picard, *Hilbert Space Approach to Some Classical Transforms* (Wiley, New York, 1989).
84. R. Picard, D. McGhee, *Partial Differential Equations: A Unified Hilbert Space Approach*, vol. 55. Expositions in Mathematics (DeGruyter, Berlin, 2011)
85. R. Picard, S. Trostorff, M. Waurick, A functional analytic perspective to delay differential equations. Oper. Matrices **8**(1), 217–236 (2014)
86. R. Picard, S. Trostorff, M. Waurick, On a comprehensive class of linear control problems. IMA J. Math. Control Inf. **33**(2), 257–291 (2016)
87. R. Picard, S. Trostorff, M. Waurick, On evolutionary equations with material laws containing fractional integrals. Math. Meth. Appl. Sci. **38**(15), 3141–3154 (2015)
88. R. Picard, S. Trostorff, M. Waurick, On maximal regularity for a class of evolutionary equations. J. Math. Anal. Appl. **449**(2), 1368–1381 (2017)
89. R. Picard, S. Trostorff, M. Waurick, Well-posedness via Monotonicity. An Overview. *Operator Semigroups Meet Complex Analysis, Harmonic Analysis and Mathematical Physics. Operator Theory: Advances and Applications*, vol. 250, pp. 397–452 (2015)
90. R. Picard et al., *A Primer for a Secret Shortcut to PDEs of Mathematical Physics*, vol. 140. Frontiers in Mathematics (Birkhäuser, Basel, 2020)
91. R. Picard et al., On abstract grad-div systems. J. Differ. Equ. **260**(6), 4888–4917 (2016)
92. R. Picard et al., On non-autonomous evolutionary problems. J. Evol. Equ. **13**, 751–776 (2013)
93. R. Picard, Evolution equations as operator equations in lattices of Hilbert spaces. Glas. Mat. Ser. III **35**(55), 1, 111–136 (2000). Dedicated to the memory of Branko Najman
94. R. Picard, Mother operators and their descendants. J. Math. Anal. Appl. **403**(1), 54–62 (2013). With an extension by S. Trostorff and M. Waurick. arXiv:1203.6762

95. J. Prüss, Decay properties for the solutions of a partial differential equation with memory. Archiv der Mathematik **92**(2), 158–173 (2009)
96. J. Prüss, On the spectrum of C_0-semigroups. Trans. Am. Math. Soc. **284**(2), 847–857 (1984)
97. T. Reis, Consistent initialization and perturbation analysis for abstract differential-algebraic equations. Math. Control Signals Syst. **19**(3), 255–281 (2007)
98. T. Reis, C. Tischendorf, Frequency domain methods and decoupling of linear infinite dimensional differential algebraic systems. J. Evol. Equ. **5**(3), 357–385 (2005)
99. R.T. Rockafellar, On the maximal monotonicity of subdifferential mappings. Pac. J. Math. **33**, 209–216 (1970)
100. R.T. Rockafellar, On the maximality of sums of nonlinear monotone operators. Trans. Am. Math. Soc. **149**, 75–88 (1970)
101. W. Rudin, *Real and Complex Analysis*. Mathematics series (McGraw-Hill, 1987)
102. G. Schmidt, Spectral and scattering theory for Maxwell's equations in an exterior domain. Arch. Rational Mech. Anal. **28**, 284–322 (1967/68)
103. R.E. Showalter, Diffusion in poro-elastic media. J. Math. Anal. Appl. **251**(1), 310–340 (2000)
104. B. Simon, *Basic Complex Analysis*. A Comprehensive Course in Analysis, Part 2A (American Mathematical Society, Providence, RI, 2015)
105. L. de Simon, Un'applicazione della teoria degli integrali singolari allo studio delle equazioni differenziali lineari astratte del primo ordine. Rend. Sem. Mat. Univ. Padova **34**, 205–223 (1964)
106. S. Simons, C. Zalinescu, A new proof for Rockafellar's characterization of maximal monotone operators. Proc. Am. Math. Soc. **132**(10), 2969–2972 (2004)
107. M. Sova, Cosine operator functions. Rozprawy Mat. **49**, 47 (1966)
108. S. Spagnolo, Sulla convergenza di soluzioni di equazioni paraboliche ed ellittiche. Ann. Scuola Norm. Sup. Pisa (3) 22 (1968), 571–597; errata, ibid. (3) **22**, 673 (1968)
109. S. Spagnolo, Sul limite delle soluzioni di problemi di Cauchy relativi all'equazione del calore. Ann. Scuola Norm. Sup. Pisa (3) **21**, 657–699 (1967)
110. A. Süß, M. Waurick, A solution theory for a general class of SPDEs. Stoch. Partial Differ. Equ. Anal. Comput. **5**(2), 278–318 (2017)
111. G.A. Sviridyuk, V.E. Fedorov, *Linear Sobolev Type Equations and Degenerate Semigroups of Operators*. Inverse and Ill-Posed Problems Series (VSP, Utrecht, 2003)
112. H. Tanabe, *Equations of Evolution*, vol. 6. Monographs and Studies in Mathematics. Translated from the Japanese by N. Mugibayashi and H. Haneda (Pitman (Advanced Publishing Program), Boston, MA, London, 1979)
113. L. Tartar, *The General Theory of Homogenization*, vol. 7. Lecture Notes of the Unione Matematica Italiana. A Personalized Introduction (Springer, Berlin; UMI, Bologna, 2009)
114. B. Thaller, S. Thaller, Factorization of degenerate Cauchy problems: The linear case. J. Oper. Theory **36**(1), 121–146 (1996)
115. S. Trostorff, A characterization of boundary conditions yielding maximal monotone operators. J. Funct. Anal. **267**(8), 2787–2822 (2014)
116. S. Trostorff, Exponential stability and initial value problems for evolutionary equations. Habilitation Thesis, TU Dresden, 2018
117. S. Trostorff, An alternative approach to well-posedness of a class of differential inclusions in Hilbert spaces. Nonlinear Anal. **75**(15), 5851–5865 (2012)
118. S. Trostorff, Autonomous evolutionary inclusions with applications to problems with nonlinear boundary conditions. Int. J. Pure Appl. Math. **85**(2), 303–338 (2013)
119. S. Trostorff, On integro-differential inclusions with operator-valued kernels. Math. Methods Appl. Sci. **38**(5), 834–850 (2015)
120. S. Trostorff, Semigroups and evolutionary equations. Semigr. Forum **103**(2), 661–699 (2021)
121. S. Trostorff, Semigroups associated with differential-algebraic equations, in *Semi-Groups of Operators – Theory and Applications. Selected Papers Based on the Presentations at the Conference, SOTA 2018, Kazimierz Dolny, Poland, September 30–October 5, 2018. In honour of Jan Kisyński's 85th birthday* (Springer, Cham, 2020), pp. 79–94

122. S. Trostorff, Well-posedness for a general class of differential inclusions. J. Differ. Equ. **268**, 6489–6516 (2020)
123. S. Trostorff, M. Waurick, Maximal regularity for non-autonomous evolutionary equations. Integr. Equ. Oper. Theory **93**(3). Id/No 30, p. 37 (2021)
124. S. Trostorff, M. Waurick, On differential-algebraic equations in infinite dimensions. J. Differ. Equ. **266**(1), 526–561 (2019)
125. S. Trostorff, M. Waurick, On higher index differential-algebraic-equations in infinite dimensions, in *The Diversity and Beauty of Applied Operator Theory*, ed. by P.S. Albrecht Böttcher Daniel Potts, D. Wenzel. Operator Theory: Advances and Applications, vol. 268, pp. 477–486 (2018)
126. S. Trostorff, M. Wehowski, Well-posedness of non-autonomous evolutionary inclusions. Nonlinear Anal. Theory Methods Appl. A Theory Methods **101**, 47–65 (2014).
127. D. Tzou, A unified field approach for heat conduction from macro-to microscales. J. Heat Transfer **117**(1), 8–16 (1995)
128. V.V. Jikov, S.M. Kozlov, O.A. Oleinik, *Homogenization of Differential Operators and Integral Functionals* (Springer, Berlin, 1994)
129. H. Vogt, J. Voigt, Bands in L_p-spaces. Math. Nachr. **290**(4), 632–638 (2017)
130. M. Waurick, A functional analytic perspective to the div-curl Lemma. J. Oper. Theory **80**(1), 95–111 (2018)
131. M. Waurick, A note on causality in Banach spaces. Indagationes Mathematicae **26**(2), 404–412 (2015)
132. M. Waurick, G-convergence and the weak operator topology. *PAMM*, vol. 16, pp. 521–522 (2016)
133. M. Waurick, G-convergence of linear differential equations. J. Anal. Appl. **33**(4), 385–415 (2014)
134. M. Waurick, Homogenization in fractional elasticity. SIAM J. Math. Anal. **46**(2), 1551–1576 (2014)
135. M. Waurick, Limiting processes in evolutionary equations - A Hilbert space approach to homogenization. http://nbn-resolving.de/urn:nbn:de:bsz:14-qucosa-67442. Dissertation. Technische Universität Dresden, 2011
136. M. Waurick, Nonlocal H-convergence. Calc. Var. Partial Differ. Equ. **57**(6), 46 (2018)
137. M. Waurick, On non-autonomous integro-differential-algebraic evolutionary problems. Math. Methods Appl. Sci. **38**(4), 665–676 (2015)
138. M. Waurick, On the continuous dependence on the coefficients of evolutionary equations. Habilitation. Technische Universität Dresden, 2016. http://arxiv.org/abs/1606.07731
139. J. Weidmann, *Linear Operators in Hilbert Spaces*, vol. 68. Graduate Texts in Mathematics. Translated from the German by Joseph Szücs (Springer, New York-Berlin, 1980)

Index

C. Seifert et al., *Evolutionary Equations*, Operator Theory: Advances and Applications 287, https://doi.org/10.1007/978-3-030-89397-2

Printed in the United States
by Baker & Taylor Publisher Services